ROCKET MAN

DAVID A. CLARY

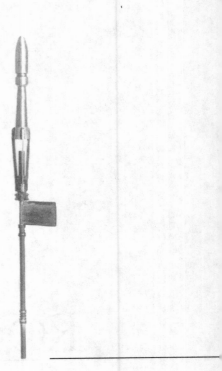

ROCKET MAN

ROBERT H. GODDARD AND

THE BIRTH OF THE SPACE AGE

AN IMPRINT OF HYPERION

NEW YORK

Library of Congress Cataloging-in-Publication Data

Clary, David A.
    Rocket man : Robert H. Goddard and the birth of the space age / by David A. Clary.—1st ed.
        p.   cm.
    Includes bibliographical references and index.
    ISBN: 0-7868-6817-1
    1. Goddard, Robert Hutchings, 1882–1945. 2. Rocketry—United States—Biography. I. Title.

TL781.85.G6 C58 2003
629.4'092—dc21
[B]
                                                    2002027321

Designed by Lorelle Graffeo

Hyperion books are available for special promotions and premiums. For details contact Hyperion Special Markets, 77 West 66th Street, 11th floor, New York, New York 10023, or call 212-456-0133.

FIRST EDITION

10   9   8   7   6   5   4   3   2   1

To Robert M. Utley, the historian's historian, and for more than three decades a boss, adviser, Dutch uncle, sterling example, and above all a friend

CONTENTS

ACKNOWLEDGMENTS

No biographer works alone. The task depends absolutely on the assistance of the many institutions that guard our documentary, material, and published heritage, and on the dedicated staffs and volunteers who make those organizations possible. I offer my deepest gratitude to the following establishments, and to those people associated with them who were especially helpful in this project.

In Worcester, Massachusetts: The Clark University Archives (especially Mott Lynn); the Worcester Polytechnic Institute Archives and Special Collections (Rodney Obien); the Robert H. Goddard Library of Clark University (Mary Hartman); the George C. Gordon Library of Worcester Polytechnic Institute; the Worcester Public Library (the helpful lady at the reference desk who declined to give her name); the Worcester Art Museum; the Worcester Area Chamber of Commerce; the Goddard Memorial Association (Barbara Berka); the Worcester Historical Museum (Joan Bedard).

In Roswell, New Mexico: The Roswell Museum and Art Center Archives (Candace Jordan); the Historical Society of Southeastern New Mexico (Elvis Fleming); the Paul Horgan Library of the New Mexico Military Institute (Jerry Klopfer); the Library of Eastern New Mexico University—Roswell (Mike Stein); the General McBride Museum of the New Mexico Military Institute; the Roswell Chamber of Commerce; the Roswell Hispano Chamber of Commerce; the Roswell Public Library; the Office of the Chaves County Clerk (Dave Kunko).

In Washington, D.C., and vicinity: The Goddard Space Flight Center, Greenbelt, Maryland (Alan Williams); the Smithsonian Institution National Air and Space Museum, Space History Division (Mike Neufeld and Frank Winter); the Smithsonian Institution Archives (Bill Cox); the National Aeronautics and Space Administration, History Office (Colin Fries), and News and Imaging

Branch; the National Geographic Society (Kristina Eide); the Library of Congress; the National Archives.

In other locations: The Golden Library of Eastern New Mexico University—Portales (Brackston Taylor); the New Mexico State Library, Santa Fe; the University of New Mexico Library, Albuquerque; the Museum and Rocket Park, White Sands Missile Range, New Mexico; the New Mexico Museum of Space History, Alamogordo; the Will Rogers Museum, Claremore, Oklahoma (Patricia Lowe).

A number of individuals also contributed advice, information, or other assistance to the project. First among them is Bill Moyer of Portales, New Mexico, who has been gathering material on Goddard for several years, and granted access to it all. He has also provided a set of Goddard patents to the Roswell Museum and Art Center. Thanks also to his father, Phil Moyer of Roswell, who combed the microfilms and reproduced all the articles on Goddard appearing in *The New York Times*, saving me considerable time. More thanks to Pearl Dean of Roswell for sharing her memories of Goddard, who did some machine work in her family business, Dean Motors; to Barbara Berka and Rodney Obien for a guided tour of Goddard historic sites in Worcester and Auburn; and to Ladislas and Barbara Berka and the board of directors of the Goddard Memorial Association for their hospitality and a wonderful seafood dinner.

Bea Clary, Jim Donovan, Mott Linn, Jay Miller, Bill Moyer, and Gretchen Young read parts or all of an early draft of the manuscript, made many useful and constructive comments and suggestions, caught some important errors, and more than once kept me from placing my foot where it would interfere with eating. Credit also to my agent extraordinaire, Jim Donovan, for making the project happen. And last but most definitely not least, my eternal gratitude to my editor at Hyperion Books, Gretchen Young, who with her assistant, Natalie Kaire, turned a pile of paper into a readable book.

And always, the fullest measure of credit belongs to my long-suffering but eternally patient Celestial Navigator, my wife, Beatriz Clary. Whenever I aim at the stars she makes sure that my feet stay on the ground. I would be lost without her.

For whatever in this book is of merit, credit belongs to all of the foregoing. For whatever falls short, blame should visit not upon those good people, but upon me.

INTRODUCTION

We are the music-makers,
And we are the dreamers of dreams,
Wandering by lone sea breakers,
And sitting by desolate streams;
World-losers and world-forsakers,
On whom the pale moon gleams;
Yet we are the movers and shakers
Of the world forever, it seems.
—Arthur William Edgar O'Shaughnessy, *Ode*

Those who knew Robert H. Goddard, America's first rocket scientist, liked him. One of his most likable traits, they all remembered, was his sense of humor. He was always ready with a joke, enlivening every situation, no matter how discouraging, with a quip. After his death in 1945, remarkably, none of his friends and associates could recall a single one of his jokes. All that was left of his humor was the memory of its former existence, empty of substance. Goddard the humorist was like the Cheshire Cat, faded away except for his smile.

So it was with Goddard the rocketeer. Between World War I and World War II he was the most famous scientist in America, the most heavily publicized in the world. The newspapers covered him more often than either Thomas Edison, who died in 1931, or Albert Einstein, who arrived in the United States the following year. He was portrayed as the dedicated physicist who all alone made rocketry respectable, who turned it into a science. He aimed to send rockets to high places, even to the moon and planets.

Because he enjoyed the support of influential scientists, philanthropists Daniel and Harry F. Guggenheim, and the famous aviator Charles Lindbergh, the world believed that Goddard's ambitions were not only feasible, but were about to be realized. When Nazi Germany launched rockets into the stratosphere and down onto Allied cities, Goddard's career in rocketry appeared also to be one in prophecy.

Unlike Goddard the humorist, Goddard the rocket man enlarged in image after his death. He was proclaimed a modern prophet, without honor in his own land until too late, who had begun life as a sickly, precocious genius, a boyhood visionary who perceived humanity's rocket-driven future in space years ahead of anyone else. He dedicated his life to realizing that vision, pressing ahead with determination and pluck despite widespread ridicule and the refusal of his own government to appreciate the importance of his work. Singlehandedly, it was claimed, he invented modern rocketry in all its details, including those used by the Nazis to bombard London.

In the aftermath of World War II, the story went, no rocket or jet plane could take to the skies without using Goddard's inventions. In 1960, the United States government confessed to stealing the man's ideas, and paid $1 million in compensation to his heirs. Three years later, a best-selling biography called *This High Man* fleshed out the Goddard the world had come to know. Monuments and medals rained upon his memory, along with institutions bearing his name. Robert H. Goddard was the "father" of modern rocketry and spaceflight.

At his moment of posthumous triumph, when he loomed larger than life, Goddard began to fade from view. First, in the interest of international good feeling, he had to share the title of "father" of space flight with two other "fathers," a Russian and a German. Then he became discounted as a man of fine accomplishments but little influence in his field. Because of his alleged secretiveness—the solitary endeavors that had made him so newsworthy during his life—it was said that modern rocketry had actually been reinvented by others. Moreover, his failures had greatly exceeded his successes.

The man who formerly had put America on the road to the Moon became a footnote in the history of technology. Again like the Cheshire Cat, he faded away until little remained but a series of juvenile biographies counseling youngsters to remain true to their dreams, to realize that persistence is the road to achievement.

The remarkable rise and fall of Robert H. Goddard's image owed to the fact that his real life was hidden behind a dense veil of legend building. He began fashioning a public persona as early as 1919, and continued it to the end of his life—one of the more surprising discoveries during this research was that the legendarily publicity-shy scientist was actually a publicity hound, who garnered reams of praise but almost no ridicule.

The legendry accelerated after his death, engineered by Goddard's widow, Harry Guggenheim, and Charles Lindbergh, culminating in the carefully scripted *This High Man*. That produced a reaction among other rocketeers and historians, who toppled the erstwhile "father" of space flight from his pedestal.

There was a real human being hidden in the shadows cast by the legends, and he was different from any of the manufactured images. Goddard the man was a complicated and often inscrutable individual, a self-contradictory person whose flaws waged a lifelong war with his virtues—a human being rather than a myth. He actually was a towering figure in the history of modern technology whose real accomplishments should have been enough to make legend building unnecessary. What he achieved was far more than his detractors in the past generation have suggested, even if less than his boosters in the previous generation claimed.

Robert H. Goddard the rocket man deserves a fair account of his life and his contributions to the history of rocketry and spaceflight. So does the question of how, beginning in his lifetime and accelerating in the decades since, he nearly disappeared into conflicting mythologies, making him either the Mount Rushmore of rocketry or a mere show horse in a field where he had little real influence. In an age when image building and spin doctoring dominate our public discourse, when the manufacture of personalities is a major industry, the life and legends of this remarkable scientist have something to teach us about what is really important in the swirl of lives and lies.

His adventure began with a dream. Once, not so long ago, seventeen-year-old Robert Goddard had a waking dream about flying farther than anyone ever had, to other worlds in the sky. He was not the first person to have such a vision, of course, but it was new to him, and it would not let go of him. In the present day, we have watched fellow humans venture into space, and walk on the Moon, and we have seen earthly machines visit other worlds in the sky. It becomes easy to take such things for granted, unless we remember how fantastic they would have seemed a century ago. To paraphrase Charles Lindbergh, we ought to ask ourselves whether Robert Goddard was dreaming then, or we are now.

CHRONOLOGY OF
ROBERT H. GODDARD

1882	5 October: Robert Hutchings Goddard born, Worcester
1883	Moves with family to Roxbury
1898	Robert's mother contracts tuberculosis; family moves to Worcester
1899	19 October: Has the cherry tree vision; date "Anniversary Day" thereafter
1901	31 March: Esther Christine Kisk born, Worcester
1903	Tsiolkovsky publishes "Exploring Space with Reactive Devices"
1904	Graduates as best male student, South High School, gives oration
1904–1908	Student at Worcester Polytechnic Institute; B.S. 1908
1908–1909	Instructor of physics, WPI, and special student in physics, Clark University
1909–1911	Fellow in physics, Clark University; M.A. 1910; Ph.D. 1911
1911–1912	Honorary fellow in physics, Clark University; also 1914–1915, 1918–1920
1912	Begins mathematical exploration of possibility of using rocket power to attain escape velocity
1912–1913	Research instructor in physics, Princeton University
1913–1914	Contracts and partly recovers from tuberculosis
1914	Awarded first two patents for rocket apparatus
1914–1915	Instructor in physics, Clark College; assistant professor 1915–1919; associate professor 1919–1920
1915	Proves experimentally that a rocket will provide thrust in a vacuum
1916–1918	Instructor in physics, Clark University
1916	17 October: Grandmother dies; buried in family plot, Hope Cemetery, Worcester

1917   5 January: Receives first grant of $5,000 from Smithsonian Institution; further grants awarded through 1934; begins work on multiple-charge, solid-fuel rockets

1917–
1918   Develops tube-launched infantry rockets under contract to Signal Corps, at WPI, Clark University, and Mount Wilson Observatory

1918   Returns to Clark University

1919   Smithsonian publishes "A Method of Reaching Extreme Altitudes"

1920   29 January: Mother dies; buried in family plot, Hope Cemetery

1920–
1923   Consultant on solid-propellant rocket devices for United States Navy Bureau of Ordnance, Indian Head, Maryland

1920–
1924   Abandons multiple-charge solid-fuel rockets, begins developing liquid-fuel rocket motors

1920–
1943   Professor of physics, Clark University

1923–
1943   Director of physical laboratories, Clark University; post is chairmanship of physics department; also assumes chairmanship of mathematics department

1923   Oberth publishes *Die Rakete zu den Planetenräumen*

1924   21 June: Marries Esther Christine Kisk

1925   6 December: Liquid-fuel rocket for first time lifts own weight

1926   16 March: First flight of liquid-fuel rocket, Auburn, Massachusetts

1928   16 September: Father dies; buried in family plot, Hope Cemetery

1929   17 July: Rocket flight alarms countryside, generates publicity; later comes to attention of Caroline Guggenheim, who alerts Harry F. Guggenheim and Charles A. Lindbergh; Nell gets her name; 23 November: Meets Lindbergh

1930–
1932   Under first two-year grant from Daniel Guggenheim, begins rocket experiments at Roswell, New Mexico

1930   28 September: Daniel Guggenheim dies; 30 December: First rocket flight at Roswell

1931   Three more flights at Roswell

1932   Tests of rockets stabilized by gyroscopes and blast vanes

1932–
1934   Guggenheim support expires; teaching at Clark University; small-scale experiments on components funded by Smithsonian and Daniel and Florence Guggenheim Foundation

1934   Annual funding by Daniel and Florence Guggenheim Foundation begins; continues through 1941; returns to Roswell; experiments resume

1935   8 March: Liquid-fuel rocket exceeds speed of sound; 31 May and 12 July: A-series rockets exceed a mile in altitude

1936   Smithsonian publishes "Liquid-Propellant Rocket Development"

1937   26 March: Rocket L-13 exceeded altitude of eight thousand feet, Goddard's highest flight

1938   9 August: NAA record flight, Goddard's only record

1941   10 October: Rocket P-36 jams in tower; last attempted flight

1941–
1942   Development of rocket-assisted takeoffs for Navy and Army Air Forces, Roswell

1942–
1945   Development of assisted-takeoff and variable-thrust rockets for Navy Bureau of Aeronautics at Naval Engineering Experiment Station, Annapolis, Maryland

1943–
1945   Engineering consultant to Curtiss-Wright Corporation, Caldwell, New Jersey

1944–
1945   Director, American Rocket Society

1945   2 June: Awarded honorary degree of Doctor of Science, Clark University; 10 August: Dies in Baltimore; buried in family plot, Hope Cemetery; Harry Guggenheim and Esther begin patent, publication, and publicity campaigns

1946   Esther begins presenting speech, "The Life and Achievements of Dr. Robert H. Goddard," and continues for thirty years

1947   21 April: Traveling exhibit of Goddard artifacts dedicated at American Museum of Natural History, New York; later donated to Smithsonian

1949   Dedication of launch tower as a memorial to Goddard on grounds of Roswell Museum

1950   First patent-infringement claim against United States filed; Wernher von Braun informs superiors that government rocket development infringes Goddard patent rights; Goddard artifacts exhibited at Institute of Aeronautical Sciences, New York, later transferred to Roswell Museum

1959   Dedication of Goddard Wing, Roswell Museum and Art Center, by Esther and Wernher von Braun

1960   Esther transfers patent rights to Daniel and Florence Guggenheim Foundation; government admits prior infringement, awards foundation $1 million for license to all Goddard patents issued since 1926

1963   *This High Man* published

1969   Reconstructed Goddard workshop and Goddard Planetarium dedicated by Esther and von Braun, Roswell Museum and Art Center

1970   *The Papers of Robert H. Goddard* published in three volumes

1971   Harry Guggenheim dies

1974   Charles Lindbergh dies

1976   Esther retires from Daniel and Florence Guggenheim Foundation

1982   5 June: Esther dies; buried in family plot, Hope Cemetery

ABBREVIATIONS USED
IN TEXT AND NOTES

AAAS	American Association for the Advancement of Science
AAC	Army Air Corps, USA
AAF	Army Air Forces, USA
AIAA	American Institute of Aeronautics and Astronautics
AP	Associated Press
ARS	American Rocket Society
ATO	Assisted Takeoff
BuAer	Bureau of Aeronautics, USN
BuOrd	Bureau of Ordnance, USN
CAC	Coast Artillery Corps, USA
CAL	Charles A. Lindbergh
CalTech	California Institute of Technology, Pasadena
CDW	Charles D. Walcott
CGA	Charles G. Abbot
CNH	Clarence N. Hickman
CUA	Clark University Archives, Worcester, Massachusetts
C-W	Curtiss-Wright Corporation
DAB	*Concise Dictionary of American Biography*, 2nd edition (New York: Charles Scribner's Sons, 1977)
Dates	Rendered in continental style, date-month-year, with three-letter abbreviations for months; for the twentieth century, "19" is omitted: 12 AUG 1888, 22 OCT 43, 3 MAY 2001
D&FGF	Daniel and Florence Guggenheim Foundation
DepNav	Department of the Navy
DepWar	Department of War
Diary	Diary of RGH, 1898–1945, RGP and abundantly excerpted in GP
ECG	Esther Christine Kisk Goddard
EGP	Esther Goddard Papers, CUA

FOIA   Freedom of Information Act

GALCIT   Guggenheim Aeronautical Laboratory, CalTech

GCR   Goddard Collection, RMAC

GE   General Electric Company

GEP   G. Edward Pendray

GP   Robert H. Goddard, *The Papers of Robert H. Goddard, Including the Reports to the Smithsonian Institution and the Daniel and Florence Guggenheim Foundation*, edited by Esther C. Goddard and G. Edward Pendray, 3 vols. (New York: McGraw-Hill, 1970). Pages are numbered serially through all three volumes, and because with two exceptions the documents are arranged chronologically, volume and page numbers appear only when necessary for clarity. The two exceptions, appearing at the beginning of vol. 1, are "Material for an Autobiography" (1927, with interpolations made 1933), and "Autobiographical Statement" (1921)

HAB   Homer A. Boushey, Jr.

HFG   Harry F. Guggenheim

HL   Paul Horgan Library, NMMI

HSSENM   Historical Society of Southeastern New Mexico, Roswell

IAS   Institute of Aeronautical Sciences

INT   Interview with

JATO   Jet-Assisted Takeoff

JPL   Jet Propulsion Laboratory, CalTech

LOX   Liquid (liquefied) oxygen

MIT   Massachusetts Institute of Technology, Cambridge

ML   Milton Lehman

MLP   Milton Lehman Papers, CUA

NAA   National Aeronautic Association

NACA   National Advisory Committee for Aeronautics

NANA   North American Newspaper Alliance

NASA   National Aeronautics and Space Administration

NASM   National Air and Space Museum, SI

NBS   National Bureau of Standards

ND   No date given

NDRC   National Defense Research Committee

NMMI	New Mexico Military Institute, Roswell
NGM	*National Geographic Magazine*
NGS	National Geographic Society, Washington, D.C.
NYT	*The New York Times*
RATO	Rocket-Assisted Takeoff
RCA	Radio Corporation of America
RDR	*Roswell Daily Record*
REGISTER	*Register of Graduates and Former Cadets of the United States Military Academy*, an annual, cumulative publication of the West Point Alumni Foundation, Inc.
REP	Robert Esnault-Pelterie
RGP	Robert Goddard Papers, CUA
RHG	Robert H. Goddard
RMAC	Roswell Museum and Art Center, Roswell, New Mexico
RMD	*Roswell Morning Dispatch*
RMI	Reaction Motors, Inc.
SC	Signal Corps, USA
SEC	Secretary
SECNAV	Secretary of the Navy
SECSI	Secretary, Smithsonian Institution
SECWAR	Secretary of War
SI	Smithsonian Institution, Washington, D.C.
TEL	Telegram
THM	Milton Lehman, *This High Man: The Life of Robert H. Goddard* (New York: Farrar, Straus, 1963)
TIAA	Teachers Insurance and Annuity Association
UPI	United Press International
USA	United States Army
USAF	United States Air Force
USAR	United States Army Reserve
USMA	United States Military Academy, West Point, New York
USN	United States Navy
USNR	United States Naval Reserve

WPI   Worcester Polytechnic Institute, Worcester, Massachusetts

WVB   Wernher von Braun

WWA   Wallace W. Atwood

WWAP   Wallace W. Atwood Papers, CUA

PROLOGUE

JUSTICE FOR BOB GODDARD

It will only spread warfare and multiply the occasions of war. In a little while, in a very little while if I tell my secret, this planet to its deepest galleries will be strewn with human dead. Other things are doubtful but that is certain. It is not as though man had any use for the moon. What good would the moon be to men? Even of their own planet what have they made but a battle ground and theatre of infinite folly? . . . Science has toiled too long forging weapons for fools to use.

—H. G. Wells, *The First Men in the Moon*

Frank Browning ceased to exist at 6:43 P.M., Greenwich mean time, on Friday, 8 September 1944, a drab and rainy evening during World War II. The street erupted as he walked down Stavely Road in London, smashing the houses on both sides and killing him instantly. Besides Browning, rescue squads found one other dead and twenty injured in the rubble of their houses, which surrounded a crater thirty feet wide and ten feet deep in the center of the concrete roadway.

The hole quickly attracted British and American military and civil defense officers, and reporters wanting to know if a new kind of German bomb had caused the destruction. "We can't tell you what it was," said a civil defense officer. "It might have been a gas main explosion." Nobody believed him. A few minutes later, another such blast occurred at Epping.[1]

So began Nazi Germany's V-2 campaign, a barrage of stratospheric ballistic missiles bearing one-ton bombs. They dropped without warning from almost straight overhead, raining by the hundreds onto London, Antwerp, and other places behind the Allied troops advancing against Germany.

Not all of them hit their intended targets; many fizzled or went astray, or their warheads failed to explode, so Allied authorities soon had their hands on a few samples. They were remarkably sophisticated machines, powered by liquid-fuel rockets and guided by gyroscopes and tail fins, or vanes. How had America's enemy developed such a monstrous contraption?

The answer was not long in coming. On 19 January 1945, the news service of the National Geographic Society in Washington, D.C., told the world that the fiendish Germans had stolen the whole idea from an American. The V-2, said the society, was a larger copy of designs worked out before the war by Dr. Robert H. Goddard, a physicist at Clark University in Massachusetts, working in isolation near Roswell, New Mexico.

"So closely do the mechanical features of the V-2 parallel the American projectile that some physicists think the Germans may have actually copied most of the design," said the report. "Most of these features were patented by Dr. Goddard between 1914 and 1932. The captured V-2 contained all the elements of Dr. Goddard's rocket, and even their arrangement inside the shell was the same." Worse, there was reason to believe that the Germans planned to bombard America with a gigantic "step rocket," a multiple-stage, long-range missile based on a 1914 patent by Goddard.[2]

It was a perfect wartime picture: strutting Nazis contrasted with a modest American inventor, pilfering his peaceful efforts to explore the upper atmosphere in order to bring death and terror to innocent Allied civilians. The story spread far and wide, and it endured. Goddard himself did not, dying the following August. His obituaries credited him with inventing nearly everything to do with rocketry, and with being the real inventor of the V-2. Robert H. Goddard became an American Prometheus.

He had been supported in his labors by Charles A. Lindbergh, perhaps the most famous human on the planet. He participated in Allied examination of the German aviation and rocket programs after the Germans surrendered in May 1945, and coauthored a scathing report claiming that with weapons such as the V-2, the Nazis had come close to winning the war. It was time for the American government to stop neglecting visionary geniuses like Goddard and start embracing the technology of the future.

To that end, Lindbergh secured a consultant's position, at one dollar a year, to advise the government in rocketry and space research. When General Carl Spaatz, chief of the Army Air Forces, asked him what he hoped to gain, Lindbergh pointed out that while the army was preparing to steal the Nazi missile systems, it should acknowledge that many vital features of them were of American origin. Lindbergh wanted only one thing in return for his work: "Justice for Bob Goddard."[3]

Who was Goddard, and what justice was owed to him?

ONE

ANNIVERSARY DAY

*I do not know whether the worlds are inhabited or not; and since
I do not know, I am going to see!*

—Jules Verne, *From the Earth to the Moon*

The two men labored for months on their fantastic project. Imprisoned in the Labyrinth of Crete—which the older man, Daedalus, had designed—they determined to gain freedom by flying away, over the sea to Sicily. Daedalus was humankind's original architect and engineer; the ancient Greeks credited to him statues and elaborate structures from Egypt through Crete and Greece to Sicily and Sardinia. He also devised a novel way to run a thread through a spiral seashell. He accomplished that by tying a thread to an ant, which obligingly dragged the fiber through the shell.

He and his fellow prisoner, his son Icarus, assembled a store of feathers and wax, and with them Daedalus constructed two pairs of wings. When all was ready, he sternly instructed the younger man: Do not fly too low, where sea spray would wet the feathers, nor too high, lest the sun's heat melt the wax. Icarus defied his father's warning, challenged the sun, and plunged into the water when his wings fell apart. The elder man continued on to Sicily.[1]

Thousands of years after Daedalus became mythology's first aviator, the dream of imitating the birds fired the imagination of an eighteen-year-old boy in Massachusetts. He intently observed the flight of swifts, and in 1901 reached a conclusion. In a letter to the editor of *St. Nicholas Magazine*, he announced that birds do not use their tails to steer themselves, but instead warp their wings. Twenty years later he used this observation—which had been made before, although he did not know that—to claim that he was the first person to conceive of this way to steer airplanes. The young man's name was Robert Hutchings Goddard.[2]

HEART OF THE COMMONWEALTH

Robert Goddard was born 5 October 1882 in the family home known as Maple Hill, in Worcester, Massachusetts. Worcester was the quintessential Yankee can-do city of the nineteenth-century industrial revolution, billing itself as the "Heart of the Commonwealth." The appellation fit, because the thriving city draped over hills and ridges just south of the state's center was also the heart of its inland economy. About 60,000 people lived there when Goddard entered the world, and their numbers increased to almost 200,000 in 1930, when he left for New Mexico.

Worcester was an industrious place filled with industrious people, the larger share of them immigrants, and they lived in ethnic neighborhoods, so there was a veritable congress of nations scattered across the town map. The Protestant Yankee population lived around the central districts, on the fringe of which stood Maple Hill. The urban landscape reflected the energy that created the place, with an earnest assertion of respectability. Brownstone, granite, and dark brick were favored for factories and public places, including banks, churches, schools and colleges, and office buildings. With New England thrift, housing stock was not discarded as it aged. There were many houses from the eighteenth and early nineteenth centuries still in use when Goddard was born, including Maple Hill.

The people of Worcester made things. Important industries included wire-making, machine tools, and shoes, along with razors, envelopes, wallpaper, corsets, grinding wheels, looms, nails, and plows. The fancy textile, grinding wheel, envelope-manufacturing, and wire-making industries blossomed after the Civil War, and by the time of Goddard's birth Worcester fencing wire was marketed all over the world.

Worcester people also invented things. Sons of the community included Eli Whitney, creator of the cotton gin and father of mass production; J. C. Stoddard, whose best-known production was the steam calliope; and Ichabod Washburn, who perfected steel-wire drawing, by which Worcester fenced the world. The tradition continued through the twentieth century, making Worcester the home of such diverse products as shredded wheat, the yellow smiley face, and the birth-control pill.

Nor was the life of the mind neglected. The city boasted six colleges during Goddard's residency, with a lively tradition of entertainment and lectures in the town's theaters. It gave birth to historian and diplomat George Bancroft, rags-to-riches novelist Horatio Alger, humorist Robert Benchley, and baseball legend Cornelius Alexander McGillicuddy, better known as Connie Mack.[3]

THE CHERRY TREE

While Goddard's hometown epitomized Yankee thrift and industry, his family reflected the very history of Massachusetts. He could trace his ancestry back to seventeenth-century immigrants from England, through an amazing fecundity and varied enterprise that produced town founders, printers, farmers, businessmen, and plain, solid citizens. Robert could count among his forebears a participant in the Boston Tea Party, and veterans of the Revolutionary War and, in the case of his paternal grandfather, the Civil War.[4]

That grandfather was Nahum Parks Goddard, who scratched out a living as an itinerant musician. Unable to support his family in Boston, in the late 1870s he moved them to the two-story, wood-frame house in Worcester. His wife, Mary Pease Upham Goddard, "Madame Goddard," was the strong-willed ruler of her household, her husband, her son, and everyone else in the place except her aging mother-in-law, who owned the house.

The move from Boston uprooted Mary and Nahum's grown son, Nahum Danford Goddard, who had been working for businessman W. B. Browne. He arrived in Worcester bearing Browne's recommendation, which described him as "a young man of excellent character and good English education . . . who by his faithfulness and integrity and energy will give good satisfaction." Young Nahum secured a job as bookkeeper for the L. Hardy Company, which manufactured machine knives for the paper and textile industries.

The business was co-owned by Henry A. Hoyt, who had a daughter, slender, doe-eyed Fannie Louise. Nature took its course, and over her father's vehement objections—he considered the Goddards to be improvident—Fannie married Nahum on the groom's twenty-third birthday, 3 January 1882. Hoyt disinherited his daughter, who moved into the Goddard house. In that environment she gave birth to Robert nine months and two days after the wedding.[5]

Nahum resigned from the Hardy firm a few months later, and moved his wife and son to the Boston streetcar suburb of Roxbury, where he purchased a large frame house on chestnut-shaded Forest Street. He went to work at another machine-knife company, S. C. Ryerson, and later he and fellow employee Simeon K. Stubbs bought the firm and renamed it Stubbs and Goddard. Nahum invented a machine knife for cutting rabbit fur, and built a market for it in the hat-making industry. He also invented the "Goddard Welder," a welding flux, and became his company's traveling salesman.

A short, dapper cigar smoker with a large mustache, Nahum was a technological adventurer. He installed electric lighting in the Roxbury house, bought an early-model phonograph, and probably aroused his son's interest in electricity. Later, back in Worcester, he would be one of the first in that city to own an automobile and a radio.

Nahum was Episcopalian, and raised his son in that denomination, but he was not often a church-goer. Republican in his politics, he was easygoing and independent. Robert remembered that he minded his own business and expected others to do the same. As the boy grew, the two enjoyed hiking, fishing, and photography in the country. Robert credited his father with making him a lover of nature.[6]

There are three sources of information on Robert Goddard's childhood—his own "Autobiographical Statement" (1921) and "Material for an Autobiography" (1927–33), and information provided to his authorized biographer, Milton Lehman, by his widow, Esther Goddard, in the 1950s. None can be impeached because Esther removed most personal and family material from both her and Robert's papers before donating them to Clark

University. What we know about Goddard's boyhood is therefore what the Goddards wanted us to know.

The picture they painted was of a precocious, inquisitive, and experimentally active youngster raised in a loving, middle-class home, where a doting grandmother provided unrelenting encouragement along with the wisdom of the ages. He was also a sickly child, who overcame that adversity along with numerous setbacks to his childish experiments. It was the story of his future life—try, fail, try again. In short, the young Goddard was a born genius, energetic despite frailty, persistent through setback or handicap, and possessed of a towering intellect. Perhaps he was, but the family account was too idealized to describe a real human specimen.

Some elements of the story are blandly conventional. Robert was purportedly a dedicated churchgoer and choir boy, habits decidedly at odds with his life record and Esther's later testimony. Madame Goddard loomed as a prescient fountain of sage advice, who told "Robbie" that a glorious new age was coming, and that he must work hard to prepare for it. On the other hand, his precocious interest in how things work and in tinkering experiments is the boyhood story of other inventive scientific pioneers, including Thomas Edison, Henry Ford, and the Wright Brothers.[7]

Goddard wanted to be regarded as a born genius. His family told him, he wrote at the age of forty-five, that he first showed interest in mechanical things when he was a few months old and spent a trolley trip to Boston "studying the bell-cord system at the top of the car. Two or three years later, it appears that I could be silenced for half a day at a time by placing the baby carriage so that I could overlook a freight yard and see the engines go back and forth."

When he was four or five years old, he said, he learned that shuffling the feet can generate static electricity, and also that batteries store electricity. He took the zinc rods from a battery jar, shuffled on a sidewalk, then jumped off a low fence, hoping these measures would make him jump higher. He stopped when his mother told him that it might work someday, and "then you'll go sailing away and might not be able to come back."

While saying that he was sickly but at the same time normal enough to play with other boys, go skating and sledding in the winter, and organize a group attempt to dig a hole to China, Goddard described his boyhood as a time of scientific ambition. He experimented with everything, especially kites and magnifying glasses. He talked his father into providing him with a microscope, a telescope, and a subscription to *Scientific American*. He also launched a scheme to build a frog hatchery that he pursued for years without result. Fascinated by textbook drawings of eyes, he tested the eyesight of his great-grandmother. She, "being a paralytic, could not get away and [her] feeble eyesight proved ideal for the purpose."

By his teen years, Goddard had tried to make diamonds from graphite, to build and launch an aluminum balloon, and to devise a perpetual-motion machine. He realized at last that his experimental reach had exceeded his uneducated grasp, and concluded wisely: "The best plan for all of us to follow is to leave our researches and investigations until knowledge and experience are attained, after which our work will either be crowned with success or buried once and for all as an impossibility."[8]

Then there was Goddard's health. Overcoming physical weakness is commonly part of the life stories of Great Men: The puny Milo carried a calf around daily until it grew into a bull and he into the strongest man on Earth. Beethoven and Edison prevailed over deafness. George Washington was sterile, awkward, and nearly toothless. Theodore Roosevelt was a nearsighted weakling, and his cousin Franklin overcame polio. Roger Banister's legs were nearly destroyed by a fire, but he grew up to break the four-minute mile.

The family account of Goddard's boyhood followed the same pattern. It does not appear that there was great concern about the boy's health until 1894, when the Goddards' second child, Richard, was born with a spinal deformity that ended his life in less than a year. Biographer Lehman said that this exaggerated the family's concern for Robert's health, with Madame Goddard becoming especially possessive.[9]

Frantic attention to the well-being of the only surviving child would be normal. It explains why Goddard was two years behind in school by 1898, because his mother and grandmother panicked at every sniffle and kept him home where they could attend to him. After 1894 Robert was mother-henned by two anxious women who catered to his every need, real or imagined. For the rest of their lives, he was scarcely out of sight of either of them, and he grew up in an environment where women made all the decisions affecting his well-being.

Another catastrophe struck the household in 1898. Robert's mother, Fannie, developed a wracking cough, which doctors diagnosed as a symptom of "consumption and complications," meaning tuberculosis. The doctors examined her thin, long-legged son for the same infection, but cleared him. Told that recurrent illness kept him out of school, the physicians opined that he might have trouble with his stomach, perhaps his kidneys, and advised exploratory surgery. His parents declined, got him a bellyband (an old folk remedy wherein the midriff is wrapped to ward off disease) instead, and again removed him from school.[10]

The tubercle bacillus (*Mycobacterium tuberculosis*) would thenceforth influence Goddard's life as significantly as did any human being. Tuberculosis, epidemic during the industrial revolution, is a chronic disease commonly affecting the lungs, although it can spread to or originate in other parts of the body, destroying any tissues or organs. The earliest sign of infection is usually a cough, followed by spitting. The disease produces debilitating effects on the whole body, including lowered blood pressure and energy, and permanent loss of lung capacity.

In Fannie Goddard's day, tuberculosis was a death sentence of a cruelly protracted and wasting sort. The best treatment physicians could prescribe was bed rest, although removal to climates where the air was dry or the altitude high also could help.[11]

Nahum Goddard sold the Roxbury house and his share of Stubbs and Goddard, and moved the family back to Worcester, where Fannie could receive constant nursing from relatives. Hoyt had sold his part of the L. Hardy Company, and Nahum returned there as shop superintendent.

The move from Roxbury was a step downward for the family. Maple Hill was heated by a coal stove in the dining room, and by a few Franklin stoves in the old hearths. An iron sink and hand pump stood in the kitchen, while outside were a well, a woodshed, and an outhouse, along with a scattering of fruit trees. Most important, the place was the domain of Robert's grandmother, now a constant presence rather than a frequent visitor. According to Esther, Madame Goddard declared: "He's going to be my boy now." Thereafter "Gram" supervised Robert while Fannie reclined.[12]

Shortly before his mother's diagnosis, the real Robert Goddard partly emerged from the shadow of his and Esther's version of his early years and started to lay down a paper trail. He began a diary early in 1898 and kept it faithfully until his death; before 1904, it is the most complete evidence of his activities and ideas. In that year, after graduating from high school, he burned the bulk of his youthful experimental notes and other writings in the dining-room stove, deciding that his work to that point had been both childish and fruitless.[13]

The early entries reveal, instead of the frail boy genius of family lore, a normal, bright, self-conscious, and inquisitive fifteen-year-old. Dated from Roxbury in 1898, the following are excerpts from typical entries:

February 16. Went to school morning and afternoon... Read Youth's Companion *etc. in afternoon.* Maine *has been blown up in Cuba. February 17. Danger of a war with Spain. February 19. Aluminum balloon will not go up. Tried to put gas in it, but could not. Aluminum is too heavy. Failior [sic] crowns enterprise. March 19. Pa has bought a phonograph today. April 23. Worked up in attic nearly all day.... I am quite sure the war with Spain has begun. May 2. Saw first bumblebee this spring this noon. May 13 ... After school went to flag raising. They read my story in the school paper this afternoon. May 17 ... May Dunn's brother has run away and come to our house. June 9. After school, stayed out in yard. Heard hurdy-gurdy play and went to ride on [Ferris] wheel in evening. June 17 (Flag Day). Fired firecrackers and common crackers ... Had rocket, Roman candles, red fire, etc. in evening. June 29. Went and graduated in morning from grammar school, and got diploma. July 4. Fired cannon, "pop," and firecrackers all day. In evening had 5 skyrockets, 3 Roman candles, 1 large pinwheel, red fire, and a Japanese match which I made.*

Nothing there evokes a sickly, scholarly youth. Nor does the first mention of rockets presage the source of his future fame. The diary does reflect the beginning of what would be lifelong habits—brief accounts of the day's activities, observations of the weather, and mention of the day's experiments or speculations. Daily reading also was a regular topic.

The return to Worcester initiated major changes in Goddard's life. He was confined by his grandmother as if he really were sickly, and he voraciously began reading books from the local library. Then he read H.G. Wells's *The War of the Worlds*, a story about a Martian invasion of Earth, serialized in a Boston newspaper in 1898.

Although its essential message—anti-imperialism with a dose of anti-vivisectionism—went over his head, the dramatic yarn set the stage for a mystical dream that would color his life to the end. This is how he remembered that experience decades later, in 1927:

[O]n the afternoon of October 19, 1899, I climbed a tall cherry tree at the back of the barn . . . and, armed with a saw which I still have, and a hatchet, started to trim the dead limbs from the cherry tree. . . . I imagined how wonderful it would be to make some device which had even the possibility of ascending to Mars, and how it would look on a small scale, if sent up from the meadow at my feet. I have several photographs of the tree, taken since, with the little ladder I made to climb it, leaning against it. It seemed to me then that a weight whirling around a horizontal shaft, moving more rapidly above than below, could furnish lift by virtue of the greater centrifugal force at the top of the path. In any event, I was a different boy when I descended the tree from when I ascended, for existence at last seemed very purposive.[14]

The content of this vision—a variation on a perpetual-motion fantasy that he pursued through his boyhood—is less important than the associated observations. Goddard expressed here a lifelong tendency to venerate his early experiences and ideas. The saw was still a souvenir a quarter century later, he repeatedly photographed the site of the event, and he remarked 19 October annually in his diary as "Anniversary Day."

He dated the beginnings of both his interest in space travel and his scientific career from this experience, and rated nearly every event of his life thereafter as "very purposive." Thus the bird-tail observation of 1901 was reinterpreted later as the first conception of the aileron.

The cherry tree experience presented Goddard with an incentive to rejoin the real world, beginning with high school. It also drew him into the most far-flung adventure of the human imagination—travel to outer space. Later he would consider how to get there.

TO FLY AMONG THE STARS

Ever since Daedalus, flight has represented freedom. His story was cited and elaborated frequently by the ancients, until in the second century A.D. the Roman poet Lucian of Samosata raised its ambitions to new heights by concocting the first known story of adventurers traveling to the Moon and beyond. It was a satire, but also the first extraterrestrial science fiction, including even an interplanetary war between the Moon and the Sun.[15]

The idea evolved along with changing notions of what was in the sky. Were there other worlds out there, what were they like, could they offer escape from the problems that plague our own planet? In the sixteenth century, when the existence of other worlds became generally accepted, the space-travel idea gained renewed currency. The first version of the Faust story appeared during that famine-plagued time. Faust sold his soul to the Devil in return for three things—plenty to eat, enough money to wear decent clothing, and the power "to fly among the stars." Historian Jacques Barzun explained the tale's enduring appeal: "To 'fly among the stars' stands for the restless discontent with mere humanness and for any aspiration so lofty that to fulfill it Man is willing to barter his most precious possession."[16]

The space-travel idea was and is the ultimate escapism. It captured human imagination from Rabelais, who sent his protagonist Pantagruel out among the stars and gods, to writers of the twenty-first century. One of Edgar Allan Poe's lesser-known tales, "The Unparalleled Adventure of One Hans Pfaall," described a trip to the Moon and back on a balloon. It inspired Poe's most avid reader in France, the "father" of science fiction, Jules Verne. In *From the Earth to the Moon* (1865) and its sequel *Round the Moon* (1870), Verne captured the boyhood imaginations of generations of would-be spacefarers.

Verne's travelers reached space in a capsule fired from a gigantic cannon, a rare scientific misstep for him. Although he designed a hydraulic shock-absorbing system for the spacecraft, it would not have prevented the cannon blast from mashing the passengers to a pulp. Nevertheless, Verne postulated some scientific principles that later proved valid. He calculated the speed required to propel a missile beyond the Earth's gravitational grip—what is now known as "escape velocity." He recognized the advantage of launching from between 28 degrees north and 28 degrees south latitude, where the acceleration donated by the Earth's rotation is the greatest. Verne also recognized that rockets can provide thrust in a vacuum—his travelers used them to steer in space, and to break their speed as they approached the Moon.

The Frenchman's yarns were plausible in terms of the science of their day. Goddard devoured them and other stories, and was ready for Wells's *War of the Worlds* in 1898. The very idea of interplanetary travel lit a fire in him. He reread Verne and Wells many times, and never shook off the sense of infinite possibilities that they inspired, in a period when Percival Lowell described signs of civilization on Mars and Wells produced yet another way to get to the Moon. *The First Men in the Moon* (1901) got there by using a secret gravity-neutralizing material. If the human mind could conceive of such things, the young Goddard wondered, could it not also figure out a way to do them?

During a visit with Gram to see relatives in Boston, a few days after the cherry-tree vision, he explained his centrifugal-force problem to his cousin, a student at Harvard. "He said it was inoperative," Goddard remembered, "but could not explain the thing in such a way as to convince me—a circumstance which is perhaps fortunate."

Later, he "started making wooden models in which lead weights were to furnish lift by moving back and forth in vertical arcs, or were to strike metal pieces as they whirled around horizontal axes. These, naturally, gave negative results, and I began to think that there might be something after all to Newton's laws...." He made experiments to test the Third Law—each action produces an equal and opposite reaction—and discovered that Sir Isaac had been correct all those years ago.

By 1901 he was bored to distraction from staying at home while his friends went away to college, entertaining himself by watching the night sky through a telescope. He realized that he had been floundering in ignorance, trying to solve problems without knowing what he was doing: "This... made me realize that if a way to navigate space were to be discovered—or invented—it would be the result of a knowledge of physics and mathematics.... I resolved forthwith that I would enter the new South High School, at Worcester, and shine in these subjects." Robert Goddard got on with his life, behind schedule.[17]

THE HOPE OF TODAY

He enrolled in South High School as a sophomore in 1901, just before his nineteenth birthday, proving as adept in English composition as in math and science. He showed no signs of whatever ailments had allegedly plagued him in recent years, and was popular with his classmates and the faculty—he was elected class president twice.

Meanwhile, he continued his experiments at home after school, investigating radio, telephony, chemistry, gyroscopes, and other gadgets. He also attended lectures at Clark University and the Worcester Polytechnic Institute. Goddard was eager to learn, and he was even more eager to do.[18]

During his first Christmas break, he wrote an essay titled "The Navigation of Space," which began: "The interesting problem of space navigation seems to be much neglected, which is not surprising considering the almost insurmountable difficulties involved." He addressed launching a spacecraft by gun, and the use of magnetic forces, each being impractical, then devoted most of his attention to avoiding meteors. His essay concluded: "We may safely infer that space navigation is an impossibility at the present time; yet it is difficult to predict the achievements of science in this direction in the distant future." He sent the piece off to *Popular Science News*, which rejected it.[19]

He penned another essay, "The Habitability of Other Worlds," which has disappeared, and over the next years wrote to distinguished authorities about his scientific speculations. His ideas included an "automatic balancing device" modeled on the semicircular canals of the ears, which he passed on to the Army Medical Museum. An army surgeon patiently explained the function of the ear canals to him without belittling his naïveté. Other correspondence covered radio waves with the Smithsonian Institution, and the amalgamation of gold by mercury with the editor of *Scientific American*. Always he received friendly explanations of the science involved in his question or proposal.[20]

Goddard's ventures became gradually less grounded in ignorance and more practical, although he retained a sophomoric streak. "During the later two years at the South High School," he recalled in 1927, "I spent considerable time considering the possibility of propulsion by a kind of machine-gun device, in which bullets were fired downward. I also experimented with gyroscopes, under the erroneous impression that the tendency for the gyroscope to remain in one plane might be used to give a resultant force." His spaceflight obsession, born in the cherry tree, had engendered two subordinate obsessions that would occupy him for decades—rocket propulsion and gyroscopes.

His experiments became more limited and practical. Witness his diary around Independence Day in 1903: "July 2. Flew kite and made rocket support in morning. Made propellers and tried them and rode bike in afternoon. July 4. Fired blanks and tried electric igniter in morning. Fired bombs and blanks etc. in afternoon. Shot two dozen rockets in evening. Fine time!" The budding scientist was still an enthusiastic boy.[21]

Goddard was slow to mature, because he had been confined under his grandmother's hand for so long, and now spent the school day among classmates three years younger than he. Like most youngsters, he was eager to grow up. Unlike most others, he could lecture himself on the ways to achieve intellectual maturity:

> *Worcester, July 12, 1903. It is a very important thing to jot down suggestions that come into one's mind from time to time, as the thoughts that are most useful do not come at a bidding. The only way to prepare the mind for the formation of suggestions is to use moderation and activity. With the former, the mind is never burdened by absorbing hope of high achievements, because these achievements may be proved fallacious dreams. If the mind is continually engrossed with one subject, other and perhaps more important suggestions will be excluded, and the means of proving the thing destined to failure may be withheld. Activity fosters growth and this furnishes suggestions.[22]*

Each day following school, Goddard sat with his mother, who continued to weaken, until his father came home. She was greatly cheered in the spring of 1904 to hear that her son had been selected to give the class oration as best male student. With guidance from his English teacher, Goddard labored on the speech. On 24 June 1904, before the graduating classes of all three local high schools assembled in Mechanics Hall, he delivered his remarks, "On Taking Things for Granted."

The last part of the last sentence made it the most frequently quoted high-school speech in history, although the full text reveals an intellect with a long way to go to reach maturity. Still, it was a remarkable performance for a high-school senior, even a twenty-one-year-old one whose schooling had been delayed by maternal anxieties.

It reviewed the history of science to show that the errors of the past were caused by taking things for granted. Three such "errors," Goddard opined, were about to be corrected by findings that there were signs of life on the Moon and Mars, by the fact that a perpetual motion machine was about to be achieved, and by concluding that the warmth of radium was due to the perpetual creation of energy.

All three notions were as fallacious then as they are today, but excusable in a youngster. The final passage redeemed the orator:

[J]ust as in the sciences we have learned that we are too ignorant safely to pronounce anything impossible, so for the individual, since we cannot know just what are his limitations, we can hardly say with certainty that anything is necessarily within or beyond his grasp. Each must remember that no one can predict to what heights of wealth, fame, or usefulness he may rise until he has honestly endeavored, and he should derive courage from the fact that all sciences have been, at some time, in the same condition as he, and that it has often proved true that the dream of yesterday is the hope of today and the reality of tomorrow.[23]

Goddard was the only member of his high-school class old enough to vote, but he was still a work in progress. Shortly after so optimistically lecturing his classmates, he found himself wallowing in uncertainty, and discouraged by his lack of accomplishment. "My own dream," he recalled decades later, "did not look very rosy. I had on hand a set of models which would not work, and a set of suggestions which I had learned enough physics to know were erroneous." In that mood he incinerated his accumulated notes.

Then his inherent optimism and determination reasserted themselves: "The dream would not down, and inside of two months I caught myself making notes of further suggestions. For even though I reasoned with myself that the thing was impossible, there was something inside me which simply would not stop working."[24]

Something else was working on him: "Practiced in Mechanics Hall in morning," he recorded in his diary on graduation day. "Spoke at graduation exercises in afternoon. Bouquet. Took Miriam to class meeting as president, and to alumni meeting in evening." It was his first date.[25]

Miriam Olmstead was South High's top female student. Goddard met her when they practiced their orations before the English teacher; her speech, "The Wealth of One Field," was a paean to nature. He fell for her at first sight, impressed with her interest in science—biology, in her case—and driven by his hormones.

They courted that summer, chaperoned by Miriam's mother, and continued their relationship as they went to separate colleges, he to Worcester

Polytechnic Institute, she to Smith College in Northampton. On Thanksgiving 1905 Robert presented her with an engagement ring. The following summer they vacationed together—still chaperoned—in New Hampshire, where Miriam took German lessons.

The whole business caused him some unease, especially as the love, like most first loves, began to fade. The affair drove him toward a near obsession with income and future security that would color his outlook for the rest of his life. He began by toying with possible inventions that could provide royalty income after marriage, but in the end he resisted her insistence on a wedding right after college. "Five years is a long time," she told him, supported by her mother. Only after he had a Ph.D. and a thousand dollars in the bank, he answered.

After graduating from college in 1908, Miriam went to Europe to study; she returned to Worcester the next year, determined to abandon the place for greener pastures. With Goddard still resistant, she moved to New York City and embarked on a career in its Department of Health. Goddard enrolled in graduate school, and Miriam disappeared from his life.[26]

It would be fruitless to draw conclusions about what this chaste affair might say about Goddard the man. But one thing is obvious: He had grown up under the wings of two women devoted to his every breath. Miriam had ambitions of her own and no desire to abandon them to cater to someone else. Though she saw him as husband material, she was not the sort of helpmeet Goddard wanted.

Before their doomed romance ran its course, in the summer of 1904 Goddard, like Miriam, aimed for higher education. While she went away to school, he saw his next step toward the stars in the form of two towers atop a high hill. He did not need to leave Gram's care to enroll in Worcester Polytechnic.

TWO

SOMETHING IMPOSSIBLE WILL PROBABLY BE ACCOMPLISHED

Why had we come to the moon? The thing presented itself to me as a perplexing problem. What is the spirit in man that urges him forever to depart from happiness and security, to toil, to place himself in danger, even to risk a reasonable certainty of death? It dawned upon me up there in the moon as a thing I ought always to have known, that man is not made simply to go about being safe and comfortable and well fed and amused. Against his interest, against his happiness he is constantly being driven to do unreasonable things. Some force not himself impels him and go he must. But why? Why?

—H. G. Wells, *The First Men in the Moon*

The two towers loomed over the Worcester Polytechnic Institute (WPI). The school began as a dream of Worcester's leading manufacturers at the end of the Civil War, self-made men educated in the school of hard knocks. Their leaders were blacksmith-turned-wire-making-magnate Ichabod Washburn and peddler-turned-tinware-manufacturer John Boynton, for whom Tech's first two buildings, each with a towering belvedere, would be named.

Washburn and Boynton believed the age of the uneducated entrepreneur had passed, overtaken by changes in science, technology, finance, and society. Worcester needed an academy offering practical education to the leaders of tomorrow, so they formed a committee with other manufacturers, raised money, and obtained a legislative charter for the Worcester County Free Institute of Industrial Science in the fall of 1865. By 1898, it had evolved into a four-year institution of science and engineering, renamed Worcester Polytechnic Institute.

When Goddard began there in 1904, it was a thriving school, attracting the best scholars to its faculty. Two had a special influence on him. One was the head of the physics program, A. Wilmer Duff, an Edinburgh-educated, very demanding applied physicist, a pioneer in fields ranging from acoustics and ballistics to electricity. His textbooks were standard on many campuses. The other was an English professor, Zelotes W. Coombs, who taught the young Goddard how to express the ideas roiling through his mind.[1]

AS USEFUL A UNIT AS POSSIBLE

According to his widow, Goddard's grandmother obtained a loan for his tuition from a grocery wholesaler. Once arrived on campus, he so impressed Duff that the physics professor hired him as a laboratory assistant, and recommended him as a tutor for other students to help pay his way through school. He continued to live at home, where Gram ensured that nothing distracted him from his brainwork.[2]

Goddard had developed a lifelong routine. He divided his day into two parts: work or school during the day, and an evening devoted to writing down his "suggestions," exploratory ideas on science and technology. Occupying part of both schedules was a drive for recognition, especially a frequent claim to have been the first to think of or do something.

Still missing from his life was an organizing principle, a clearly conceived goal. When Coombs ordered the freshman English students to produce essays with the theme "Who I Am and Why I Came to the Institute" in November, Goddard reviewed his years of intellectual floundering and confessed to bouts of laziness and changing interests as a youngster. He was still unsure of his footing: "In this predicament it is not surprising that there is sometimes the temptation to choose an elective course, where several interests can be further cultivated; for, after all, practical education, from an economic standpoint at least, should tend to develop what is best in a man and to make him as useful a unit as possible in the community."[3]

Four years of rigorous instruction helped Goddard to sharpen his mind, and by the time he graduated as a general science major in 1908 he was dedicated to solving the challenge of space travel. Although he did not yet know how to reach that objective, he was prepared to address it intelligently.

He had also matured as a social being, in an environment where he was surrounded by bright minds. In January 1905 he was elected vice president of his class ("Missed president by 5 [votes]," he noted in his diary), the next month he pledged Sigma Alpha Epsilon fraternity, and a year later he became class president. He displayed a whimsical sense of humor and a taste for practical jokes, and interrupted lectures with questions whose points were not always clear.

Among the marks he left at WPI was the school song, the music and words of which he wrote in 1908. "In the symbol of our life," it went, "the hammer in the iron hand; in sacred comradeship, with memories all intertwined; in men throughout the land, whose highest aim is usefulness, the brotherhood supreme; there may Tech spirit ever live!" The chorus was on the same artistic level. It reflected a lifelong interest in music, for which his enthusiasm greatly exceeded his talent.

Besides holding virtually every class office, Goddard was elected editor of the yearbook, a member of Sigma Xi engineering fraternity, class representative to and speaker at the 1908 annual banquet, and the "brightest senior." He and the two other boys who graduated in general science were described as "all good-natured pleasant fellows even though the average standing of their division was very high." Graduation as "brightest senior" earned him a check for $75.

"Bob (pronounced Borb, with the accent on the second syllable)," read his biography in the yearbook, "came to us from the South High. The word 'shark' fails to convey any idea of his appetite for knowledge, for he fairly revels in the weirdest of physics and kindred stumbling blocks to the less fortunate of us. As a relaxation he has made a study of the theory and application of the gyroscope, and wrote a treatise on the subject. On account, probably of his extreme age and the venerable air which his bald pate has given him, Bob has been continually before the class in official capacities."

Elsewhere in the yearbook appeared an anonymous quip: "At school I knew him—a sharp-witted youth, grave, thoughtful, and reserved among his mates, turning the hours of sport and food to labor, starving his body to inform his mind."

By graduation day he was a balding, lanky young man with deep-set, dark eyes and a placid expression. His senior thesis, "On Some Peculiarities of Electrical Conductivity Exhibited by Powders and a Few Solid Substances," caused Duff to arrange his appointment as a physics instructor at WPI for the 1908–09 school year, which paid $850. With that pending, Goddard revised the thesis and some later work into a paper about exceptions to Ohm's law to be found by passing current through electrolytic powders. That line of work later gained him a patent for an electronic oscillator tube.[4]

While Goddard the student advanced through Tech, Goddard the speculator and diarist kept his evenings occupied. The diary recorded commonplace events interspersed with observations and aphorisms, especially at the end of the year. "Anything is possible with the man who makes the best use of every minute of his time," he closed out 1904. "There are limitless opportunities open to the man who appreciates the fact that his own mind is the sole key that unlocks them." A few months later he said: "If there is no law against it, why—then 'twill happen some day. We should not be original for originality's sake."

His mind continued to sharpen: "Experience enables one to apply things that one knows; theoretical science enables one to apply things that have never been dreamed of," he wrote in July 1906. Late in 1908, "Pa has bought an auto" preceded an unattributed quotation and his own lament: " 'The years forever fashion new dreams, when old ones go.' God pity a one-dream man."[5]

That Goddard had become a one-dream man was revealed in his greater evening project, recording his "suggestions" in five green notebooks, filled before 1915. He claimed in 1927 that they included the original formulation of important ideas and inventions. They also recorded fruitless speculations and plain dead ends. As he summarized their content:

The suggestions were very diversified, and concerned the possibility of using the magnetic field of the earth, shooting material to a "spaceship" by means of electric and other guns, an airplane operated at high speed by the repulsion of charged particles, artificially stimulated radioactivity, artificial atoms of great energy consisting of moving positive and negative charges, propulsions in space by repulsion of charged particles, reaction against displacement currents in space, repulsion of highly heated material particles at the focus of parabolic mirrors, the use of solar-energy by light devices on a "spaceship," the idea of the multiple-charge rocket, the use of liquid propellants, and several other plans. A summary of twenty-six methods, involving means in space, means taken with the apparatus, and means sent from earth, was written December 28, 1909.[6]

By that time he had reached a preliminary conclusion that it would be possible to get an object from Earth into outer space, and that the way to do so would be by using rockets. It had not been an easy road to that end, as the notebooks filled up with speculation, questioning, doubting, rejection, and fresh approaches. Early entries wandered the field, broadly exploring the problems of space travel. Often he digressed into such matters as the dangers posed by meteors, how to decelerate and land on another planet, and where and how to obtain food and air.

He also addressed survival during space travel through suspended animation, postponement of death, and ways to achieve immortality, which reflected concern for his mother's health. A chemistry professor showed him they would not work.

Goddard refocused on the basic item in space travel, the vehicle. Balloons and propellers would not get the "car" into space, he decided early on: "At present the thing is impossible. There may be some trick of applying things that are known, or there may be something that is not known, that may change this, but at present the thing is impossible." In March 1906 he nearly gave up, writing in his diary that he had "decided today that space navigation is a physical impossibility." He reverted to mathematics, calculating how much energy would be required to elevate a given weight a given distance. By 1909, he had decided that some form of rocketry would do the trick.[7]

His days filled with schoolwork, his evenings with his diary, "suggestions," and reading, Goddard also expressed the formerly isolated boy's need for attention. He later developed a reputation for being closemouthed, and his authorized biography claimed that he kept his space-travel speculations to himself for fear of ridicule. His life and his paper trail contradict both notions.

Early attempts to gain recognition for his ideas involved several themes for English class that he marketed for publication as soon as Professor Coombs handed them back. One was an essay written near the end of his freshman year, "The Use of the Gyroscope in the Balancing and Steering of Airplanes." It combined his bird-tail observations of 1901 with a proposal to accomplish the same end in airplanes with gyroscopes. Duff liked it, and urged him to send it to *Scientific American Supplement*, which accepted it in 1907 and paid him ten dollars. He later had it reprinted in WPI's *Journal*, and in 1927 he claimed that "So far as I have been able to learn, this was the first suggestion of a gyro-stabilizer for aeroplanes."[8]

Goddard's ego flared with a speculative fiction assignment Coombs gave his composition class in January 1906, on traveling in 1950. Goddard handed in a story called "The High-Speed Bet." Various editors rejected it, but then in November 1909, *Scientific American* accepted it as a condensed and unsigned piece called "Limit of Rapid Transit." Its essential idea was a "tubular railroad," a train traveling in a vacuum tube, suspended by electromagnets and capable of reaching speeds of up to twelve thousand miles per hour.

He would not let it go. He persuaded the WPI *Journal* to publish the full text in 1914, after he had seen an article about a Frenchman, Emile Bachelet, offering a similar proposal. He wrote to the editor of the *Boston Sunday American* to make two points: (1) Goddard had the idea first, beginning with the English theme, and (2) the train could go faster than Bachelet proposed. Near the end of his life, he applied for patents on an "Apparatus for Vacuum Tube Transportation" and a "Vacuum Tube Transportation System." He received them posthumously.[9]

As for Goddard's being shy about his interest in space travel, another theme for Coombs in October 1907 demonstrated otherwise. "On the Possibility of Navigating Interplanetary Space" revived the old high-school essay with material lifted from the green notebooks. He focused on the three problems of sustaining life in space, avoiding accidents, and finding a means of propulsion. Quickly disposing of the provision of air and food and of dodging meteors, Goddard turned to propulsion. That, he said, involved finding a source of energy to accelerate the vehicle to 4.01 miles per second. Combine that with solar energy, "reaction against the ether," and atomic energy, and learn how to control it all, and one could navigate in space. "Thus something impossible will probably be accomplished through something else which has always been held equally impossible, but which remains so no longer."

This mishmash and throwback to his graduation speech was a step backward intellectually. Nevertheless, along with the two other themes, he pointed to it with pride when he was forty-five years old, despite his lack of success in marketing it. He sent it to *Scientific American*, which rejected it as both too long and as omitting subjects the editor thought important.

Still game, Goddard sent it on to *Popular Astronomy*, whose editor replied: "The possibility by your showing is so remote is it worthwhile to publish it? The speculation about it is interesting, but the impossibility of ever doing it is so certain that it is not practically useful. You have written well and clearly, but not helpfully to science as I see it."[10]

Goddard did learn the chief rule of good writing, which is that it is mostly rewriting. He also demonstrated the persistence that would be a hallmark of his adult life. Moreover, his four years at WPI gave him a comfort in and fondness for that institution that he would not find in any other. In the following years it proved difficult to tear himself away from the place.

Yet leave he must. If he were to realize his spaceflight dream, he required further preparation, through a doctoral program in a first-class graduate school.

Across town, his academic record had been recognized at Clark University, where the redoubtable Arthur Gordon Webster chaired the physics department. Webster wanted Goddard in his graduate program, and early in 1909 offered him a place as his student.

Goddard wrote down the pros and cons of leaving or staying at WPI, and talked it over with his father and with Professor Duff. Webster clinched it in March by telling him he could get a fellowship and finish his Ph.D. in one year.

That did it. "Saw Duff in afternoon," he wrote in his diary. "Told him about giving up position." Duff approved, and advised following graduate school with further study in Europe. Goddard rejected that, because it would not pay enough.[11]

THE GENERAL NOTION OF A PLAN

He had already concluded that he knew how to travel much farther than Europe. "The general notion of a plan of raising apparatus without employing the air," he wrote in 1918, "came to the writer while trimming a cherry tree October 19, 1899, in the afternoon in Worcester, Massachusetts. The idea of the rocket action as having possibilities—on January 24, 1909."[12]

He referred to a series of notebook entries beginning on that date. The first sentence said: "If an explosive or deflagrator is burned in tubes in such a way that all its energy is converted into kinetic energy of the particles expelled and the body propelled, it is, theoretically, possible to obtain propulsion." There followed calculations on the basic principles of rocket propulsion. He soon fixed on the idea of a series of "guns" or rockets, firing in sequence.

The whole business turned on two problems—overcoming the resistance caused by the missile's friction with the atmosphere, and finding the most efficient fuel source. In early February he addressed the possibility of "an arrangement of H [hydrogen] and O [oxygen] explosive jets." In June he focused on achieving "continuous propulsion produced by liquids under pressure," possibly by ignition "just before they reached the throat" of a combustion chamber. By the end of the year he had boiled it all down to twenty-six steps toward space.[13]

Goddard later claimed to have been the first to conceive of using rocketry for high flight, and most other details of practical rocketry, including the "step" or multiple-stage rocket. Neither history nor legend supports his claim.

In the case of rocket-propelled human flight, stories going back to the early sixteenth century tell of earlier travelers. The Chinese adventurer Wan Hu, according to a tale from about 1500, took off in a chair propelled by forty-seven rockets mounted around it, aiming to reach heaven. Whether he got that far was not apparent, because he vanished in an explosion a short distance above the ground.

A century later a Turk named Larari Hasan Calebi shot into the sky on a rocket to celebrate the birth of the sultan's daughter. He returned with greetings from the Prophet Jesus, and was richly rewarded.[14]

Both stories are myth, but they reveal the idea as being as old as rockets themselves. A rocket, like a cannon, is an explosion contained in a tube closed at one end. The force of the blast diverts toward both ends of the tube, expressed at the open end by the blast, at the other by the recoil of the tube—the equal and opposite reaction of Newton's third law. The difference between a cannon and a rocket is that in the former the blast expels a projectile from the tube, which recoils with equal force. In the latter case the tube *is* the projectile, propelling itself by the recoil effect on the closed end.

Over two thousand years ago, Chinese alchemists observed the explosive quality of a mixture of sulfur, saltpeter, and charcoal. By between 200 B.C. and A.D. 200, they were setting off firecrackers, which saw their first use on battlefields to frighten horses. By the tenth or eleventh century, Chinese artisans had applied black powder to the first rockets—a "fire arrow" and a six-foot-long "flying fire lance." The former was reported to have dispersed a 100,000-man army in 994, while the latter was described as a two-stage rocket.

By the end of the twelfth century rocket-powered arrows and lances had appeared in a widening variety that included multiple-rocket launchers mounted on wheelbarrows. By 1300 some rockets had become birdlike vehicles with wooden wings and ranges of 1,500–3,500 feet. Within another hundred years the Chinese deployed two-stage rockets with motors mounted in tandem and firing in succession. The range of these missiles was reportedly over a mile, and near the end of their trajectory their warheads burst to release swarms of arrows onto enemy troops.[15]

Along with gunpowder, rocket principles spread to the West by the thirteenth century, and European rockets first flew in battle in Silesia in 1241. They received heavy use in war-torn Italy, where they acquired the name *rochetta* (from a Latin word for "spindle," on account of their shape). Rockets had spread over all of Europe by the fifteenth century, but went into decline because of improvements in firearms and artillery.

Their possibilities did not disappear from view, however. An unpublished manuscript around 1425 proposed, with drawings, a rocket-propelled car powered by a secret gunpowder-based fuel. Seven decades later, Leonardo da Vinci suggested extending the range of a cannon by firing a rocket from it.

In 1529 Transylvanian artillerist Conrad Haas observed that the major obstacle to long range was that as the fuel burns, what remains must carry the dead weight of the missile. He designed a multiple-stage rocket of successive paper cylinders, each one below at the end of its own burn igniting the one above. They would propel what he called a "payload"—an explosive charge of gunpowder for war, or a little house for riding into space.[16]

After that, not much changed in rocketry until the late eighteenth century. In the later Mysore Wars of 1781–99, British troops were bedeviled by Indian bamboo rockets that inflicted many casualties. In response, William Congreve developed an iron rocket fueled by packed powder. Dragging a stick as a stabilizing tail, it boosted a variety of warheads over a range of about two thousand yards. They were in use to shell cities by 1805, but the British deployed them mostly aboard ships—hence the "rockets' red glare" during the attack on Fort McHenry, Maryland, in 1814.

Congreve rockets were wildly inaccurate. Early in the nineteenth century William Hale replaced Congreve's stick with three small vanes in the rocket exhaust, slotted so as to induce spin and thereby increase accuracy. Hale's major customer was the American army, which employed his rockets in the Mexican War and the Civil War.[17]

Hale's rockets were only relative improvements over what came before, however, and inadequate compared to contemporary advances in rifled artillery. They were retarded by their fuel, still that old mixture of sulfur, saltpeter, and carbon known as black powder or gunpowder. It was a limited source of energy.

Shortly before the Civil War, chemists discovered that treating cotton fiber with nitric acid yielded a controllable explosive with much greater power—nitrocellulose, also called guncotton, cordite, or "smokeless" powder. That presaged further advances in explosive nitrate chemistry for propellants, and the first "high explosive," dynamite. Humans now had within their grasp explosive energy of unprecedented power.

The potential of rocket propulsion suddenly appeared unlimited. In 1891, German engineer Hermann Ganswindt published a plausible design for a spaceship, and spent the next decade perfecting it. He proposed using dynamite for motive power, but his design was flawed because his calculations were only approximations.

Before reaching space by rocket could be proved feasible, rigorous mathematics had to address the problem. That, Goddard recognized, was the major challenge facing him in 1909. What he did not know was that another mathematically inclined physicist had beaten him to the punch.[18]

He was Konstantin Eduardovich Tsiolkovsky, born the son of a Russian forester in 1857, in a remote village. He grew up in a happy home, taught by his mother, but by the age of thirteen his mother was dead and he was nearly deaf. He was a voracious reader—Jules Verne was among his favorites—and an inventor of flying devices and machinery in his boyhood.

After he had consumed the village library he moved to Moscow. There he practically lived in the public library, concentrating on higher mathematics, attending lectures, and subsisting on brown bread. He became a village schoolteacher at nineteen, and soon an impoverished family man. In 1881 he fell under the wing of Dmitry Mendeleyev, the inventor of the Periodic Table of the Elements, who taught him chemistry.

During his time in Moscow he inhaled the weirdly mystical atmosphere of the public library. That institution was dominated by the librarian Nikolai Fyodorov, who influenced, among others, Madame Blavatsky, founder of Theosophy. The Theosophists believed that Earth was not the human race's natural home, and that humans more properly belonged in the cosmos. Tsiolkovsky agreed. Finding a way into space became his life's work, which generated hundreds of publications in science, science fiction, and philosophy.

In 1895 he began working on rocketry. He concluded that no powder fuel provided enough energy, and that liquefied hydrogen and oxygen, mixed and ignited in a chamber, offered the best source of propulsive force. The combination of two parts hydrogen and one part oxygen that produces water is the most energetic chemical reaction in nature, and liquefying them would concentrate them. In practical terms, liquid hydrogen being scarcely available, liquid oxygen and any combustible liquid might do.

The fruits of his work appeared in 1903 as "Exploring Space with Reactive Devices," in the Russian journal *The Scientific Review*. The paper presaged everything to come in rocketry, including how to calculate the thrust of the rocket motor, and the velocity required to escape Earth's gravity. He proposed a multi-stage rocket as the most practical design, to shed dead weight as it went. This was all presented with hard facts and rigorously calculated numbers.

Tsiolkovsky had proved that reaching space was not only possible, but feasible. Since his aim was to disperse humans over the cosmos, he also considered using solar energy to provide power to colonies in space. But this appeared in a journal read by few in Russia, and almost no one elsewhere.

Tsiolkovsky remained obscure, despite publishing supplements to this work in 1911, 1912, and 1914, and after Goddard became famous a revised and updated version in 1926. The Communists had no use for his mysticism, but some of his ideas were congenial to their plans to change the world, so in 1919 they appointed him to the Soviet Academy of Sciences. That gave him a pension and the freedom to explore his ideas, mostly in fiction.[19]

Goddard was unaware of his Russian competition, and since he could not know that he had been trumped, he soldiered on. Near the end of the spring semester in 1909, he was assured of the next stage of his life. The fellowship at Clark would give him a living stipend, and the graduate education would be first-rate.

Now twenty-six years old, he was determined to appear more grown up. Before arriving at Clark he grew a thick, dark mustache, compensating for his increasingly glabrous crown. A photograph of him in September 1909 reveals a spare young man, prematurely bald, with fierce mustache, hooded eyes, and jug ears. There was something else: He had acquired the habit of cigar smoking, inherited from his father. Puffing on dark New England cheroots would become a large part of his life.[20]

AND EVERYBODY CONNECTED WITH YOU

Like WPI, Clark University reflected the philanthropy of a Massachusetts boy who had made good. Jonas G. Clark built a fortune in California during the gold rush, returned east in 1864, multiplied his wealth in New York City real estate, and became a famous book collector. In 1878 he moved to Worcester, where he liquidated his enterprises and prepared to found a college bearing his name.

His motives were considerably different from those of WPI's founders. They had wanted to provide practical instruction; Clark planned to combine the best European and American examples of higher education. Clark University, founded in 1887 and opened in 1889, was the second graduate institution in the United States, preceded only by Johns Hopkins in Baltimore.[21]

The institution of dignified brick buildings standing among shade trees became world-famous from the outset, because its founder wanted it so and could afford to foot the bill. He hired G. Stanley Hall away from Johns Hopkins to serve as president, a post Hall manned until 1920. He was a founder of the discipline of psychology, and remained a seminal figure in that field decades after his death. He founded the American Psychological Association, and was a patron of William James. In 1909 he sponsored a famous visit to the United States—which began with lectures at Clark—by Sigmund Freud and Carl Jung.

Hall was oriented toward research and specialization, European-inspired departures from the older ideals of a liberal education. He was also an arbitrary and dictatorial man who left behind a legacy of internecine struggles that plagued Clark's faculty culture for decades.

Moreover, he could be devious. Jonas Clark proposed to build a university for undergraduates and graduate students both, but Hall wanted to preside over only a research institution and graduate school. He declined to establish an undergraduate college until the place had produced enough Ph.D.s to staff it. He kept stalling until 1902, when Clark's will provided money to establish Clark College. Until 1920, Clark University and Clark College existed under the same roof, with separate presidents.[22]

Hall spent Clark's money liberally to hire the best in every field. Nowhere was that more evident than in the physics department, whose first head was Albert A. Michelson, who later became the first American scientist to win the Nobel Prize. He left in 1892, and was succeeded as department chairman by Arthur Gordon Webster.

Born in New England, Webster had studied physics and mathematics in Paris, Stockholm, and Berlin, and was universally regarded as one of the truly brilliant minds in his field. He was a mathematical physicist, but also an outstanding experimentalist, a pioneer in the physics of sound, among other interests. Founder of the American Physical Society, he, like Duff, was an authority on ballistics. The author of many textbooks, Webster delivered polished, comprehensive lectures without notes. He eschewed written examinations but worked his students hard in the laboratory.

During three decades as the head of physics at Clark, he accepted only twenty-seven doctoral candidates, one of whom was Goddard. Thanks to Webster's guidance, Goddard had his understanding of classical and modern physics revised and completed, his previous errors corrected, and his skill in applied mathematics sharpened.[23]

Webster built on a foundation laid down by Duff at WPI. He thereby became the second in a series of guiding influences in Goddard's professional life—a patron, in the fullest sense of that term. Just as his mother and grandmother had looked after the details of daily living so efficiently that he never learned to fend for himself, Goddard the scientist required guidance from someone more worldly wise as he pursued his scholastic interests. Duff and Webster got him through World War I, to be succeeded by three equally important advisers—Charles G. Abbot, Charles A. Lindbergh, and Harry F. Guggenheim. Throughout his life Goddard depended on others to lead him through the real world.

He arrived at Clark at a momentous time in September 1909, a five-day celebration of the university's twentieth anniversary. The affair attracted a gathering of eminent scientists from all over, including Nobel laureates Michelson and Ernest Rutherford. Goddard attended lectures delivered by those and other luminaries, including one by Mars visionary Percival Lowell on the planet Venus. It was a heady intellectual brew surpassing anything he had encountered before. When Freud, Jung, and James lectured later that fall, however, Goddard was not interested. Physics, not psychology, was his life.[24]

His scholastic work continued on lines laid down at WPI. With a thesis called "Theory of Diffraction," he received his M.A. in June 1910. After he defended his dissertation, "On the Conduction of Electricity at Contacts of Dissimilar Solids," a year later, Webster told him: "You did yourself proud and everybody connected with you." Similar compliments came from other professors on his committee. On 15 June 1911, he received his Ph.D. at graduation, the only one there who obtained his degree cum laude. The next day he waxed ecstatic when he was addressed as "Doctor" for the first time.[25]

Then he seemed at a loss as to what to do next. The University of Missouri offered him a job at $1,000 per year, but he turned it down; the salary was not enough to draw him away from home. There, everything seemed to be looking up except for his mother's health. Nahum—contradicting the story of tuition borrowed from a grocer—had prospered at Hardy Company. After Robert's graduation he moved Fannie into a new house he had built for her on Bishop Avenue. She now suffered from arthritis as well as pulmonary tuberculosis, so he designed the structure to accommodate her afflictions, and bought a car modified to carry her wheelchair. Goddard remained at Maple Hill with his grandmother, spending the summer relaxing, working on his notebooks, and at the end of the season catching his first sight of an airplane.[26]

Despite his aimlessness, Duff and Webster looked out for him. In August he went to see Duff about returning to WPI as an instructor, but his mentor could offer him nothing more than tutoring. A month later, he telephoned Goddard about a position at Columbia University paying $1,600. After he bought a new hat and went to interview for the job, Goddard turned down the offer when it came—he said it involved too much teaching and not enough time for research. Webster, meanwhile, secured him an extra year at Clark as an honorary fellow in physics, which would keep him employed until the spring of 1912.

Goddard was not lazy, merely a homebody who resisted leaving the family nest that had sheltered him for three decades. His advisers kept after him, and almost reluctantly he came around. When Webster persuaded him that it would be a good way to make contacts in his field, Goddard attended the December 1911 meeting of the American Association for the Advancement of Science (AAAS) in Washington, D.C. It was his first overnight trip alone away from Worcester.

Webster next encouraged him to deliver a paper, "On Mechanical Force from the Magnetic Field of a Displacement Current," at the American Physical Society meeting in Cambridge, Massachusetts, in April 1912. When he got home, he learned that Duff could offer him a teaching job at WPI, while continuing to recommend him to other universities. Webster went after a fellowship for him at Princeton University, a nursery for the sons of privilege then transforming itself into a leading research institution.[27]

While Webster was pursuing opportunities on his behalf, Goddard started work on his first patent application, and began a long relationship with Charles T. Hawley, a partner in Southgate and Southgate, one of Worcester's major patent law firms. In August he gave Hawley the materials on a "Method of and Means for Producing Electrical Impulses or Oscillations." The Patent Office issued him Patent No. 1,137,964 on 4 May 1915. Although the first applied for, it became the third patent in his name.

What it covered was a concept and design for a charged-particle generator or oscillator. He conceived it as a way of measuring high-frequency oscillations, but it was also an alternative to the vacuum tubes developed successively by Thomas Edison, Arthur Fleming, and Lee DeForest that culminated in the Audion. That signal-generating foundation of the radio industry would support a fiercely defended claim to monopoly on the part of the Radio Corporation of America (RCA).

Goddard sold his interest in the patent to the Westinghouse Company in 1920, and forgot about it until it reentered his life in the middle of the next decade. When it reappeared, it changed the history of electronics in America.[28]

AN INDELIBLE IMPRESSION ON MY MIND

While the patent application was underway, Webster and Duff in concert persuaded Princeton to offer Goddard an instructorship. Actually, it was that only in name, as the university intended him to do more research than teaching. W. F. Magie, head of Princeton's Palmer Physical Laboratory, invited Goddard to pursue whatever research appealed to him, although he was expected to continue in electronics.[29]

"Princeton University," he remembered later, "with its Palmer Physical Laboratory, its beautiful buildings and spacious campus of old trees, created an indelible impression on my mind, as did also the quiet town and the exclusively college atmosphere." Not only was the place congenial, but he would be free of homesickness there. He wrote his mother immediately on arrival 10 September 1912. The next day he went to a boarding house and rented two rooms—one for himself, the other for Gram. She would take care of him wherever he was.[30]

Goddard spent busy days at the Palmer Lab, and nights exploring rocket science. He devoted equal concentration to both, but rocketry was uppermost in his mind. When he visited his parents in their new home over Thanksgiving, he dropped in on an old classmate and told him little of Princeton but a lot about rockets. "If you can send a rocket off the earth," he was gratified to hear from his friend, "it will make a sensation. Remember, Jules Verne used a gun."[31]

Goddard had other means of propulsion in mind. Among possible sources of energy, he even considered atomic power, then still a fragmentary area of science, and concluded: "If none of these are possible, solar plus chemical energy gives the only possibility."

The notebooks continued to absorb ink. "During the evenings of this year," he said later, "I worked on the theory of rocket propulsion, assuming that with smokeless powder, and hydrogen and oxygen, an efficiency of 50 percent could be secured, which I later found by experiment to be true." The "efficiency" that concerned him was a measure of the proportion of energy released by fuel combustion that could be translated into kinetic energy moving the rocket. Energy not so employed was wasted.

On 19 October 1912, Anniversary Day, Goddard outlined what lay ahead: "Order [of further work]: air resistance theory; calculate for guncotton; calculate shape of jet from entropy of perfect gas, and the proportion that is steam; design feeding mechanism and cartridges." Every evening he entered notes, ideas, calculations, arguments, and digressions on how to deal with spaceflight problems. Subjects included protection against meteors; control of acceleration and its physiological effects; use of barographs and thermographs during altitude trials; using cameras from orbit around other planets; returning spacecraft to earth; attaining high altitudes in general; and electrical, solar, solid-chemical, and liquid-chemical propellants. By February 1913 he believed that he had worked out the mathematics to show that attaining high altitudes by rocketry was indeed possible.[32]

His days at Princeton were interesting and productive. On 3 March

1913, the faculty and students saw the president of Princeton, Woodrow Wilson, off at the depot, on his way to become the president of the United States. Ten days later Goddard demonstrated the "Production of Light by Cathode Rays" to the faculty. Nights were consumed by the "theory," or as he also called it, the "problem."

Just before spring break began on 19 March, he did something else. Although he had not sorted his ideas out yet, he began work on a patent application for a rocket.

Then his world, quite literally, nearly came to an end.[33]

THREE

SOME FIRST-RATE WORK YET

I am thrilled,
and yet my mind
trembles with fear
at seeing
what has not been seen before.
—*The Bhagavad-Gita* XI:45

Young Robert Goddard, chafing at confinement, had climbed a
cherry tree and let his mind soar among the planets. Now, in 1913,
he found himself confined again. Never physically adventurous, somehow
he summoned up enormous determination and beat off an enemy that
would dog his trail to the end. What happened to him then, and how he
overcame it, forged the last elements of Goddard the man.

HE WILL HAVE THE VITALITY TO PULL
HIM THROUGH

Goddard suffered a series of bad chest colds during the winter of
1912–13. He visited his parents in March, coughing heavily. Nahum summoned Dr. Edgar A. Fisher, the family physician, who called in Dr. George
N. Lapham, a tuberculosis expert. They agreed that he had come down with
the disease that was killing his mother, and as Goddard learned later, they
gave him two weeks to live. He wrote to Gram in Princeton and asked her
to pack up his papers and bring them to him, because he feared that he
might die without leaving behind proof of his ideas. Nahum engaged a nurse
to tend to him.[1]

Goddard's boyhood penchant for demonstrating the truth of "impossibilities" drove what happened next. He engineered his own regimen, devised experiments in deep breathing, and spent most of his time on his parents' veranda, heavily wrapped while filling his lungs with the frigid winter air.

He was visited by Dr. John C. Hubbard, a physics professor in Clark College, and Dr. Harold F. Stimson, Webster's research assistant. The trio became close, as Goddard exercised his offbeat sense of humor. Observing their respective initials, he declared that they constituted a Holy Trinity. Goddard was God, Hubbard was J.C., the Son, and Stimson was H.S., the Holy Spirit. "Underneath, Bob was a free thinker and childish," Stimson remembered years later.

The memory also endured for Hubbard, who wrote after Goddard's death in 1945: "I shall never forget the smile and twinkle from the depths of a hammock, when there were several inches of snow on the ground and the frosty air was around zero. The thing that affected me most was that so far as we know he had no medical authority for his action (living on the veranda). It seemed to be his own idea entirely and he had in no uncertain terms absolved his family and his doctor from any responsibility."[2]

He improved steadily, and by late March was back at work on his rocket theories. By May he was eating with the family, and completing the materials for a patent application. Two decades later, Goddard entered the following in his diary: "Note for Autobiography: Dr. Lapham, T.B. expert, who examined me when I was sick in March 1913, examined me again in June 1913, and said he never expected to see me as well as I was...but that I should never do any more research, but live a good deal out of doors. I told him there were some things I just couldn't help working on, and would try them."[3]

He took long walks in the country beginning that summer, and by the spring of 1914 his illness was almost over. All that was left of his hair was a fender around the sides and back of his head, his weight was 134 pounds and his clothes were baggy, and he was hollow-chested and stoop-shouldered. Gram also showed the strain, worn down from worry and her years. She refused Nahum's offer to take her in at his house, so Robert hired a housekeeper for her, and visited Maple Hill daily. He also negotiated through Hubbard with Edmund C. Sanford, president of Clark College, for a part-time position.

Dr. Lapham examined him again. "If you go five years you will probably never have a recurrence," the doctor told him, "and will be in better condition than ever before." Lapham told Dr. Fisher that "what happens to Mr. Goddard will depend entirely on his past history. If he has been a man of good habits and has led a clean life, it will be worth a million dollars to him, for he will have the vitality to pull him through."[4]

Lapham might have addressed Goddard's conduct in the future, but he assumed that someone with a history of pulmonary disease would have sense enough to avoid irritating his breathing apparatus. Defying good sense, however, Goddard resumed his previous habit of smoking cigars, not because he was foolish but because he could not help himself—he was addicted to them.

The signs were all over him. Except for the weather, in his diary he mentioned no other subject as often as his smoking. Those who worked with him in the 1920s and 1930s all remembered the frequency of his lighting up. And witness the way he smoked, in the words of Paul Horgan, who knew Goddard in the 1930s:

> [H]e always smoked cigars. He loved cigars and would smoke them down to the last pinch. I never saw anybody smoke a cigar making every draft on it a conscious drawing. It was a curious and funny thing—he loved smoking cigars so that to watch him do it was almost to do it yourself. And yet I always felt that this had something to do with the curious sound of his voice and the trouble in his chest. I think he knew that cigars weren't good for him, and he used to recite a funny poem about "tobacco is a filthy weed, I like it."

"His voice," Horgan continued, "was very husky, a light voice, and he seemed to be speaking through a filter or fog in his throat, which I traced to the TB. He did have an effortful way of speaking. He had to come clear with his words through a foggy trachea in some way."

Goddard deflected mention of his addiction through humor and the poem Horgan could not quite remember. It was old doggerel composed by Dr. Benjamin Waterhouse: "Tobacco is a filthy weed, That from the devil does proceed; It drains your purse, it burns your clothes, And makes a chimney of your nose."[5]

This was a behavior he could not control, and which he aggravated by being both a social and a solitary drinker.[6] Goddard accordingly redoubled his efforts to govern everything else about his life, or at least those parts of his life that the women and his patrons did not take care of for him.

THE PROBLEM OF RAISING A BODY

On Anniversary Day (19 October) 1913, fourteen years after the vision in the cherry tree, Goddard outlined the next steps he had to take in his reach for outer space: "Complete patent application if necessary of nozzle and plurality; take out application on reloading feature; also complete application for electric pump; repeat calculation carefully, for smaller intervals; look up [Sir George] Darwin's theory of the lunar motion; and look up meteors. Also try a jet." He still could not focus his mind on a single problem, or organize a systematic approach.[7]

His first application for a rocket patent had already gone to Washington, and he was at work on another. The first, issued 7 July 1914 as Patent No. 1,102,653, "Rocket Apparatus," covered everything to do with rocketry more sophisticated than simple fireworks or a ship rocket. It established his claim, in American patent law, to priority for the idea of a "step" or multi-stage rocket, wherein propulsive charges fire in succession to attain the desired velocity at the end—the principle that carried men to the Moon.

This idea was not original to Goddard. A Belgian patent had covered the same thing in 1911, and fireworks makers had dabbled with successive propulsion for centuries.

Goddard accomplished something more fundamental, however. His most significant achievement in the first patent was to adapt a nozzle to the rocket. He selected the De Laval steam-turbine nozzle, a cone that used the expansion of exhaust gases to do work. It allowed the gases to exert pressure against the cone after they had left the chamber, increasing the power of the rocket.

It also redefined the rocket motor as a combination chamber and nozzle, rather than just a charge in a tube. Goddard's idea presented, in the judgment of technology historian Peter Alway, "not only the most efficient rocket to date, but the most efficient heat engine to date," a heat engine, broadly defined, being any device that converts heat (as from combustion) into mechanical or electrical energy.

Efficiency is essential to long-range rocketry. Goddard realized that there were two factors involved in making a rocket practical. One was raising the fraction of mass that was propellant (conversely, reducing the proportional weight of the unfueled rocket). The other was raising the speed at which exhaust gases were ejected from the rocket motor, increasing the reactive force propelling the rocket.

The more combustion energy converted to kinetic energy (energy of motion) driving the rocket—as opposed to energy wasted as heat or noise—the higher the efficiency of the rocket. The most efficient rockets of Goddard's day converted about 2 percent of their energy into thrust. His chamber-and-nozzle combination alone increased that nearly tenfold.[8]

The second patent, also for a "Rocket Apparatus," went back and forth between Goddard, patent attorney Hawley, and the Patent Office through the fall, winter, and spring of 1913–14. At first Goddard emphasized the design of a breechblock mechanism to load successive solid charges into a firing chamber, making the rocket a flying machine gun shooting blanks.

On 26 March he had another idea: "Thought of adding fluid part to Rocket Patent in morning," he recorded in his diary. He added to his application a description of a complete liquid-fuel system for his rocket motor. Others had considered the possibility in broad terms before, but Patent No. 1,103,503, issued 14 July 1914, made Goddard the "father" of liquid-fuel rocketry.[9]

"These two patents are worthy of special attention, in passing," Goddard said in 1927, "for they give as nearly as possible an answer to the question as to what the 'Goddard Rocket' is." That boiled down to "the three broad principles covered in the claims of these two patents, namely the use of a combustion chamber and nozzle; the feeding of successive portions of pro-pellant, liquid or solid, into the combustion chamber, giving either a steady or a discontinuous propulsive force; and, lastly, the use of multiple rockets, each discarded in succession as the propellant it contains is exhausted."[10]

At the time he wrote that, he had succeeded only twice in getting a rocket off the ground, and it had not used the third element—which he never conceded was not original to himself anyway. The three principles by that time enjoyed wide currency in rocket literature, especially in Europe. Goddard was saying clearly: "I had it first!"

Twelve years later, he proclaimed: "It was quite evident that these prin-ciples would have to be followed if practice were to catch up with theory. In 1914 I obtained two patents in which these principles were set forth, *indicat-ing that extensive work along these lines had not previously been undertaken* [emphasis added]."[11]

Goddard's lifelong emphasis on claiming priority for his ideas—not just through patents, but by notarized and dated entries in his notebooks—bothered many of his colleagues, understandably, because science has always emphasized exchange of ideas through publication. His motives were mixed, and they evolved.

Patents go back to the beginnings of the scientific revolution. They encourage invention by rewarding inventors for their creativity, via their exclusive claims to profits or royalties from use of their ideas. They also encourage further progress by limiting the claim to a certain period, so that others can improve on the original without fear of an infringement chal-lenge.[12]

A patent is the opposite of secrecy, as it is a publication of an invention. At the same time, it grants the inventor the right to assert control over his creations. That points to the simplest explanation of Goddard's focus on patents: He was mercenary, as when he rejected Duff's suggestion that he study in Europe because nothing there would pay enough. Goddard was a Massachusetts Yankee. Ever since Emerson had talked about building "a better mousetrap," his was a culture that encouraged invention and the garnering of honest profits. Any idea that had the remotest possibility of paying off was perforce patented.

There was also that once-isolated boy's desire for recognition. A patent was another way of saying "I had it first!" and requiring others to agree. This narcissistic compulsion joined with practical considerations. Goddard, who claimed to have invented the aileron, observed the Wright Brothers' defense of their patent on wing-warping aircraft through a succession of lawsuits. The flagship case, against Glenn Curtiss, ended with a secret out-of-court settlement rumored to exceed a half-million dollars. Goddard regretted not patenting the idea instead of writing a letter to a magazine all those years ago.[13]

Patents were not the only things on Goddard's mind in 1913 and 1914. There was also his convalescence. That he had a way to go was demonstrated by a project that occupied most of September and early October 1913, an essay he called "Outline of Article on 'The Navigation of Interplanetary Space.'"

Drawing upon a prediction by astronomer Sir George Darwin that the Moon would someday crash into the Earth, Goddard began: "From an economic point of view, the navigation of interplanetary space must be effected to insure the continuance of the race; and if we feel that evolution has, through the ages, reached its highest point in man, the continuance of life and progress must be the highest end and aim of humanity, and its cessation the greatest possible calamity."

What followed was less mystical, but still sophomoric. The paper drew upon his childhood daydreams and student term papers to address the entire problem of spaceflight. Much of it was a rambling consideration of various means of propulsion based on his earlier speculations. Goddard thought an ion or atomic-energy propulsion system might derive from his oscillator tube. Then he thought a guncotton-driven rocket might prove feasible, because liquid hydrogen and oxygen were for the present impractical. The Moon, he said, offered a place to make a soft landing, to manufacture hydrogen and oxygen, and to stage further voyages to the planets. Finally, solar energy was a possible source of power for planetary flights. He closed with brief attention to the problem of providing air and food, and the need to dodge meteors.

This exercise in confusion suggested that Goddard still suffered the enervating effects of tuberculosis. Emphasizing space travel rather than how to get there, he did not focus on just what would be involved in getting an object off the planet. He speculated on what kind of a rocket would work, but not whether the whole idea was physically possible. As he retitled it, this essay was just an "Outline" and not a solution to his "problem." That would come soon enough.[14]

First came practical details of life. He continued taking long walks, which improved him physically and gave him time to organize his thinking. During those excursions he worked out the details of an article for the *Physical Review*, after which he had time "to think over my main problem."

Despite his physical improvement, he was still a homebody, and reluctant to leave Worcester with his mother and grandmother ailing. So he burned his bridges at Princeton, writing to Magie about the state of his health and his inability to return and perform hard work.

Magie was understanding. He expressed pleasure at Goddard's recovery so far, "though my delight is tempered with regret that you feel yourself incapacitated from taking up intensive work again. I felt that you had a great future before you as an investigator. Maybe by working patiently but not long at a time you may turn out some first-rate work yet."

Magie joined Webster, Hubbard, and Stimson in urging President Sanford to give Goddard something at Clark matching his reduced stamina. In May Sanford did—one year at $500. "I wish to say," Goddard told Sanford, "that the terms of this appointment appeal to me as being very fair to all concerned." Magie let him know that he could return to Princeton whenever he could carry a full load.[15]

Goddard's instructorship would begin in the fall of 1914. It gave him the prestige of affiliation at a top-ranked university, with minimal demands on his time, and the additional advantage of laboratory space and some research money. It was a sinecure in the guise of occupational rehabilitation, and he could remain at home.

Merely having it at hand seemed to energize him, and to clear his mind as well. Walking through both the countryside and his "main problem" that summer, Goddard worked his way out of the earlier confusion, to produce what would become his magnum opus, under the title "The Problem of Raising a Body to a Great Altitude Above the Surface of the Earth," drafted between 10 and 21 August.

It began: "The problem is concerned with the practicability of doing two things: The raising of apparatus, such as recording instruments, to a great altitude, and letting it fall back to the ground by suitable parachutes; and second, the sending of apparatus to such great distances from the earth that the apparatus comes under the influence of the gravitational attraction of some other heavenly body."

That was an amazing turnaround from everything Goddard had written before. An essential truth had hit him like a thunderbolt—it was fruitless to talk about roaming space until he figured out how to get there. For the first time, he had placed first things first.

Now Goddard was the consummate applied physicist. He briefly mentioned the value to be gained by finding some better way than balloons to study the upper atmosphere, or telescopes to study other planets. He marshaled a logical progression of known facts and principles along with graphs and tables to determine what total mass of rocket would raise one pound of payload to given altitudes. He considered everything from Newton's laws and the conservation of energy, to coefficients of friction, atmospheric pressures, and the efficiencies of potential fuels. He assembled a mathematical proof that *it can be done—a rocket can make it into space.*

Perhaps a startling epiphany—some striking vision like that of the cherry tree—had suddenly turned Goddard from mystic, dreamer, and fantasist into the complete physicist. There is no record in his writings of such an event. Nor is there evidence that anyone showed him how to organize his approach to his "main problem." Perhaps it was a product of his recovery from the tuberculosis.

Whatever the cause, he somehow learned to sweep away the nonsense and settle down to business. The clarity, logic—indeed, elegance—of his treatment of the subject was reflected in the subheads of his text:

IMPORTANCE OF THE SUBJECT; METHOD TO BE EMPLOYED; STATEMENT OF THE PROBLEM; DISCUSSION OF THE EQUATION OF MOTION; INTEGRATION OF THE EQUATION OF MOTION, AND CONDITIONS FOR A MINIMUM INITIAL MASS; SOLUTION BY APPROPRIATE METHOD; VALUES OF THE QUANTITIES OCCURRING IN THE EQUATION; DIVISION OF THE ALTITUDE INTO INTERVALS; CALCULATION OF MINIMUM MASS FOR EACH INTERVAL; MINIMUM MASS NECESSARY TO REACH VARIOUS ALTITUDES WITHIN THE EARTH'S ATMOSPHERE; MINIMUM MASS REQUIRED TO RAISE ONE POUND TO AN "INFINITE ALTITUDE" WITH MODERATE ACCELERATION; INITIAL MASS REQUIRED TO SEND ONE POUND TO THE SURFACE OF THE MOON; MINIMUM INITIAL MASS FOR LARGE FINAL MASSES.

Goddard boiled the whole question down to determining how big a rocket would be required to raise, after shedding mass as combustion products, one pound. But this was all theory. The next problem would be finding fuels efficient enough to do the job, and a design that would work:

The calculation ... has been carried out for two propellants: guncotton and a mixture of hydrogen and oxygen. The former has been chosen owing to the fact that, mixed with small amounts of other substances, it forms nonexplosive but rapidly burning substances of high heat value; and the latter because it affords the greatest heat energy per unit mass of all chemical transformations. If this were employed it would have to be liquid, owing to air resistance due to the large volume that would be occupied by gaseous hydrogen and oxygen. Means of using liquid hydrogen and oxygen are, however, not herein discussed.

Goddard also briefly addressed a two-stage rocket, the second smaller than the first. Despite his claim in 1939, this now familiar form of rocket was not so apparent in the first two patents. In this paper it took recognizable form as succeeding mechanical stages bearing fuel charges, as opposed to the broader idea of succeeding charges. Finally, having proven that sending a projectile as far as the Moon was theoretically possible, Goddard considered how to verify success. He suggested sending up enough flash powder to be observable by telescope from Earth.[16]

This performance exceeded anything he had accomplished before. Despite his extravagant claims to originality on bird wings and gyroscopes, this time he really had produced something original and scientifically important. He had replaced fantasy with science.

He had the manuscript typed in multiple copies and handed it to Webster and Duff to review. While he awaited their reactions, he began his part-time work as an instructor, and started turning rocket theory into practice.[17]

TAKE CHANCES, AND DO WHAT WE CAN

His employment was as a physics instructor in the undergraduate Clark College. He taught a course on electricity and magnetism for three hours per week, and met with Dr. Hubbard one hour weekly to discuss his laboratory work. "This gave me," Goddard remembered, "considerable leisure time, and besides walking, I worked out the theory and calculations for smokeless powder and for H[ydrogen] and O[xygen] completely, and began experiments on the efficiency of ordinary rockets."

This academic featherbed did not relieve him from the sense that doom waited around the corner, a lasting effect of the tuberculosis. "It's appalling how short life is," he lamented in January 1915, "and how much there is to do one would like to do. We have to be sports, take chances, and do what we can."[18]

Once again he lost focus and struck out in several directions. The most important new emphasis was a return to a subject partly elided in his paper "Problem of Raising an Object"—the energetic efficiency of hydrogen and oxygen. Checking his calculations with Webster and Duff, he concluded that the most energetic chemical reaction in nature offered the highest potential for rocket propulsion.

"Curiously enough," he recorded in 1927, "the initial mass to send one pound to infinity, for hydrogen and oxygen at 50 percent efficiency—namely 43.5 pounds—was closely that estimated roughly (45 pounds) on January 31, 1909." Always reluctant to concede the possibility of youthful error, he meant that his figures had been correct from the beginning.[19]

He could not yet tackle gas or liquid fuels, for want of background and budget. Instead, he experimented on solid-fuel rockets, an area of applied science that he would pioneer. First he must determine whether he was starting out on the right track. In January 1915 Webster was not entirely persuaded. "He couldn't find a flaw in it," Goddard recorded in his diary, "but seemed to think it unreasonable. Suggested talking to Duff."

His WPI patron was more enthusiastic, and asked Goddard to address a physics department colloquium on his findings. Goddard talked about sending cameras and instruments to high altitudes with his inventions, earning himself an effusive response in the local press, which predicted: "A veritable new science will be opened up if the proposed machine proves to be as practicable as the inventor hopes."[20]

Thus encouraged, he assembled a collection of commercial rockets ranging from fireworks and signal flares to "ship" rockets used to pass lines between vessels at sea. He talked DuPont Chemical into sending samples of its smokeless powders, the Maxim Silencer Company into providing its new powder, and various others into lending their wares. Tests of rockets and propellants were underway at the Clark physics lab by early 1915, raising smoke and noise to the consternation of all around. When Reed College, in Oregon, recruited Goddard in March, he turned the offer down, claiming that he did not want to interrupt his research.[21]

Almost every day in 1915 and 1916, Goddard conducted an experiment of some sort. Mostly, he tinkered, measuring the explosive power of various powders, adjusting the size and shape of nozzles, poking around the physics of rocketry. The most practical work involved building the first of several "ballistic pendulums," hanging weights with scales to measure the thrust of his small rockets when secured to the pendulum. In April 1915 he tested a "gas-gun" behind the barn at Maple Hill and concluded: "Need better fuses." On the tenth he "went to Clark and tried 3 ship rockets . . . scared janitor."

He recorded everything meticulously and discussed his results with physics and chemistry faculty at Clark and WPI. He also subjected Webster, still skeptical about the point of it all, to demonstrations of his work. He could not resist the fun part, however. On 16 June: "Went with [Stimson and a student] to Coes Pond and tried rockets for altitude in afternoon. One #9 went right across the pond—alt. = 486 ft. approx."[22]

The joy of pyrotechnics did not distract Goddard from patenting his ideas, and he completed the materials for five successful applications from fall 1914 to summer 1917. The first, once he renewed his attention to hydrogen and oxygen, was for a device to produce them by electrolysis. The next three covered further advances in motor design, one a "magazine-rocket." The last of the five was for a way to produce electrified streams of gas to drive spacecraft. He now held nine patents, pending or granted, in electronics and rocketry.[23]

More important than the patents were experiments during 1915 that proved that a rocket would provide thrust in a vacuum. In the simplest form, Goddard set up a .22-caliber revolver loaded with blanks, mounted on an arm swiveling around a spindle, all in a bell jar from which the air had been withdrawn. When the pistol was fired by pulling a string attached to the trigger, the whole assembly twirled around. Just when he first devised this simple demonstration is not now apparent, but he repeated it often during the 1920s.

Science required rigorous measurements and impeccable experimental design to prove the point, so Goddard devised a large vacuum chamber with a rocket (at this stage, he usually called it a "gun") mounted inside, and fired when the chamber had been evacuated. Gauges measured the thrust caused by the rocket's lifting of the top of the chamber. Next, to prove that the motion observed was not the result of exhaust rebound, the experiment was repeated in a large circular vacuum tube.

More than fifty tests revealed that rockets provided about 20 percent more thrust in a vacuum than in air. More important, in Goddard's words, the work proved "that the phenomenon is really a jet of gas having an extremely high velocity, and is not merely an effect of reaction against the air."

This should have been obvious, but it was not, even to some physicists. Proving that a rocket could operate in a vacuum was a stellar achievement. Demonstrating that rockets could operate independent of their environment was of immense importance for the future of the space age.[24]

IT WAS ALMOST IMPOSSIBLE TO TURN HIM DOWN

The space age lay in the unperceived future. More immediately, there was a living to be earned, and Goddard persuaded President Sanford that he could carry a heavier load. In May 1915 Sanford promoted him to assistant professor, his pay "proportional to the amount of work which you do up to a total of $1,125." A year later, with Hubbard departing, Sanford told Goddard that he would recommend him as acting head of the college physics department. He still ranked assistant professor, but his pay rose to $1,500 for one year, "provided that your health permits you to carry full work."[25]

Teaching was always a secondary occupation for Goddard, but in the beginning he was good at it. Percy M. Roope arrived at Clark in 1916 uncertain about where his interests lay, and sampled what the faculty had to offer. As he recalled a half century later, "after the first day's class work I knew what I was going to do. I knew whom I must study under; his name was Robert Hutchings Goddard."

Goddard's classroom style combined physical demonstrations with Socratic questioning to make the students think, and that ignited a fire in some of them. "There's no question but that he was the greatest teacher that I have ever known," Roope said, "and I am quite certain that many others would agree that he was the greatest teacher they had known."[26]

Such extravagant praise reflects the solicitude that Goddard always showed to his subordinates, whether students or assistants. He never failed to answer a question or explain some theoretical or technical point. He may have struck his contemporaries as an oddball, but always he was courteous and solicitous, especially to youngsters. So recalled Hugh L. Kennleyside, who earned a master's at Clark in 1921:

> *In person he was mild, courteous, introspective, and almost wholly absorbed in his professional studies. He was generally looked upon as being slightly but harmlessly mad, with his absurd ideas of designing rockets that could travel to the moon as an initial step toward his ultimate objective, Mars. It is hardly necessary to add that at the time I knew Robert Goddard I had no remote conception of the range and precision of the amazing mind and imagination concealed behind the façade of the pleasant professor who, in spite of the fact that I was not one of his students... never failed to greet me with friendly courtesy. He remembered me and frequently expressed interest in the progress of my studies and my response to life at the university.[27]*

Nils Riffolt, a Swedish immigrant and aide to Webster who became one of Goddard's technical assistants, observed other facets of the man's personality: "He was always in a hurry. He worked hard at whatever he was doing. For instance, I never saw him walk slowly. Hot weather bothered him, he perspired freely, but it never slowed him down. On the other hand I did not perceive any sign of impatience. He appeared to take ups and downs calmly. My most vivid impression of Goddard was his sense of humor, his ready laugh, and his ability to be elated like a youngster over small incidents."[28]

Riffolt was a talented instrument maker who could turn out nearly anything on a lathe. Operating on a shoestring budget in a complex technical area, Goddard persuaded Webster to lend Riffolt's services on the cheap.

That reflected something else about the rocketeer's personality. Growing up in an environment of extreme solicitude, on account of his supposed frailty, Goddard learned to manipulate other people to serve his own ends. It was a talent that made him an effective scrounger.

L. C. Leach, who owned an industrial laboratory in Worcester, recalled the scrounger at his best:

It was almost impossible to turn him down. Once he brought us two dozen of his gunpowder mixtures, and talked us into running thermal tests in our bomb calorimeter [a device for measuring the heat content of substances]. We used it mostly for testing coal samples and didn't much care for testing explosives, which seemed likely to blow up our equipment and ourselves as well. But Bob lectured us on the relative forces of coal in oxygen as compared to gunpowder. As he said, coal turned out to be more explosive. So we ran off his tests, not knowing what he was up to, but feeling sure that he did.[29]

Goddard retained his mystical streak. On 19 July 1915 he reread H. G. Wells's *The First Men in the Moon*, which he read again ten days later. Then, on 8 August: "Dreamed at 6:15 A. M. of going to moon, and interested, going and coming, on where to land respectively on moon and earth. Set off red fire at a prearranged time and place so all can see it. Was cold, and not enough oxygen density to breathe—got into chamber for a while. Saw and took photos of earth with small Kodak while there—two for stereoscopes—and glimpsed earth once during return—South America? Not enough oxygen when I opened my helmet to see if so."[30]

Such visions, including breathing difficulties normal to his tubercular history, did not deter him from rocketry, and by late 1915 he was dedicated to realizing the ideas expressed in his patents. He would perfect solid-fuel rocketry by discovering the optimum fuels and motor design, and he would devise a multiple-charge, solid-fuel mechanism. Over the next year he pursued his experiments, with a general organizing principle not always apparent. "Tried gun in large tank with black-powder primer in afternoon," he recorded 25 February 1916. "It went! Vel[ocity] = 5890 ft/sec, efficiency 39%." Four days later, "Tried gun with double fuse wire, 23 volts 5 gm, *no* black powder primer. It doesn't go."

In April, he presented to the Clark faculty an abstract, "Jets of Gases Having High Velocities." According to his diary, "Webster said, 'Very ingenious idea, and you've done a lot of work on it!'—Didn't believe the average person realized how small the efficiency of the ordinary rocket was—said he didn't either."

His overriding concern remained rocket efficiency, to be achieved by the design of combustion chambers and nozzles. In September he claimed an efficiency—combustion energy turned into motive force—of 63 percent, compared with that of a reciprocating steam engine at 21 percent and a diesel engine at 40 percent.

Also pertinent was finding the appropriate explosive, not to mention figuring out how best to install it in a rocket. He asked the Maxim Company for advice on making nitrocellulose cartridges safely. Hudson Maxim answered patronizingly that if the charge was ignited in a steel cylinder, "you would be able to confine the combustion to one end of the combustible, and by choking the exit orifice, could burn it as fast as you liked, and actually at the rate of combustion of smokeless powder in a gun, which we call explosion."[31]

He was equally interested in igniting a flash on the Moon, as mentioned in the "Problem of Raising" paper. Beginning in December 1915, he ran a series of experiments on the visibility of burning magnesium, by quantity and with admixtures, at various distances. Tests in the laboratory yielded indifferent results.

One night in the fall of 1916 he sent two graduate students to the truck farm of "Aunt Effie" Ward, a distant relation, in Auburn about two miles from Gram's house, where he stationed himself. As he watched, they ignited various flash powders. After measuring the brightness, Goddard went from feeling optimistic that a few pounds of the stuff on the Moon could be seen from Earth, to the less-hopeful conclusion that he would need about one hundred pounds of it.[32]

That was not his only discouragement that fall. His grandmother died at home on 17 October, at the age of eighty-three. After a funeral in her parlor two days later, she was buried in the family plot at Hope Cemetery. Goddard walked from the graveyard to the town center, then "Walked over by lane—saw cherry tree." It was Anniversary Day.[33]

FOUR

EXTREME ALTITUDES

Then is it like, they will set forward to invade the territories of the Moon, whence, passing through both Mercury and Venus, the Sun will serve them for a torch, to show the way from Mars to Jupiter and Saturn.

—François Rabelais, *Pantagruel*, IV:51

In January 1915, Goddard's patron Duff had reacted with enthusiasm to his paper "Problem of Raising an Object," and urged him to seek financial support. During the next two years Goddard approached the Aero Club of America and the National Geographic Society, without result. He needed to sharpen his skills as a salesman.

I REGARD THE SCHEME AS WORTH PROMOTING

In September 1916 he wrote a long, illustrated letter to the secretary of the Smithsonian Institution, the distinguished paleontologist Dr. Charles D. Walcott. It began: "This communication I had intended sending a little later, but I feel that it would not be desirable to delay any longer. Incidentally, I think it would be best not to make it public."

He then dangled the bait: "For a number of years I have been at work upon a method of raising recording apparatus to altitudes exceeding the limit for sounding balloons; and during the last two years I have tried out the essential features of the method at the laboratory of Clark University with very gratifying results. These experiments are now completed, and I feel that I have settled every point upon which there could be reasonable doubt."

Goddard knew that the Smithsonian Institution would be interested in a way to put "recording apparatus" up where balloons could not reach. He refrained from describing how he would do that. Instead, he employed the hard sell, saying that his device would be useful in warfare, but even more useful to science. If circumstances impelled its conversion to weaponry, that would occasion a great loss to researchers.

That touched a nerve at the Smithsonian, where Walcott's predecessor, Samuel P. Langley, had pioneered heavier-than-air aircraft, only to be upstaged by the Wright Brothers and then to see further development dominated by the military.

Now he got to the point, describing the history of his rocket activities, and his proof that rockets could reach high altitudes. He reviewed his experiments, and described his chamber and nozzle designs, his tests in a vacuum, and his prediction that with multiple charges he could start most of a rocket's mass as propellant.

He concluded by asking the Smithsonian to assemble a committee of experts to review his ideas. If the committee found merit in them, "would the Smithsonian Institution be willing to take upon itself the recommending of a fund, sufficiently large to continue the work, either from a society such as the National Geographic Society or from private individuals?" Having established that he was not just another beggar for research grants, he ended: "I realize that in sending this communication I have taken a certain liberty; but I feel that it is to the Smithsonian Institution alone that I must look, now that I cannot continue the work unassisted."[1]

Goddard benefited from fortunate timing. Walcott was away hunting fossils when the letter arrived, so it went to the acting secretary, Dr. Charles Greely Abbot, who was renowned for his research on solar radiation. Long-lived (1872–1973), he had been at the Smithsonian Astrophysical Observatory since 1895, and served as its director from 1907 to 1944, retaining the post while secretary from 1928 to 1944. Tempting such a man with an invention that would place his instruments high in the sky was like waving candy before a hungry child.

"When I had read Goddard's paper I took it to Secretary Walcott recommending it as the *best* presentation of a research I had ever seen," Abbot recalled four decades later. Actually, he advised Walcott at the time: "I consider the method probably sound." He alerted Dr. Edgar Buckingham, a mathematical physicist at the National Bureau of Standards, who was enthusiastic. Abbot outlined for Walcott the important research questions that a rocket could answer in the upper atmosphere, and wondered what it would cost.[2]

Walcott sent Goddard a short reply on 11 October, asking about costs. "We are greatly interested in a number of problems that possibly might be solved by the use of your method," he closed. On the day of Gram's funeral, Goddard outlined for Walcott the work ahead—selecting firing-chamber size, developing a reloading mechanism, more tests to reduce the weight of the chamber. If the highest altitude was the goal, he would need a "secondary" rocket atop the first. "Now," he concluded, "as to cost: I do not think that the work I have outlined could possibly be done within a time as short as one year for less than $5,000." Once perfected, he believed his rockets could be produced for around $250 per copy.[3]

Walcott circulated Goddard's letter to physicists in the Washington area. They agreed that before they could advise on making a grant they would need to see Goddard's theoretical work. As Walcott told him, they also wanted a description of his "mechanical arrangements, particularly the reloading device of which you give no description." He promised an early response if Goddard provided that information.[4]

Walcott got more than he bargained for, because Goddard had revised his 1914 paper, "Problem of Raising a Body," now called "A Method of Reaching Extreme Altitudes." It was essentially the earlier paper, with new material describing the work conducted since 1914, in particular the proofs in a vacuum.

He gave the report a fine binding and enclosed it in a handsome wooden box. His cover letter, dated just five days after Walcott's, said that his manuscript should answer all questions, and offered to make a presentation in person. "Such a procedure," he closed, "will not, however, be really necessary because everything I am sending is self-explanatory."[5]

It was overwhelming. Walcott passed it on to Abbot, who checked the mathematics and the patent specifications, and forwarded the package to Buckingham. Abbot told Walcott: "I believe the theory is sound, and the experimental work both sound and ingenious. It seems to me that the character of Mr. Goddard's work is so high that he can well be trusted to carry it on to practical operation in any way that seems best to him. I regard the scheme as worth promoting." Buckingham chimed in: "I hope the Smithsonian Institution will see fit to help Mr. Goddard in developing his invention, and I shall be glad if my expression of opinion contributes to such a decision."[6]

Funding Goddard would help Walcott solve a problem. He was on the board of directors of the Research Corporation of New York, which administered a bequest to the Smithsonian called the Hodgkins Fund. Thomas George Hodgkins, born in England in 1802, arrived in the United States in 1830 and spent decades studying relations between the atmosphere and human well-being, while amassing a fortune in business. In 1891 he gave the Smithsonian $200,000, provided the first $100,000 went to increase knowledge of the atmosphere. Spending the first half would gain Walcott's institution access to the remainder.[7]

On 5 January 1917, Walcott granted Goddard $5,000 from the Hodgkins Fund to do the work outlined in his October letter, enclosing a first payment of $1,000. He advised that Abbot and Buckingham would constitute the Smithsonian's review committee for the grant, and invited Goddard to consult with them at any time. He must submit at least one progress report a year, along with statements of expenses with receipts.[8]

Goddard was elated. "Read letter to Ma and Pa," he recorded in his diary. "Pa said, 'You certainly put it up to them in wonderful shape.' Ma said, 'I think that's the most wonderful thing I ever heard of. Think of it! You send the Government some typewritten sheets and some pictures, and they send you $1,000 and tell you they are going to send four more.'"

Duff was even more impressed, but most rewarding was Webster's reaction. After taking the news to President Hall, he arranged for Goddard to address a faculty meeting. Webster held his grant proposal up as a model for others to follow, "although you may not get $5,000." In a day when such grants were rare, Goddard's was unusually large. It made him a local celebrity.[9]

At Duff's urging, WPI president Ira N. Hollis offered Goddard the use of the old Magnetic Building on the corner of campus, along with the services of C. D. Haigis of the physics department, assisted by a student. The building was a small, stone, Romanesque structure with a corner turret, built decades earlier for studies of electromagnetic fields. It had been rendered useless by the construction of an electric trolley line on the street in front of it. Goddard soon filled it with tools, machinery, and drafting boards.

There he worked in what WPI students regarded as "mysterious isolation," but not entirely. Frequently he walked up the hill to the main campus to discuss his research with Duff and other faculty. Many of his parts were manufactured in the WPI machine and forge shops under the direction of the shop chief, Carl Johnson.[10]

He was hard at work by March, testing a new "gun" with Haigis, but not getting enough acceleration from his powders, which bulged his gun when ignited. He sent an optimistic report to the Smithsonian, mostly about the generosity of WPI and the Winchester Repeating Arms Company, which had given him some nickel steel. He closed by stating: "It seems likely, in view of the long range and ease of transportation of the apparatus, that it may be useful in warfare; but to be of the greatest use, the applications along this line should be known to the United States Government."[11]

He was still a scrounger. "The chambers, or guns, have been made at the L. Hardy Co.'s shop," he told Walcott in July. He was by then fully staffed. Besides Haigis, his crew included WPI student and machinist Carl S. Carlson, Nils Riffolt (borrowed from the Clark physics workshop), who did fine machine work, and Louis T. E. Thompson (also from the Clark physics department), who did the same. He had bought blankets to cover the Magnetic Lab windows "to deaden the sound of firing—at the request of President Hollis." Noise there was: "Gun blew up—was hardened—with 42 gm of powder," he recorded on 9 July.

DuPont donated sheet pyronitrocellulose for his experiments, while he continued to improve his combustion chamber. By August 1917 he believed that he had eliminated the bulging problem. Still redesigning the reloading mechanism for his multi-charge rocket, he optimistically reported that he was a third of the way to producing a device useful for warfare. On the strength of that, the Smithsonian sent him an installment of $500 from his grant.[12]

In fact, he was floundering again, jumping from problem to problem without a clearly defined aim. He had a goal—a multi-charge rocket with some kind of reloading and firing mechanism—but it was elusive. He had one advantage in dealing with his grantors—the country had been at war since April, and that gave him an excuse for making little progress.

In December he claimed that he was hampered because his machinists were all part-time, and because arms manufacturers were too busy with war work to help out. He had redesigned his breechblock again, but still had not produced one that worked. Now he predicted that it would be six months to a year before he could perfect a self-reloading, multi-charge rocket.[13]

It was dangerous work sometimes, as Riffolt recalled: "The charge was taken out, but the primer was still there. In other words, I was drilling holes all the time in projectiles with the [primer] charge in." The efforts went nowhere, but that did not bother Goddard. He presented a positive face to his benefactors at the Smithsonian, excusing lack of achievement on account of insufficient help, competing demands of the war, or a bout of the "grippe" in December.[14]

Goddard was not so much at a dead end as in a technical morass. He was trying to build something highly impractical—a flying machine gun shooting blanks—but would not admit it was a hopeless cause. He tried one thing after another, and hoped that funding would continue until he stumbled onto something that worked. Devoted to the multi-charge idea conceived in his youth, he could neither organize his approach to the challenge nor, least of all, consider alternatives.

Goddard needed a break. When he was a boy, he dreamed of flying away to Mars. He was a man now, so that option was closed to him. He could, however, go off to war.

CAN'T YOU SEE THAT THERE'S MILLIONS IN THAT THING FOR YOU?

Despite his protests in his grant application, Goddard was not averse to using military money to support his project. During two world wars he capitalized on the mood of the moment to tap into national-defense resources. He desired only two things—that he control his own work, and that the money be green.

Even before World War I broke out, he regarded the military as a funding source. On 1 April 1914 he noted in his diary: "Thought of aerial torpedo in morning. Drew element diagram in afternoon." In July he sent copies of his patents to Secretary of the Navy Josephus Daniels, suggesting that they offered a way to shoot down airplanes. He also offered to design a wireless rudder control to guide his rockets.

His patents, he averred, had been developed to support high-altitude research, and if the navy adopted them he would retain the right to use them for the original purpose. Daniels asked for more detail, as the patents were too vague to evaluate his proposal. "It would be still better," said the secretary, "if you would submit actual samples of your rockets for test." He also wanted to hear more about wireless control.[15]

Goddard answered that he hoped to perform meteorological experiments with rockets in the next year or two, when their value to aerial warfare would become apparent. As for wireless control, he extended it to marine torpedoes, but offered no details.

When Acting Secretary of the Navy Franklin D. Roosevelt invited him to demonstrate his device, Goddard was caught up short. The vague ideas that had dominated his thinking were something less than the solid objects military bureaucrats wanted to examine. "I beg to say," he told Roosevelt, "as I have already intimated, that it would be out of the question for me to give a demonstration at the present time. I should like to take advantage of your offer sometime later if it is a possible thing."[16]

Unable to put up, Goddard shut up, for the moment. On Anniversary Day 1915, he recorded another list of things to do: "Get patent on breech-block reloading mechanism. Finish vacuum experiments. Have the Navy or Army Department develop it for coast defense etc. provided they will allow research at high altitudes etc. with it, under government control and assisted by foundations, if necessary."[17]

The American declaration of war against Germany in April 1917 opened new opportunities. Goddard told Walcott that rockets offered longer ranges than artillery, as well as ease of transportation and firing, without fixed positions for the enemy to target. If the War Department got into rocketry, it was "essential" that Goddard direct it. Abbot advised Walcott that the War Department ought to be urged to grant $50,000. "I believe," he said, "he would succeed in perfecting means to propel large bombs or shells to distances of say 100 miles from the point of firing."

The more practical Walcott told Goddard that "until demonstration has been made that the apparatus can actually be used," there was nothing on which to base a proposal to the War Department. "Can you not make a small rocket that will serve to put your theoretical work into practice?"[18]

Goddard felt that the march of history was leaving him behind, without any military money in his pocket. He told Walcott that his main problem was in constructing a reloading mechanism for his multi-charge rocket. "For this reason," he said, "I think it best to make the apparatus I had originally planned"—by which he meant an improved, single-charge rocket—"which I am sure *will* be operative, and to have a War Department representative present when the trial is made. A great range having once been demonstrated, it should be easy to arouse interest among the proper authorities."[19]

For one of the few times in his life, Goddard stopped pursuing a long-held idea—the multiple-charge solid rocket. He was not willing to admit that it would not work, but otherwise he could not tap into the military budget. In August he approached the army directly, writing to the chief of ordnance to offer general specifications for a military rocket carrying an explosive shell.[20]

At the army's invitation, in August he traveled to West Point to present his ideas to Captain Halsey Dunwoody. Since graduating from the Military Academy in 1905, Dunwoody had transferred from the Artillery to the Coast Artillery. He had learned how to fly and now sought transfer to the Air Service of the Signal Corps. He was from an inventive family—his father, also a career officer, had invented the crystal radio.

Hearing Goddard out, Dunwoody had a startling first reaction: "Can't you see that there's millions in that thing for you?" He proposed preparing a supply of small rockets carrying warheads, firing lots of them, and seeing where they hit.[21]

In November, Goddard traveled to Washington, where Dunwoody was lobbying to enter the Signal Corps. He visited the officer several times over three days, and the news was not encouraging: "Plans to adapt it for airplane gun with percussion firing," Goddard recorded in his diary. "If the Government takes it [Goddard's rocket], they don't give much for it." He was discouraged, because if the military adopted his invention, it would involve larger firms in its manufacture.

Dunwoody was more upbeat: "He agrees in thinking it best to manufacture one ... and then suggest that ... some others make a dozen. ... If Government helps (not probable) it will control it. Therefore all he can do is to help indirectly or urge that a manufacturer give us facilities and protection—for a consideration of course." This Goddard did not like at all. A large company could take his rockets away from him once the government had a claim on them.[22]

He returned to Worcester believing that Ordnance had given him a runaround. He also had an idea about how he could avoid losing control of his rockets to a big firm. It was to collude with a local business that could manufacture his rockets without taking their design out of his hands. He spent December negotiating with George I. Rockwood, president of the Rockwood Sprinkler Company of Worcester, with the help of attorneys.

However, the two parties could not come to terms over royalties, licensing, and rights to existing patents, shares in profits, and in particular Goddard's insistence that the product be called the "Goddard Rocket."[23]

There were limits to Goddard's ability as a salesman, beginning with his failure to determine the interests of his potential customers. In wartime, military authorities were inundated with crackpot schemes to furnish the "ultimate weapon." Goddard's proposal of what-might-be was sufficient to persuade a research institution, the Smithsonian, to support his development to the point of practicality. Generals and admirals needed something that already worked.

Whatever Goddard thought of rocketry's potential, at a time when rockets were mostly holiday entertainment, few other people had any idea what he was talking about. His mercenary appetite had been whetted by Dunwoody's talk of "millions," but he was out of his element when dealing with practical people inhabiting the workaday world. His patrons had to go to bat for him.

INDICATIONS ARE THAT SPIES ARE ACTIVE

Webster went to the plate first, drawing upon his connections with the navy. On 16 January 1918 he wrote a letter of recommendation for Goddard to take to his friend Rear Admiral Ralph Earle, of the navy's Bureau of Ordnance. Goddard left for Washington the next day, and became lost in the naval bureaucracy.

He went to the Smithsonian to meet with Abbot, Buckingham, and Walcott. They decided that, to support any military work for him, they needed a committee of physicists and engineers who would consult with military experts "to decide whether or not the importance of the thing outweighs the difficulties that are to be expected." Or so they told Goddard.[24]

In fact, Abbot and Buckingham, eager to get their instruments to "extreme altitudes," were too impatient for committees. They prepared a report on Goddard's "successively propelled rocket," saying that alternative developments were still required to perfect the thing. Either he should continue the "slow way as at present" of trying one idea after another, or pursue all alternatives at once. Such practical choices, Walcott understood, were just the sort of solid options to which a bureaucrat could react. And he knew just who might give them a fair hearing, and who had money to spend.

It was his good friend Major General George Owen Squier, another former artillerist who had gone into the Signal Corps, in which he served as chief signal officer from 1917 to 1923. An 1887 graduate of West Point, he obtained a Ph.D. at Johns Hopkins in 1893, and became a pioneer in cable and radio communications, as well as in military-aviation doctrine. Walcott and Squier were kindred spirits, and Walcott was the chairman of the Military Committee of the National Research Council—his opinions carried weight in the War Department.

Backed by Abbot and Buckingham's report, and supported by S. W. Stratton, director of the National Bureau of Standards, Walcott approached Squier. He asked the Signal Corps to add $10,000 to the project. The four civilians offered to serve as a "Supervisory Committee" for the effort.[25]

Goddard returned to Worcester with the Smithsonian's promise to lend him a lathe formerly used by Samuel P. Langley in his aircraft experiments. The news that the Signal Corps would fund his work arrived just before the lathe, in early February 1918. The money allowed him to improve his laboratory space, get Haigis relieved from teaching so he could work full-time, and hire a chemist, Dr. Henry C. Parker, to work on powders.

By March he had remodeled the Magnetic Building, adding an attic and lights, had installed a fully equipped machine shop, and had also hired two toolmakers, a carpenter, a draftsman, a part-time janitor, and a night watchman. The Smithsonian reimbursed him and obtained recompense from Squier's office.[26]

Abbot visited Worcester in the middle of March, and after Goddard had fired a shot for him said: "You certainly have made progress. I don't see any reason why this won't work now." The next day, Goddard's doctor sent him home to recover from German measles.

Unaware of that, Abbot reported to Walcott that the work was going well, thanks to the energy of Goddard and the "drive" of Haigis. "I believe from what I have seen that the rocket will be ready to try within a month," he predicted, then warned: "People are very curious about the work, and indications are that spies are active. Drunken men ring the bell, etc., etc., but disappear quickly when the watchman brings his repeating Winchester shotgun. The building is of thick stone with but one door. I recommended to wire all windows and the door so as to alarm Mr. Haigis, who rooms nearby, if an attack should be made."[27]

Goddard had a burglar alarm installed, and at least one person was in the building round the clock. His more pressing problems were homegrown, because he had reverted to the multiple-charge rocket. His chamber and breechblock became more complex, and they sprouted a forest of springs, but still they did not work. He larded his reports to Walcott with photographs of his sophisticated mechanisms, which impressed the Smithsonian head, despite the fact that Goddard had spent $8,000 so far without producing a working model.

Walcott told Squier that "it is apparent that rapid progress has been made, and a successful outcome of the development is probable." He recommended that the Signal Corps allot another $10,000, "which I hope will bring us to the stage of trial of the rocket."[28]

In May Goddard told Walcott that he proposed to abandon his old breech design in favor of a new one, but he felt that Haigis was so devoted to the old way that he was no longer useful. Then the rocketeer had an attack of practicality—he would test a single-charge rocket "which promises to be a very great improvement on all rockets and trench mortars now in use." Meanwhile, he asked clearance to go to Washington to talk with Abbot and Walcott about "certain matters."[29]

Those matters concerned Haigis. On 8 April he had threatened to quit unless he had a partnership in Goddard's work. The next day he said he wanted just a salary but more freedom to use his own judgment. Goddard, who always avoided confrontations, agreed to that so long as he had final approval on Haigis's plans. "I will confess that Mr. Haigis's work has been giving less and less satisfaction," he told Walcott.

He visited Abbot 15–16 May. Following his patron's advice, when he returned to Worcester he laid Haigis off till the end of the month, when the man's employment would cease—firing him without facing him. Learning that Haigis had plans of his own to compete in rocketry, Goddard asked a Department of Justice agent to talk to the former employee, "and the latter gave his word that he would say nothing outside concerning the work."

No sooner had he reported that to Walcott, and learned that the army would give him another $20,000 (twice what Walcott recently proposed), than Goddard went into a panic that produced a frantic telegram to the Secretary: "MY FOREMAN AFTER DISCHARGE NOW EMPLOYED BY CONCERN HERE DEVELOPING INFRINGEMENT ON APPARATUS. HIS EMPLOYER REPRESENTS THAT AT REQUEST OF COLONEL SHINKLE OF ORDNANCE DEPARTMENT HE IS TO MANUFACTURE APPARATUS. IS THIS REQUEST TO MANUFACTURE AUTHENTIC?"[30]

Before he had determined what Haigis and his new employer—who turned out to be Rockwood, the sprinkler man—were up to, Goddard began a defense of his property rights in rocketry. Or more exactly, he asked Walcott to wage the fight for him, sending a long peroration on ways of sorting out patent rights, licenses, and royalties between himself and the government.

Goddard was out of his league again, because the issues he raised were beside the point. As Walcott and Abbot realized, the immediate challenge was to make sure that nothing interfered with progress. Ownership and royalty questions could be addressed at a more appropriate time.[31]

As events unfolded, it came to light that Haigis had gone to work for Rockwood immediately after realizing that he was out of a job, probably on 18 May. Three days later Rockwood, who evidently believed that his earlier negotiations with Goddard gave him a claim to the rocketeer's products, wrote the Ordnance Department asking for a letter requesting that he produce for the army the rocket that Goddard had designed for the Smithsonian. Colonel Edward M. Shinkle refused "because of the action already taken by other Departments of the Government," meaning the Smithsonian and the Signal Corps. His office, he said, was now consulting with the other departments and hoped to know soon whether Goddard's device was practical or not. That, it turned out, was just a statement for the record, as the colonel began to pull strings to get himself into a position to profit by abetting Rockwood's efforts.

A 1901 graduate of the Military Academy, Shinkle had followed a career from Artillery to Coast Artillery to Ordnance. An ambitious bureaucrat, his curiosity aroused by Rockwood, he arranged for Brigadier General C. M. Saltzman of the Signal Corps to ask the chief of ordnance to send him to Worcester. He showed up on 29 May unannounced. He called Goddard and asked to see him at a downtown hotel the next morning; Goddard declined. Since he had no instructions from Squier on this matter, he called Abbot, who consulted the chief signal officer, then told Goddard to say nothing.

Shinkle next appeared in Webster's office at Clark with Rockwood and Haigis in tow, but following Abbot's instructions Goddard dodged them. The colonel called Goddard's father, Nahum, who telephoned Abbot to say that Shinkle had threatened to close the laboratory down unless Goddard saw him. Abbot told Nahum that Squier would see Shinkle back in Washington.[32]

This episode illustrated how dependent Goddard was on his patrons. He was not the sort to confront troublesome people directly, and realized that his friends at the Smithsonian were better prepared to deal with a bureaucratic buccaneer like Shinkle than he would ever be.

Abbot now emerged as Goddard's chief guardian angel. When he heard what was going on in Worcester, he happened to have a visitor, Dr. George Ellery Hale, director of the Mount Wilson Observatory near Pasadena, California. He briefed Hale on the situation, and asked whether he might have room for Goddard to do his work, far from prying eyes. Hale agreed.

Abbot took the story to Secretary Walcott, who called General Squier. Because important bureaucratic issues must rest on paper, on 31 May the secretary sent the general a strong statement. He implied that Shinkle had improperly connected himself to Rockwood, then intervened in Goddard's work without authority.

"Owing to the unfortunate interference that has taken place," Walcott fumed for the record, "it appears that two weeks or more of delay has been caused by the Rockwood-Haigis-Shinkle intervention, and the secrecy which was considered to be so important a matter would have been practically thrown away if Dr. Goddard had permitted Colonel Shinkle and Haigis, his former foreman, to obtain full knowledge in relation to the development of the rocket."

Walcott knew that what he wrote would be passed on to the Ordnance Department, and so was helping Squier fight off an invasion of Signal Corps turf. Shinkle was transferred to France.[33]

OFFICIAL INERTIA ONLY OBSTACLE

Abbot worried that the events following Haigis's defection threatened a breach of security that could reveal a secret weapon to the enemy. Goddard reacted only to the potential threat to his proprietary interest in rocketry. The experience made him cautious about the danger renegade employees might pose to his personal and financial stake in the field. From that point onward, he swore his assistants to secrecy.

This new secretiveness was reflected in what he did on returning to Worcester from Washington, forewarned by Abbot that he would move to California. He sealed a manuscript in an envelope labeled "Special Formulae for Silvering Mirrors," and gave it to Thompson to hide in a safe pending his return. The document was a reprise of his 1913 "Outline" on the navigation of interplanetary space. This time it was called "The Last Migration," a discourse on the "use of solar energy, hydrogen, and oxygen, avoidance of meteors, atomic energy, and the use of the sun in increasing the speed during interplanetary flight, but also speculated as to the last migration of the human race."

It offered far-fetched ways of storing humans during space travel by means of cold storage of "granular protoplasm," of forcing the evolution of the species, and of waking the space pilot at intervals of ten thousand to one million years. It was one more aimless rhapsody originating in his youth, but a decade later he pointed to it with pride as preceding anybody else's speculations on space travel.[34]

He left this behind, he said, "in order that some word would remain, no matter what circumstance occurred during the war." He was not bound for the battlefield, however, merely for Pasadena. It was farther from the shooting than Massachusetts, but it was the farthest he had ever been from home. He was reassured, before he left, to learn that the Clark trustees had renewed his position for another year.[35]

At Abbot's urging, on 1 June 1918 Walcott asked Dr. R.S. Woodward, president of the Carnegie Institution of Washington, for permission to move Goddard's establishment to Mount Wilson, which his organization funded. Woodward agreed, upon which Walcott directed Goddard to move "to a place of which Dr. Abbot informed you." He sent him $2,000 to pay for the trip.[36]

On the same day, Goddard tested a single-charge rocket weighing about two pounds, one-eighth of it propellant; it flew about a half mile. "This looks like a good start on the single-charge device," he crowed to Abbot, "and also gives an indication of what is to be expected when the propellant is one-fourth or one-third entire weight." This news drew an immediate "Congratulations!" from Abbot.

They next corresponded about the move to the West Coast, especially Goddard's idea of carrying his propellants in suitcases on the train, violating safety regulations. They both thought the plan perfectly safe, but Goddard wanted "assurance of backing, if any difficulties should arise." Abbot, who was not that naïve, telegraphed: "DESIRED AUTHORIZATION IMPRACTICABLE BUT PLAN SHOULD WORK WELL," meaning we cannot tell you to do it, but do it.[37]

Goddard and his crew left Worcester on 5 June and arrived at Mount Wilson five days later. Much of his material was loaded onto a truck carried on a rail car, its license plates covered up; two of his crew rode with the truck. Nahum crated the heavy shop equipment and shipped it by American Express.

Reaching Pasadena, Goddard decided to concentrate on a single-charge rocket as his best hope of early success. By the thirteenth he had assembled a three-inch rocket and went looking for a place to test it. In the nearby desert at Arroyo Seco ("dry ditch" in Spanish), the first launch took place on the twentieth. The rocket, which had stabilizing vanes inside its nozzle, turned back thirty feet from the launch site and charged at him, inciting a mad scramble for cover on the part of the rocketeer and his aides.[38]

George Hale, his host at Mount Wilson, was a world-renowned astronomer, and he placed his organization almost entirely at Goddard's service. The observatory shop was his to command, except that it reserved a quarter of its effort to completing a one-hundred-inch telescope then under construction. This was not all dedication to national defense, however. Hale wanted to take high-altitude photographs of the sun from Goddard's rockets.[39]

Hale's enthusiasm lagged quickly, as Goddard cut into his shop's time too much, and he was surrounded by astronomers who wanted the big telescope completed. Goddard gave a demonstration launch for Hale and the others, and regained their support.

Goddard also focused on producing something militarily useful. Scrapping the vane-in-exhaust idea, he adopted launching tubes to aim his rockets, testing tubes both closed and opened at the breech. With the first small success he grandly expanded his predicted applications to include aerial warfare and even the replacement of trench mortars. Yet he chafed at being unable to work on his multiple-charge idea.

Goddard was not about to let military necessity divert him entirely from the goal of "extreme altitudes." Nevertheless, by mid-July he had obtained respectable ranges and accuracy. He had discovered that a rocket was more accurate when its head end was heavier than the rest of the missile, an elementary principle of ballistic stability known to archers for millennia.[40]

Besides his inability to concentrate on multi-charge designs, Goddard was denied his earlier freedom to tinker with this and that in a meandering approach to his work. The government imposed a time limit—show results by midsummer. He suspected that he would go beyond that time, and in mid-July alerted President Sanford that he might ask for a leave of absence for the coming semester. "The work is going well," he said, "and we have evolved several new and apparently important applications. We have so far done better than I expected; but it is no easy task to develop and perfect a device within a very limited time."

He felt ready to demonstrate his results to the army, and asked his supervising committee to arrange for an inspection. Stratton passed that on to Squier, who asked the chief of ordnance to send an officer to Mount Wilson. No one had showed up by August. On the fifth, still with no inspection, Goddard asked Sanford for leave, and Abbot and Stratton begged the army to send someone out.

Goddard feared that his funding would run out before an inspection. Hale, now a firm ally, sent a strong telegram to Abbot, who replied that there was still $5,000 left, the holdup was in Ordnance, and: "FEEL CONFIDENT FURTHER SUPPORT WILL COME. OFFICIAL INERTIA ONLY OBSTACLE."[41]

Goddard was eager to show off his work, because he had achieved results that exceeded anything he had done before. As he told Abbot early in August: "The single-charge device has been developed so that it shoots straight, can be fired with about the rapidity of a hand grenade or Stokes trench mortar, and has a single impact-exploding device which has been tested. A firing tube which can quickly be converted from open to closed, and vice versa, has been made and tested, and also a single shock absorber. Further, a firing tube in which the opening can be varied gradually has been designed." Goddard had developed a portable infantry rocket, offering a way to toss explosives farther than grenades or trench mortars, and in several calibers.[42]

While he had Parker working full-time on designing projectiles, he diverted some of his funding to put C. N. Hickman, a Clark graduate student, to work on the multiple-charge problem. Hickman had been recruited before the move to California; he was a brilliant physics student with hobbies as a clarinet player, archer, and magician. He knew that the multiple-charge reloader was impossible, but he gave it a good try.

In early August, a cartridge exploded while Hickman was opening it, taking off the thumb and two fingers of his left hand and half a finger of his right. He emerged with cuts on his face and chest, but his eyes were all right. He was still game—he continued his work after two more accidents.

Goddard next began to emphasize a "Roman candle" multiple-charge rocket, with charges aligned in the chamber firing in sequence. It was simple, with few moving parts, but he could not get that to work, either.[43]

At Hale's urging, Goddard took Sundays off, visiting the Hales at their mountain home and going on nature-watching expeditions to the mountains or Catalina Island with Hickman and his wife and others. He had the others photograph him cavorting in parks and on trails, and he developed an interest in young women. He was too shy to make advances, but he enjoyed watching and photographing them.[44]

Back at Mount Wilson and Arroyo Seco, there was no time for such indulgences. With Parker, Goddard continued to refine the tube-launched single-charge rockets and their warheads, while with the wounded Hickman he soldiered along on reloading mechanisms. Meanwhile, letters and telegrams shuffled between Goddard, Abbot, Hale, Squier, and various officers, seeking an inspection and some promise of future funding. August wore away into September, and still no inspectors arrived.[45]

It was especially frustrating because for the first time Goddard had produced working rocket systems. He had three different tube-launched rockets in 1-inch, 1.75-inch, and 3-inch calibers, with the smallest one shoulder-mounted. They carried explosive warheads, and they did what they were supposed to do. That demonstrated to Abbot and others that Goddard could accomplish something real, as opposed to theories and extravagant predictions.

Without warning, not one but two ordnance captains showed up on 13 September to spend two days watching Goddard's inventions at work. After his 1-inch model lobbed a pound of explosive 180 yards and hit the target, they were impressed, offering reactions that he recorded joyfully in his diary: "Do you know that that stuff you've got there is going to revolutionize things? That's wonderful stuff you have got there. We thought it was going to be some kind of toy." Goddard provided the captains a plan and budget for further work, and hoped for the best.[46]

Ordnance asked for a demonstration at Aberdeen Proving Grounds in Maryland. Telegrams flew back and forth regarding the dates and Goddard's needs for special materials before the demonstration. At last a date was scheduled, and a supervising officer assigned, but it was rescheduled in late October after Goddard left California for Maryland.[47]

With Hickman doing the actual demonstrating, Goddard put on his show at Aberdeen on 6 November. He was not satisfied with the performance, partly because his audience of army and navy officers came and went throughout the day, but mostly because his multiple-charge rocket blew up. The single-charge, tube-launched weapons all worked well. Abbot, among others, was especially impressed when Hickman fired a launcher resting on a pair of music stands, which did not fall over. But the war ended five days later, and with it any need to pursue military rocketry.[48]

It has become part of the Goddard legend that he invented the "bazooka"—the portable anti-tank rocket launcher of World War II—during 1918, and demonstrated it to an obtuse military that failed to appreciate its importance. The actual story was not that simple. The Goddard tube-launched rockets of 1918 and the bazooka were distinct contraptions; their most important common element was C. N. Hickman, who worked on both.

The utility of Goddard's rockets in the trench warfare of World War I is doubtful. That was mostly an artillery war, and machine guns, barbed wire, and entrenching tools dominated the local landscape. The trench mortar, which could drop projectiles into ditches, was the only squad artillery with much use at all. Goddard did work on a rocket-powered mortar, but it was his tube-launched projectiles that he showed off at Aberdeen. That idea would have use in the future, but there were few people in 1918 who could predict the armored vehicles and tactics yet to come.

With the end of the war the army and navy canceled all procurement commitments. The Smithsonian could offer him no more money, so Goddard went looking for the $10,000 he estimated it would take to finish his work, and Abbot tried to talk the Signal Corps out of at least half that amount. Goddard prepared drawings and specifications for airplane-mounted rockets for the Ordnance Department, which could not pursue the work. He was invited to resubmit his proposal once the War Department had resumed peacetime status.

The rocket man lobbied the army, the navy, the Smithsonian, and the Weather Bureau, but got nowhere because the wartime budgets had evaporated. In late March 1919 Abbot advised him to give up. The army had not repaid the Smithsonian for the funds it had advanced for the Aberdeen tests. Until those bills were paid, the Smithsonian could go no further with rocketry.[49]

Goddard was not ready to throw in the towel. He had had a taste of military funding, and like the proverbial tiger that once chews human flesh, he was determined to hunt up more of it. He was not to be turned off by the budgetary realities of the postwar government, nor was he ready to give up on the multiple-charge, solid-fuel rocket.

Determined as he was on chasing both goals, he did not realize that life still offered surprises, sensations that he had not yet enjoyed. One was celebrity. Another was love.

FIVE

PROFESSOR HAS PERFECTED
INVENTION FOR EXPLORING SPACE

There will be sojourners come from the Earth, who, longing after the taste of the sweet cream, of their own skimming off, from the best milk of all the dairy of the Galaxy, will set themselves at table down with us, drink of our nectar and ambrosia, and take to their own beds at night for wives and concubines, our fairest goddesses, the only means whereby they can be deified.

—François Rabelais, *Pantagruel*, IV:51

Goddard was back at Clark by December 1918, preparing for another semester of teaching physics to undergraduates. He was the best-dressed man on campus, favoring fine suits and excellent haberdashery, crowned by a crisp homburg. He could afford to dress well, as he lived at home and his family was prosperous. Pictures of him in the classroom show an animated face, that of a man who enjoyed science and passing his enjoyment on to youngsters.

Yet his heart was not in professorship. He had spent months in shirtsleeves working on rocketry, breathing easily in clean desert air, and he had used government money to pursue a solid-fuel rocket that fired charges in succession. This was impractical if not impossible, because the mechanism would have to be complicated and heavy. But he could see no alternative; using liquid fuels had been an afterthought in his patents, and mostly forgotten since.

I WAS ENTHUSIASTIC OVER THE WAY I HAVE BEEN TREATED

Goddard had a well-equipped machine shop, shipped back from California. He was not a craftsman, but with skilled mechanics he could produce anything. The nearest modern equivalent to his workshop is that of a professional racing-car team, which can manufacture from scratch any precision part or indeed an entire engine and drive train. Goddard's establishment could do the same in rocketry.

In February 1919 he went to Washington to seek funding. He lobbied Abbot and Walcott at the Smithsonian, and visited Dr. W. J. Humphreys, a physicist at the Weather Bureau; John C. Merriam of the Carnegie Institution; and C. N. Hickman, now at the Bureau of Standards. None had any money. There was still some available from his Hodgkins grant at the Smithsonian, but because of postwar budgetary confusion the institution could not release it until 1920.

In the interim, Abbot put Goddard in touch with F. W. Baldwin, head of the boat-building department of Alexander Graham Bell's laboratory in Nova Scotia. He had already produced the fastest boat on Earth, a hydrofoil, and was interested in rockets as a possible way to break his own records. But Goddard had no rockets to offer.[1]

He next tried enlisting public support by telling the newspapers about his accomplishments. That produced sensationalistic coverage. WORCESTER MAN INVENTS A DEADLY WAR ROCKET, screamed the *Worcester Evening Gazette*, which described his "terrible engine of war." The story spread to the Boston papers, then by the Associated Press (AP) to others around the country and overseas.[2]

Goddard resumed his money hunt in 1920. He visited Abbot in March to tell him that he had a sixty-charge rocket ready for trial. His Smithsonian funds would pay for that, he said, but he needed more cash for further development. Abbot, however, had become restless about the time that had elapsed without a high-altitude rocket taking to the air. "I told him," he advised Walcott, "that I felt that a successful trial was the indispensable preliminary to any efforts to obtain additional support, and he quite agreed with this view."[3]

The rocketeer continued trying to raise funds by generating publicity, giving frequent talks to groups in New England. In April he went to Chicago to address the American Association of Engineers and the Aviation Club, securing pledges of support, then on to New York, where he visited Admiral Richard Byrd, the aviator. There followed an address to the National Academy of Sciences in Washington, "The Possibilities of the Rocket in Weather Forecasting." The speaking tour continued into 1921, when the Smithsonian's annual report predicted that a demonstration of the multi-charge rocket would take place soon.[4]

Goddard's publicity campaign, together with his actual work on rockets, kept him out of the classroom often. Over the next decade he had a tardiness and absenteeism rate that would not have been tolerated in a student. But Clark had a new president in Wallace W. Atwood. Believing that fame in a faculty was a fund-raising asset, Atwood urged Goddard to ask the Clark trustees for $5,000 to complete the multi-charge rocket. Once demonstrated, Goddard told the board, the rocket would attract the higher funding needed to produce it in quantity. Walcott supported the proposal with a letter asserting his faith in its scientific value. Suitably impressed, the trustees granted $2,500, followed by another $1,000 in 1922.[5]

That was small change compared to what Goddard believed the military should give him. Accordingly, he paralleled his campaign for civilian money with a hard press against the armed services. He was most successful with the navy, which by early 1920 was taken with his proposal for a rocket-launched antisubmarine depth charge. In late April, Rear Admiral Ralph Earle of the Bureau of Ordnance—Worcester native, an old acquaintance of Duff and Webster, and in 1924 president of WPI—gave Goddard a consultant's contract. He provided preliminary designs and other assistance, for which he received a retainer of $100 per month, plus direct expenses and $15 per diem when traveling.

Goddard journeyed often to Indian Head, Maryland, to supervise tests of propellant powders. He commented in his first report: "I was enthusiastic over the way I have been treated at Indian Head, and at the way the force there has placed data concerning pressures and rates of burning of various powders at my disposal." In March 1921 Goddard reported that he and the navy men had successfully fired a 725-pound depth-charge rocket, and he proposed to build a cannon-launched rocket projectile. The navy renewed his contract in June, extending it to cover the cannon rocket. That continued until 1923, when the navy's money ran out.[6]

He had gone after the navy instead of the army because he believed that the latter had shot its budgetary wad. The army, however, had not forgotten Goddard. In May 1920 Brigadier General Amos A. Fries, head of the Chemical Warfare Service, approached him about the use of rockets to project gas, incendiary, and high-explosive charges. The rocketeer provided separate figures for single- and multiple-charge rockets. "What you write is of the utmost interest to me," the general replied. In April 1923 Goddard gave a presentation in Fries's office, just as postwar demobilization hit bottom. The army had no money, but expressed a desire to remain in touch.[7]

THE IDEAL TYPE OF ROCKET IS, HOWEVER, THE LIQUID-PROPELLANT ROCKET

Goddard was not running out of money so much as out of ideas on how to make a multiple-charge, self-reloading, solid-fuel rocket work. He had the remains of his Smithsonian grant, another from Clark, the navy contract, and his share of the physics department budget, and he used them to keep Nils Riffolt and others at work.

He also had, as always, boundless optimism. In September 1919 he sent Abbot a report on his latest version of a reloading mechanism, a simple apparatus he called "inertia-loading." His tests of it, however, revealed a need for "adaptations."

Riffolt attached a sheet-metal shed to the Clark Science Building to conduct static tests without filling the building with noise and smoke. When Goddard felt ready to launch one of his devices, he took it to the truck farm of Aunt Effie Ward, at nearby Auburn. His most successful demonstration there was of a piston-loading cartridge system, with four cartridges, which rose about sixty feet.

Other experiments were recorded in his diary: "Tried a single cartridge and a long follower in the morning. It fed, without jamming.... Tried multiple in afternoon, with cartridges shellacked at both ends and a layer of filter paper over cap, held with rubber cement.... It fired five shots, and the tappet for the sight dropped off.... Tried 4 shots in afternoon; the fourth blew up just after being in central tube, sounded like a single, third explosion. Probably due to catching fire."

This constant improvisation was discouraging, because of both the difficulty in making the thing work and the looming prospect that a youthful idea had been wrong. Anniversary Day 1920 marked four years since Gram had died. He walked over to her house at Maple Hill and "went down back of barn, ate apples, leaned against cherry tree. Looked at other apple trees and inside of house and photos." Memory was comfort.

He revived his Roman candle idea from Mount Wilson. "The chief advantage of the Roman-candle rocket," he recalled in 1927, "lies in the absence of mechanism; and the chief disadvantage lies in the fact that every cartridge must be contained in a tube strong enough (and therefore heavy enough) to withstand a black-powder discharge."

It took that long for him to admit that, no matter how hard he flogged his multi-charge dead horse, it would not get up. Finally, in 1922, when Walcott asked Goddard for an accounting, he sent what turned out to be his final report on the multiple-charge rocket.[8]

He had a position to fall back on, and fortunately for his ego it was one he had addressed when younger. It was the use of liquid propellants, which he had considered in his notebook in 1909 and mentioned in his 1914 patent. He returned to hydrogen and oxygen. Each element was readily obtained in compressed gaseous form, but gases occupied too much space with attendant container weight to be practical for rockets. Liquefied hydrogen and liquefied oxygen (LOX) were theoretically ideal, as being the elements in their most condensed form.

To become liquids, the gases must be deprived of most of their native heat while under pressure. This was most difficult with hydrogen. It was still practically unavailable in 1920, and the first liquid-hydrogen-fueled rockets would not take off until the 1960s. Liquefied oxygen was another story, because it is a by-product in the manufacture of gases for oxyacetylene welding. But it boils at around 20 degrees above absolute zero, which is about 460 degrees below 0 degrees Fahrenheit. That made it difficult to store.

Moreover, although LOX was available, few people had experience with it. As G. Edward Pendray recalled: "Even in the 1930s," he said, "engineers told us of frightful things that would happen if we used the stuff. They said if it got on our clothes and we happened to light a cigarette, we'd go off like a torch. Mixing and igniting LOX with an inflammable substance like gasoline seemed a guarantee of catastrophe. Goddard's first test was a tremendously courageous thing."[9]

Goddard began to consider hydrogen and oxygen in January 1919. He was back in his best form, addressing the theoretical possibilities of known processes. A year later he offered the Smithsonian a report updating his thinking about the goals of his work.

Instead of concentrating on liquid-fuel rocketry, however, the "Report to the Smithsonian Institution Concerning Further Developments of the Rocket Method of Investigating Space" focused on using rockets to drop devices on alien planets, inscribed with geometric figures and diagrams of the constellations. In 1929 he revised this to advise destruction of the rockets, "to avoid promiscuous use of rockets by the inhabitants of another planet."

Goddard was lost in space again. Hydrogen and oxygen as propellants were obscured among speculations about ion propulsion, solar mirrors, the problems of talking to extraterrestrial aliens, and other irrelevancies. This continual revisitation of his youthful fantasies reconnected him to his ultimate objective, but it also distracted him from the practical question of how to get there.[10]

In early 1921, Goddard began to think seriously about trying liquids, making arrangements to acquire LOX from a local source. Eight years later, he explained his delay in getting to this stage by saying: "The ideal type of rocket, however, is the liquid-propellant rocket, first suggested June 9, 1909.... Such a development had not been attempted, for the reason that the handling of such liquids ... did not appear so easy as the handling of powder cartridges."[11]

A METHOD OF REACHING EXTREME ALTITUDES

By the time he wrote that, Goddard was world-famous as a prophet of rocketry on the road to outer space. The origins of his fame lay in his decision to talk to the press about his work in March 1919. When the *Worcester Evening Gazette* proclaimed on 28 March in a subhead INVENTS ROCKET WITH ALTITUDE RANGE 70 MILES, the story said that his "terrible engine" had been developed under the patronage of the War Department, the Smithsonian, Clark University, and WPI.

The story appeared in the *Washington Star* on 30 March, raising alarms in the War Department and the Smithsonian. Abbot told the army that the information did not originate in his office, and passed on this message from Goddard: "MATTER IN PAPERS CONTRARY TO MY WISHES. STATEMENT SHOULD NOT BE MADE UNTIL YOU RECEIVE INFORMATION WHICH I AM SENDING." That information reached Abbot on April Fools' Day. Goddard said that a *Gazette* reporter had inquired about his work, and possessed considerable information from the paper's Washington correspondent. He claimed that he had tried to kill the story, but the editor rebuffed him.[12]

The rocketeer was being disingenuous, because he had clearly talked to the reporter. Moreover, he owed his clients a degree of confidentiality, especially where military research in wartime was concerned. Goddard should have let Abbot or the War Department know that he had been approached by someone who knew details of his work, and he had not done so. He was protected by the fact that War Department proprietary information—the Aberdeen test results—had been disclosed, and suspicion thus turned from him to others. The uproar subsided.[13]

Webster now urged Goddard to publish his results in an appropriate outlet, and threatened to do so himself if the younger man did not. Thus pressured, the rocketeer proposed combining his earlier versions—the 1914 "Problem of Raising," a 1916 "Method of Sending Recording Apparatus to, and beyond, the Highest Levels of the Atmosphere," and parts of a 1918 paper called "Results on a Method of Reaching Extreme Altitudes"—into a new text to be published by the Smithsonian. Abbot obtained clearance from Ordnance to put Goddard's war rockets on display at a Physical Society meeting and permission to publish his work. Ordnance had paid its bills, leaving $3,000 available from the Hodgkins grant, which could be used to pay for publication; Walcott approved.

The rocketeer prepared a lecture for the Physical Society meeting, summarizing the Smithsonian paper. Publicity about his lecture at the meeting attracted the notice of French and British military authorities, who asked for copies of his remarks. His publication had an audience awaiting it.[14]

He spent most of May 1919 preparing the final text, which ended up being substantially what he had sent to the Smithsonian in 1916. "There's one good thing about the work," he reflected in his diary; "you don't see very many monuments to those who opposed the airplane about twenty-five years ago. The physical principle of this is just as sound as that of the airplane." Since his youth he had regarded himself as the pioneer of scientific rocketry, and now he saw glory just around the corner. The manuscript was published in December, and Goddard became a famous man.[15]

The paper, published in the *Smithsonian Miscellaneous Collections* as "A Method of Reaching Extreme Altitudes," established Goddard's claim to being the first (Tsiolkovsky was still unknown) to publish a mathematical proof that by means of rockets escape from Earth's gravity was possible. It was without question the first published proof that rockets would work in a vacuum.

The theory was supported by experiments and demonstrations, as well as untested descriptions of parachutes to recover a rocket. As with the 1914 and 1916 papers, "Method of Reaching" based its calculations on solid pro-

pellants, and just mentioned liquid fuels. More important, the fifteen foot-
notes he added included one on "distribution of mass among the secondary
rockets for cases of large total initial mass." This staked his claim to being the
author of multiple-stage (as opposed to multiple-charge) rocketry.[16]

The majority of Goddard's paper comprised a mass of calculations,
tables, demonstrations, and logical explication to prove that his theory was
sound. His overriding point was that rockets offered research value as a way
to carry scientific instruments to "extreme altitudes."

As a thorough scientist he could not stop there; he had to address reaching an "infinite altitude." So he did, near the end:

*Theoretically, a mass projected from the surface of the earth with a velocity of
6.95 miles/sec would,* neglecting air resistance, *reach an infinite distance,
after an infinite time; or, in short, would never return. Such a projection
without air resistance, is, of course, impossible. Moreover, the mass would not
reach infinity but would come under the gravitational influence of some other
heavenly body. We may, however, consider the following conceivable case: If a
rocket apparatus such as has here been discussed were projected to the upper
end of interval s_8, with an acceleration of 50 or 150 ft/sec²,* and this acceleration were maintained to a sufficient distance beyond s_8, *until the parabolic velocity were attained, the mass finally remaining would certainly
never return.*[17]

There followed several more paragraphs equally dry and academic,
larded with mathematics that the layman would not understand. To the
physicist or engineer, they demonstrated something that to others would be
utterly amazing—that it would be possible to rocket from the Earth into
outer space. They were followed by a section on using flash powders upon
impact with the Moon, and an appendix that declared that dodging meteors
was not a problem.[18]

Finally, the last part of Goddard's summary returned to the "infinite altitude" issue: "Even if a mass of the order of a pound were propelled by the

apparatus under consideration until it possessed sufficient velocity to escape the earth's attraction, the initial mass need not be unreasonably large, for an effective velocity of ejection which is without doubt attainable. A method is suggested [flash powder] whereby the passage of a body to such an extreme altitude could be demonstrated."

The *Smithsonian Miscellaneous Collections* carried technical papers for limited audiences. Goddard's piece belonged there, for only a few physicists and engineers would wade through its arid parade of facts and figures. Rarely did a paper in the *Miscellaneous Collections* rate much attention at all, let alone a press release.

To a publicist at the Smithsonian, however, Goddard's work promised adventure. On 11 January 1920, he issued and AP picked up a release announcing the publication. Its text emphasized the value of rocketry to high-altitude research, but it ended with this:

An interesting speculation described in the publication arising from Prof. Goddard's work is on the possibility of sending to the surface of the dark part of the moon a sufficient amount of the most brilliant flash powder which, being ignited on impact, would be plainly visible in a powerful telescope. This would be the only way possible of proving that the rocket had really left the attraction of the earth, as the apparatus would never come back once it had escaped that attraction. While this experiment would be of little obvious scientific value, its successful trial would be of great general interest as the first actual contact between one planet and another.[19]

Goddard picked up his copy of the *Boston Herald* of 12 January 1920 and jumped at the front-page headline: NEW ROCKET DEVISED BY PROF. GODDARD MAY HIT FACE OF THE MOON, followed by the subhead: CLARK COLLEGE PROFESSOR HAS PERFECTED INVENTION FOR EXPLORING SPACE—SMITHSONIAN SOCIETY [*sic*] BACKS IT.

As he recalled a few years later: "From that day, the whole thing was summed up, in the public mind, in the words 'moon rocket'; thus it happened that in trying to minimize the sensational side, I had really made more of a stir than if I had discussed transportation to Mars, which would probably have been considered as ridiculous by the press representative and doubtless never mentioned."[20]

He conveniently forgot that he had spent the past decade addressing ways to explore outer space. To be sure, his writings on that subject had received limited circulation, and when he did publish them, it was in an academic outlet and in academic terms. Nevertheless, he had conceived himself to be the intellectual harbinger of space travel, and now he was.

It was sensational. Witness the headlines from around the country, all on 12 January: AIMS TO REACH MOON WITH NEW ROCKET (*The New York Times*); SAVANT INVENTS ROCKET WHICH WILL HIT MOON (*San Francisco Examiner*); SCIENCE TO TRY SHOOTING MOON WITH A ROCKET (*Chicago Tribune*); MODERN JULES VERNE INVENTS ROCKET TO REACH MOON (*Boston American*). By the thirteenth the story had generated cartoons and editorials, mostly of a gee-whiz sort. The story was then all over the country, and had begun appearing overseas.[21]

It is difficult to appreciate the storm of publicity that Goddard's Smithsonian paper generated, or that the uproar continued for the next two decades, or the impression it made on Goddard, validating his self-image as the world's first rocketeer. By March the story was generating predictions that the moon rocket was about ready to go. That attracted newspaper accounts about and letters to Goddard from people who wanted to ride along. By April the papers had Goddard about to leave for Mars.

It all might have died down then, except that the National Geographic Society announced, on no apparent authority, that Goddard would make his first test flights that coming summer, and might reach the Moon. That generated scores of new articles and cartoons. The newspaper noise ratcheted up again in September with more reports that the rocket was about to fly. This time, Goddard put word out that his rocket was small, and that he needed an angel to finance it. But the following January he told *The New York Times* that he should make the first trials of his rocket, ultimately aimed at the Moon, the following summer.

No sooner had the fuss this stirred up died back down than the Smithsonian announced in July that it expected Goddard's rocket to reach outer space. That earned Goddard a visit from the AP, which in August circulated a long article about his plans. Printed all over the country, this piece, as had been some earlier ones, was illustrated with fanciful science-fiction pictures. Not surprisingly, it attracted more volunteers wanting to ride to the Moon. So it went for years to come.[22]

Goddard was bemused by the fuss over his moon rocket. He was an academic, through and through. As such he was put off by erroneous press accounts of his science, although he was not offended by them. Instead, he reveled in his newfound celebrity.

He recognized that the daily press worked from the Smithsonian news release rather than from his actual publication. Accordingly, he tried to set the facts straight, beginning with his own press release on 18 January 1920, which declared that there had been too much speculation. "Any rocket apparatus for great elevations," he continued, "must first be tested at various moderate altitudes. Also a knowledge of densities at high levels is essential. Hence from any point of view an investigation of the atmosphere is the work that lies ahead." He ended by proposing to raise, by public subscription, $50,000 to $100,000 to support this work.

Perhaps he thought such a dry pronouncement would moderate the enthusiasm in the press, as if academic mumbling would sell more newspapers than extraterrestrial fantasizing. Sending a copy of the release to Walcott, he said that "people must realize, nevertheless, that real progress is a succession of logical steps, and not a leap in the dark, and hence it is very important that, for whatever reason interest is taken in the work, adequate support and interest should be given the preliminary investigations." If the Smithsonian provided more money, he could get on with his multi-charge mechanism and put a rocket high into the sky.[23]

Goddard freely granted interviews to newspapermen over the years, aiming to replace sensationalism with scientific reason. He was dull copy in his own words, but exciting stuff when a reporter's imagination ventured into space with him. When he issued his first press release, he persuaded his mentor Webster to offer a statement in his support, but his adviser proved equally dry: "No one in the history of heat engines," he said, "has ever increased the efficiency of an engine in the same degree [as Goddard]. This is triumph enough. Never mind whether it hits the moon."[24]

Goddard next approached the scientific press. "I am beginning to appreciate the difficulty," he said in a letter published in *Science* in early February, "of making oneself understood in a statement where matters are suggested rather than explained in detail, and where a critical attitude is urged until a result is actually verified by experiment, even when one feels perfectly confident beforehand what the result will be." He affirmed that development of his rocket could go forward so long as he obtained funding, and reaffirmed that its range need not be limited to the atmosphere.[25]

Only those sharing his interests wanted to hear that kind of talk, as opposed to fanciful speculations on where he was headed. When the *Inventors Bulletin* asked for an "Autobiographical Statement" describing who he was, what he had done, and what his goals were, he provided one.

With the general press, where he hoped to replace fantasy with conservative science, he could be his own worst enemy. He told *The New York Times* early in 1921 that he would have the first "workable model" of the rocket that would eventually reach the Moon ready for trial soon. That caused the paper to lead with: "Professor Goddard declared today that he was more convinced than ever of the possibility of making a rocket that would reach to the moon, designed in such a manner that it would give off a flash when it struck that could be seen from the earth."[26]

The *Times* ran that piece under the headline MOON ROCKET READY SOON; PROF. GODDARD EXPECTED TO TRY OUT INVENTION NEXT SUMMER. It set off another hoorah in papers around the country. Goddard then turned to the lay scientific press, publishing a long statement in *Scientific American*, 26 February 1921, refuting the "misconceptions" that were in the air. But he reiterated the practicality of getting to high altitudes and even the Moon. His experiments, he said, were going slowly because his funds were short. He appealed for help in raising money.[27]

He hit the lecture circuit, playing on his notoriety to drum up cash. "I am not going to say much about rockets this afternoon," he told one group. "I am doing very little with it because it costs so much to do anything these days. Patience is all right, but you must have money. The Weather Bureau is much interested in my invention, but it has no money, even to carry on its own investigations." He told another group about his plans for high-altitude weather research via rockets, and his need for funding, but press reports emphasized his saying that a rocket could go faster in the vacuum of outer space than in the atmosphere of Earth.[28]

Goddard's authorized biographer claimed that, throughout his life, the rocket scientist tried to conceal his interest in space travel. Nothing could be further from the truth. He continually reminded the world of where he wanted his rockets to go; he just wanted people to understand that it would not happen at once. His most repeated theme was that he needed money to take the first steps toward the Moon.[29]

The publicity made Goddard famous around the world. It also ignited bands of rocket enthusiasts, who in 1920 formed the first rocket clubs in Russia and the United States, followed by scores of others in many lands. He always gave such groups a stiff arm, believing that rocketry was his area of work, where he wanted no interlopers. That was a serious mistake, because amateurs could have offered him support for his funding.[30]

Other interest sparked by the publicity was more welcome. When the British scientific journal *Nature* asked him to summarize his work, he complied gladly. He was similarly responsive to observatories and fellow scientists.

Toward others his reaction was mixed. Robert Esnault-Pelterie, known as REP, of France was the most notable. He had started building airplanes in 1904, and was a famous aviation pioneer. He developed an interest in rockets by 1912, and a decade later founded the Hirsch Prize for achievements in astronautics. After the Smithsonian report came out, he began a long correspondence with Goddard, who was responsive to someone who could talk his language, albeit put off by the notion that somebody else had been into rocketry as long as he had. They exchanged copies of their respective publications.

When REP suggested that Goddard's plan to hit the Moon with flash powder was conservative—REP wanted to fly a camera around the Moon and back—the latter became standoffish. He spurned the Frenchman's invitations to address conferences in Europe and later to submit entries for the Hirsch competition. Nevertheless, REP remained a great fan of the American.[31]

Theodore von Kármán arrived in the United States from Hungary in 1926. He was an aeronautical expert, who would go on to found the Jet Propulsion Laboratory and become one of the great pioneers of American aviation and rocketry. Three decades later, he recalled that before he left Europe he was aware of three outstanding figures in the science of rocketry. They were Goddard, Tsiolkovsky—and Hermann Julius Oberth. By the late 1960s the three would be acclaimed as the triple "fathers" of modern rocketry, but Goddard and Oberth presented a peculiar dichotomy.[32]

Oberth was born 25 June 1894 in Transylvania, a German area of Austria-Hungary. He enrolled in medical school, but had already become fascinated with rockets. In combat during World War I, he proposed to the German government that a long-range rocket could win the war; he received no answer. In 1919 he left medical school to study physics at Heidelberg. When his dissertation on spaceflight was rejected, he had it published privately in Munich in 1923. It was dry and turgidly academic, but when translated into more readable language by enthusiasts including Max Valier, a Munich astronomer, and Willy Ley, a German science writer, it spread the rocket-into-space idea around Europe, especially in Germany. His work as a technical adviser to Fritz Lang during the making of the 1929 movie *Die Frau im Mond* (released in America as *Woman in the Moon*), which featured spectacular special-effects rocketry, established him as the "father" of rocketry in Germany.[33]

When Goddard's "Method of Reaching" appeared, Oberth wrote the American, saying that he also had been working on the problem of getting out of the atmosphere. He was nearly done writing his book, and was surprised to learn that he was not the only one interested in this line of research. He asked Goddard for a copy of his Smithsonian report, and Goddard complied.

When he received it, Oberth promised to address Goddard's work in an appendix to his book. *Die Rakete zu den Planetenräumen* (*Rockets into Planetary Space*) appeared in 1923, and Goddard read the appendix as an unfair comparison of Oberth's work with his own. When he heard from Oberth thereafter, his replies were terse.

When *Nature* published a favorable review of Oberth's book in 1924, Goddard dashed off a rebuttal. He was fiercely determined to establish his own priority, and to refute what he saw as Oberth's implication that Goddard lacked the vision to see the spaceflight possibilities. He not only suggested that Oberth had purloined his ideas, but dismissed him as a mere theorist. Only Goddard was experimenting, he averred, and he would work out getting to the Moon in due course.

It was hate at first sight, and it endured. Nils Riffolt recalled that throughout the 1920s Goddard was obsessed with Oberth and what he was up to. Goddard was convinced that the German had stolen his ideas.

Three decades later, Oberth claimed that the only influence on his early work had been Jules Verne, and that he had as much as never heard of Goddard. Even later, Oberth's biographer suggested that it was actually Goddard who had first written to Oberth, from whom he stole his information, and he damned the American for having no vision of rocketry's potential. The Goddard-Oberth relationship poisoned the history of rocketry, introducing personal partisanship into what should have been an effort to record events.[34]

The moon-rocket fuss could have left a foul taste in Goddard's mouth, which might be attributed to the bad blood between him and Oberth, but it did not. Having the chairman of the chemistry department, Dr. Benjamin Merigold, greet him daily with "Well, Robert, how is your moon-going rocket?" might have put him off, but there is no evidence on that score either. In fact, Goddard enjoyed all the attention.

He did not wince even when *The New York Times*, in an editorial on 13 January 1920, offered the following remarks: "That Professor Goddard, with his 'chair' in Clark College and the countenancing of the Smithsonian Institution, does not know the relation of action to reaction, and of the need to have something better than a vacuum against which to react—to say that would be absurd. Of course he only seems to lack the knowledge ladled out daily in high schools." That might have infuriated some people, especially since it was the *Times* writer who had the science wrong.[35]

But Goddard took the statement in stride, because as a whole the commentary was effusively enthusiastic about his rocketry. He paid the editorial no more attention than any of the other news coverage. He had seen scientific errors in the press before and had not taken them personally.

Years later, however, he showed a copy of the *Times* piece to a visitor, and vented bitterness at the way the press had treated him. Something in the intervening period had changed his outlook. He had gotten married.

SHE WOULD CERTAINLY SEE THAT HE WAS CONSTANTLY ENCOURAGED

Fannie Goddard died on 29 January 1920. She had been declining for some time, tended by her live-in nurse, Jennie Ward Messick, a widow. Robert was deprived of the dearest women in his life, first Gram and now his mother. On 1 February, he and visiting relatives attended the funeral, after which Robert, Nahum, and an uncle spent the evening smoking cigars. A decent year later, Nahum, age sixty, married Mrs. Messick, who was fifty-two and had continued to reside in Nahum's house. Robert was their best man.[36]

His father provided for him, and offered advice and support whenever he needed them. But such attention did not substitute for the feminine care and protection Robert had known all his life from his grandmother and his mother. Nor did Jennie take their place.

Without womanly protection, he showed some signs of independence after his California adventure. In the spring of 1919, he took driving lessons and a few months later bought his first car, an Oakland roadster. This was intended to serve a new venture he had organized with L. T. E. Thompson, Nils Riffolt, C. N. Hickman, and chemist Rowland C. Beebe—Industrial Research Laboratories, a consulting firm.

It failed for want of business, and Goddard was left with the automobile. "He had a little car," Riffolt remembered, "a little coupe, wasn't much of a car. He wasn't much of a driver, either. One day when he backed out of the garage, the engine fell out." The man who would build machines to conquer space was mechanically inept.[37]

Driving a car was a famous way to meet girls, and he did take a few for drives. In October 1919, he met Esther Christine Kisk, a typist in the Clark University president's office. Just eighteen years old, she had graduated from South High School the previous spring and worked at Clark to earn money for college. Her parents, immigrants from Sweden named August and Augusta Kisk, were a hardworking couple. He was a foreman in a wood-working shop, while she ran a lunch stand at another factory. As Esther recalled, they did not at first approve of her taking up with a man nearing forty, but they warmed up to him.

Esther was both intelligent and physically striking. When she attended Bates College, she earned straight As in English, French, German, Greek, history, and public speaking. After she married Robert she took summer classes at Clark, with similar results again in German and English. She was a handsome blue-eyed blonde, slender without being willowy, and her high brow, straight nose, and strong chin gave her an unmistakable resemblance to Goddard's mother.

She was also strong-willed and energetic in going after what she wanted, and was not above turning on a spunky sort of charm as if with a faucet. She decided that she wanted Robert, turned on the charm, and he was smitten. Before long she was riding beside him in his jalopy.

The cause of their meeting was his frequent need for a typist, and Esther took in typing to earn extra money. Soon he was a regular visitor. "He was a tall, slender man," she remembered, "but had a very gentle and kind way with him. He used to come to my home, which was near the University, to read over the material, and frequently he would bring a little punch along with him.... My mother... was very neat, and it would annoy her that the little round bits of paper from this punch would be scattered on her rug." He played the piano for her, took her for walks and ice cream, and wrote gushy letters.[38]

Esther persuaded Robert to take breaks with her on weekends. She also encouraged him to paint. "He was profoundly aware of the beauties of nature," she said, "and on Sunday afternoon walks liked to photograph in color a lovely cloud, a colorful tree, a bursting milkweed pod. When time permitted, he painted in oils, usually landscapes. He played the piano by ear only, but with a physicist's precision in harmony, and could drift along for an hour, harmonizing."

Perhaps because playing the piano was one of the first things he did in her presence, Esther ensured that Robert's authorized biography emphasized that activity. In fact, while his acquaintances remembered his painting, almost none was aware that he played the piano.[39]

As for Esther, a Clark biologist remembered that "she was as different from most girls as he was—scholarly, intellectual, reserved, also on acquaintance very cute, interested in cultural things, new books, theories—not a common person. During the early years she ran fast to keep up with him." He recalled that she read a lot, and was "an interesting talker," but "plainly dressed, very prim, no lipstick, naïve about marriage."[40]

In 1922, they announced their engagement. Calling herself "old-fashioned," she dropped out of college and returned to her typist's job to save up for her trousseau. When she was ready, on 21 June 1924 they were wed in a noontime ceremony, with no attendants, at St. John's Episcopal Church in Worcester. Though they had both been raised in that church, neither visited it often. "We didn't go to church regularly," Esther said of their life together.

The event attracted attention on account of Robert's notoriety. ROCKET TO MOON WORKER MARRIED, said the *Boston Globe*. The *Worcester Evening Post* described the ceremony and Esther's "gown of white georgette crepe made over Canton crepe and a corsage bouquet." She left for their honeymoon, an auto trip through the White Mountains of New

Hampshire, in a "roshanara crepe gown with a gray wrap and hat to match." She was twenty-three; he was nearly forty-two.[41]

They were extraordinary people, so their marriage was unusual from the outset. They were both photographers, but they returned from their honeymoon with only two pictures of themselves. The images were almost identical, showing a mountain waterfall, with a small figure in the lower left corner—Esther in one shot, Robert in the other. There were no views of them together.[42]

There was a problem with intimacy. To satisfy her parents, Robert underwent a physical examination before the wedding, and the doctor gave him a clean bill of health. It is likely that the physician also gave them a consultation together about Robert's tubercular history. He would have advised them to restrict, perhaps entirely avoid, physical contact. Mouth-to-mouth contact would be forbidden absolutely, so as not to endanger Esther. Throughout their marriage, Robert and Esther slept in separate beds, with a barometer hanging on the wall between them.[43]

Esther put her considerable talents and energy into arranging their life. Robert had inherited the Maple Hill house from Gram, and they moved into it after the honeymoon. It needed an overhaul, and Esther planned and directed one. She put in all new utilities, replaced all windows and sash, ripped out a partition to create a large living room, installed a fireplace, and added a wraparound porch. From roof to cellar, the old place became modern and comfortable.[44]

She next turned her attention to her bridegroom. He needed guidance, in her view. "Most of the time," she remembered, "he did not know whether or not he had any money in the bank, or in his pockets. Fifteen minutes after dinner, he could not have told what he had eaten. Clark students told a story about his walking along a corridor in the main building, still holding an open umbrella over his head!"

Esther tucked Robert under her wing, as had Gram. She said later that it was "the social and recreational side of his life that was curtailed." She dragged him to campus parties and picnics, and to faculty bridge parties. He hated bridge, and during games sat in a corner puffing his cigars.[45]

He was too shy to become a social butterfly, so Esther did not push him too hard. She believed that he needed protection, however. Although he never showed resentment about the moon-rocket commotion, she confessed to Robert's biographer that she did, and her attitude slowly affected his. She therefore embarked on what Paul Horgan called her "job," which "was to see that he was effective and happy; and in her lifelong modeling of Bob's façade as that of a Great Man, she would certainly see that he was constantly encouraged about his own achievement and his qualities, his stature. She was indefatigable about it, in fact."[46]

Robert had been trying to fly his rockets high for years, without success. Now the Goddard team took on the challenge.

SIX

THIS FLIGHT WAS SIGNIFICANT, AS IT WAS THE FIRST

Bring me my bow of burning gold,
Bring me my arrows of desire,
Bring me my spear—O clouds, unfold!
Bring me my chariot of fire!
—William Blake, *Milton: Prefatory Poem*

One of Goddard's favorite occasions during the 1920s was "Sub-Freshman Day," when prospective members of the next freshman class visited campus, and he gave them flashy demonstrations of physical principles. Sometimes he quoted the 1920 commentary in *The New York Times*, then fired off his blank-loaded pistol in an evacuated bell jar. As the weapon twirled, he said: "So much for *The New York Times*." It was not an expression of resentment, but a demonstration of the error of common assumptions about science. Students who attended Clark, he promised, would learn better.[1]

The trustees appointed Goddard a full professor beginning in the fall of 1920. "I trust that you will see in this," Sanford told him, "an appreciation on the part of the Trustees of the valuable scientific work which you have been carrying on." Goddard next learned that Sanford was retiring, and wrote to express gratitude for all that he had done for him. "I can say, finally," he closed, "that I can remember no one with whom I have ever been associated with whom it has been a greater pleasure to work."[2]

The rocket man was becoming more comfortable at Clark, but it was not WPI, where he did not reoccupy the Magnetic Building; Tech was in a period of rapid changes, and needed the structure for other uses. Moreover, Duff was not as paternal as before. He had a habit of forgetting names, and after he returned from service as a ballistics expert with the navy, he may not have recognized the younger man.

Goddard wanted a dependable atmosphere around him, and Sanford's departure promised uncertainty. Fortunately, he still enjoyed the patronage of Dr. Webster.

HE'D GO AWAY FOR A FEW DAYS OR A WEEK AND HIS CLASSES WOULD BE CANCELED

Goddard was a tall but stooped man, thin and lanky, shy and meek and scholarly. His patrons were small, avuncular, and benign. Duff, Webster, and Abbot were all slight of frame, bushy of mustache, and soft of speech. Sanford was a model of the quiet, unworldly academic who governed by collegial consensus. G. Stanley Hall was exceptional among this crew, a square-shouldered empire-builder with a high brow and senatorial beard. But he had headed the university, not the college, so Goddard had never answered to a superior who even approached Hall in formidability. A peaceable man governed by peaceable mentors, the rocketeer was about to meet a redoubtable new presence in his life.

Sanford and Hall retired in 1920, and the trustees lured Wallace W. Atwood away from the Department of Geography at Harvard to replace both of them. They folded the college into Clark University, and gave Atwood the sole presidency.

He was a big bear of a man, with a square, clean-shaven face, iron jaw, and the cold, hard eyes of a Gilded Age robber baron. He was fifty years old when he took command at Clark, a giant in geomorphology and physical geography. One of the last in his field to emphasize the empirical tradition of the nineteenth century, between 1909 and 1948 he spent every summer in the Rocky Mountains gathering data. From 1909 until his death in 1949 he was a geologist for the U.S. Geological Survey, and for years advised the National Park Service on how to interpret nature to the public. His popular

The Rocky Mountains and his landmark *The Physiographic Provinces of North America* remained in use a half century after he was gone.

As Hall had been on a mission to establish psychology as a discipline, Atwood arrived at Clark determined to create a graduate school of geography. He accomplished that by terminating several other doctoral programs, although physics withstood his assault. He gained a reputation as a dictator—in his first year he caused a student uprising by turning out the lights on a guest speaker who annoyed him. In Percy M. Roope's opinion, Atwood was fair if hard as an administrator, but he showed no interest in students.[3]

Atwood began pressuring Webster to retire as early as 1921, to squeeze him out of the faculty and terminate the physics doctoral program. Whether because of that or some other demon tearing at his soul, in May 1923 the sixty-year-old Webster shot himself in the physics lab. Goddard delivered a eulogy at his funeral, and succeeded him as head of the physics program. He became head of mathematics as well, and chaired the two departments until 1943. He was not an effective administrator, and his nearly constant absence after 1930 allowed Atwood to kill the physics graduate program.[4]

It appeared at first that Goddard and Atwood would get along well, despite their competing ambitions. The new president helped the rocketeer secure a grant from the Clark trustees, and in 1921 gained him a raise in salary to $2,500 per year. Further raises followed through the decade, with Goddard earning $4,000 annually by 1928. Atwood also instituted sabbaticals for Clark faculty. When Goddard's term came, he consulted with W. F. Magie about a fellowship to study overseas, then begged off on account of the demands of his research and his father's declining health.[5]

The first sign of trouble between the two men appeared in 1926, when the president discovered that the professor was double-dipping in retirement plans. "It appears that we are purchasing an annuity in cooperation with Dr. Goddard for his estate," Atwood told the trustees, "but from my records it appears that Dr. Goddard is covered by the Carnegie Foundation. He was appointed instructor in the fall of 1914, and has been continuously in service since that time."

The Carnegie Foundation for the Advancement of Teaching, to provide security for college professors, had established a national plan in 1906 to offer them retirement annuities. This program was replaced by the Teachers Insurance and Annuity Association (TIAA) in 1918, but continued to cover enrolled members. It encouraged universities to establish their own pension or annuity plans. Clark's plan was not intended to include those already covered by Carnegie, however. In Atwood's view Goddard was feathering his nest at the expense of the system, but there was nothing the trustees could do about the matter.[6]

Goddard was also indifferent to his administrative responsibilities. One duty of faculty was to sponsor student extracurricular activities, and in 1920 Goddard organized a "Wireless Club" at Clark. The organization began by listening to crystal radios, but shortly the members raised money to establish a broadcasting station, which was on the air by 1921. By 1925 the club operated a licensed station, WCN, broadcasting daily with a full-time employee, along with experimental broadcasts under two experimental licenses.

Goddard had lost interest, and the club ran itself as well as unsupervised youngsters might be expected to do. Once, when Atwood was away from campus, the acting president complained that—besides singing, piano playing, and other commotion at inappropriate times—the members had set fire to the roof of the Science Building.[7]

Then there was the omnipresent Esther, who began managing Goddard's life even before their marriage in 1924; she remodeled the physics department office when he succeeded Webster. Becoming an active participant in his work, she boosted him to all who would listen, aiming to win adherents through her own grace and considerable charm. She was a photographer, often seen around the university taking pictures of Goddard in the classroom and in the laboratory, urging him to spend money from the department budget for photographic equipment and supplies. When she learned to drive in 1926, there was no escaping her.[8]

Esther was no more peculiar than her husband, whose chief interest struck most colleagues as fanciful. "No question in my mind," biologist

William Cole remembered, "in those days the idea of a rocket ship was just foolishness. Anyone interested in it, we thought, must have been a little off. I felt the faculty respected him as a physicist...but still [thought he was] queer." As time went on, "He got more and more mysterious about what he was doing. He'd go away for a few days or a week and his classes would be cancelled."[9]

Roope, who assisted in Goddard's experiments, understood how he mystified others. "Most of his time outside the class was spent in the laboratory," he explained, "and they didn't realize that rocketry was a combination of theory and experiment, and Goddard, being the leader in the field, had no one to turn to. Every advance had to be done by experiment. This involved a lot of gadgetry, a tremendous amount of machine work."[10]

THIS HAS BEEN DISTINCTLY AN AMERICAN PIECE OF WORK

Goddard filed seven successful patent applications between World War I and the time he left Worcester late in the summer of 1930. None, as it turned out, had any commercial value.[11]

He filled his notebooks with his ideas, then formalized them as affidavits that he buttonholed other faculty members to sign as witnesses. According to Cole, they never understood this obsession: "When the rest of us had something to say, we tried the professional journals, but Bob wanted patents and affidavits instead. I never knew why." Patent attorney Hawley left his own observation: "What he really wanted was to protect his priorities."[12]

Goddard's secretiveness was a legend that grew up later. He was not averse to publicity; he just wanted it to convey his message in his terms, so he talked with reporters and generated his own news stories when it suited him. Stung by the Haigis-Rockwood affair, he took steps to ensure that only he did the talking, and that his employees did not speak out of turn.

He insisted that they sign secrecy pledges at the start of employment, of which the following was typical: "I, Henry Sachs, of Bladensburg, Maryland, hereby agree not to divulge the nature of any of the work which will be done by me, or will come to my knowledge, in the rocket investigation upon which I am to work under Dr. R. H. Goddard, of Worcester, Massachusetts." When they left, they signed a similar statement.[13]

Confidentiality commitments have an ancient place in science, going back to the early days of the scientific revolution. They prevented assistants or students from claiming a share of the master's credit for the work they had participated in. Far from defying tradition, the egoistic Goddard embodied it.[14]

More troubling to fellow academics was Goddard's failure to publish his results in a peer-reviewed scholarly journal. If asked about that, he said he was reluctant to put his work out for review before he had perfected all details, a stage difficult to reach in his field. When he did publish, it was often in popular periodicals, which was considered an academic sin in the twentieth century; witness the collegial cold shoulder earned by recent "popularizers" such as Carl Sagan and Stephen Jay Gould.

If Goddard's reticence was interpreted as secrecy, then it contradicted an ancient tenet of science, which held that all scientists gained from a free exchange of information. Hence the emphasis on publication, to support the verification of experiments and results. The claims of science, it was believed, were higher than those of the person, politics, creed, race, or country.[15]

Goddard was trained as a scientist and aware of the ethics of science. He was also a product of his own life. As the matter of the annuity demonstrated, and his lust for patents reflected, he wanted to profit from his work whenever possible. But that was only part of it, because Goddard was even more hungry for recognition of his originality than for money. He would not place the claims of science higher than those of person.

Nils Riffolt remembered him routinely saying in the 1920s, "Let's keep this under our hat." He remembered also that the professor was free with information among his crew, and only somewhat less free with the press. Regarding a certain other party, Goddard was on guard: "Oberth was working in Germany, and Goddard didn't want to disclose too much."[16]

Goddard was outraged at Oberth's claim to independent invention, asserting that the German's work plagiarized his own 1919 Smithsonian paper. In 1923 Goddard prepared a long, detailed report on all his tests, and an additional report asserting his priority, dating from 1899. He refuted Oberth's claims and alleged statements that Goddard's system could not reach into space. "I do not wish to open the question of priority and thrash out all the phases of the matter in public," he told the Smithsonian. Then he proceeded to do just that, pointing out flaws in Oberth's scheme for space travel, and evincing resentment of competition for first place in rocketry that, in the case of Oberth, bordered on paranoia.[17]

Goddard addressed the annual meeting of the AAAS at the end of 1923. There he reviewed his work since 1909, refuting Oberth's contention that Oberth had been the first to consider hydrogen/oxygen propulsion in 1912. Goddard pointed out that only he held patents in rocketry, and reviewed the flaws in the German's "purely theoretical" approach, contrasted with his own actual tests.

He concluded: "This has been distinctly an American piece of work; it originated in America, as the writer's own interest and endeavors date back to 1899; the first theoretical work was published and the first experiments performed in America; and it seems very desirable that enough support be had to enable the work to be completed at an American laboratory."

Here boldly were Goddard's two driving urges—reaching back to his boyhood daydreams for proof that "I had it first!" and seizing every opportunity to ask for research money.[18]

This speech prompted the editor of *Monthly Weather Review* to request a progress report on rockets, and Goddard complied promptly. He did not mention Oberth by name, but emphasized that he had been the first to suggest using liquid fuels and that he had been testing them since 1921. He also said, yet again, that a rocket could reach the Moon, even if that was a long way off. "The object of the work is, however," he remarked, "much more than the performance of some single spectacular stunt. It is the development of a new method."[19]

The AAAS presentation kicked off another round of "moon rocket" commotion in the press. Goddard had gotten his message out—part of the commentary now focused on the need for the United States to maintain a lead over foreigners. Simultaneously, Oberth's book appeared on this side of the Atlantic and was reviewed in American papers, which sent a succession of reporters to Worcester to ask Goddard about rocketing to the Moon, resulting in articles illustrated with photographs of Goddard posing with his inventions. The whole world expected his rocket to take off for outer space very soon.[20]

In 1925, *The New York Times* reviewed the work of Max Valier, who, it claimed, was about to send a rocket to the Moon that was based on the work of both Goddard and Oberth, which he had "devoured." Goddard was quoted as saying that any ideas Oberth had he had stolen from his own Smithsonian publication.[21]

Years later, apologists for the German rocket program would maintain that the Germans were mostly unaware of Goddard. In fact, he had a wide and avid audience in Germany during the 1920s, and although the American despised Oberth, he did not harbor ill will against Germans in general. He was receptive to honest inquiries from Germany; he was informative when he wanted to be, and guarded when he did not want to help the competition.

In the fall of 1925 a German technical journal asked him to write an article describing his work, which the editor said had drawn wide interest in his country. Goddard answered that he was working on a model, and until he had completed it he did not feel he could say much. He would, however, keep the request in mind, "and in the meantime I wish to say that I feel honored because you have written me upon this subject, and also because the German public is interested in the work." He said the same to the editor of a Berlin newspaper.

When writer Otto Willi Gail sent him a copy of his latest science-fiction novel in April 1926 and asked for help getting it published in America, however, Goddard declined, because "it so obviously puts the American contribution to the subject in so unfavorable a light that I doubt if it would be very acceptable to the American public in its present form."[22]

Also in 1926 he assented to a request from a German publisher who wanted to translate and publish his 1919 "Method of Reaching" paper. But when the German commercial attaché in New York asked for an update of his work, he responded: "I have not published anything of the sort since this was published in 1919."

Then in December 1926, he asked Abbot to make sure that nobody released any of his letters or reports, "for the reason that persons abroad are trying every possible means of getting information as to what I am doing. Their latest move has been to write to the German Consulate in this country, which has written to me, but which does not give the name of the inquirer." Where the German official simply relayed a request for information, his failure to say who had originated the question became in the rocket man's view evidence of a secret plot against him.[23]

Throughout the 1920s Goddard's fame spread across Europe, and he received inquiries from editors, technicians, science-fiction writers, and spaceflight volunteers in countries from the Mediterranean to the Baltic Seas. He was most pleased by the opinions of Nicolai A. Rynin, Professor of Aerial Transport at Leningrad University, who accused Oberth of plagiarizing the American's work. Rynin, a formidable aeronautical scientist, was starting a second career in theoretical astronautics that would make him a prolific author of books, encyclopedias, and exhibitions. He and Goddard became friendly correspondents, exchanging publications while Rynin kept the Clark professor informed about the coverage of his work in the Soviet press, where there was no love lost for the Germans. Goddard provided material for Rynin's *History of the Idea of Flight to the Moon.*[24]

The occasion for all this foreign attention was the continuation of the moon-rocket publicity. PLANS GIANT ROCKET TO EXPLORE SPACE, *The New York Times* headlined an account of the AAAS speech. The article referred to Goddard's earlier plans to strike the Moon, but focused on his high-altitude objective. The story spread, and papers predicted that, once again, Goddard's rocket was about to take off. The volume of coverage tended to rise during the midwinter months, but to the end of the decade it never abated entirely.

In 1925, the moon rocket gave way partly to features on the transmission of radio signals through space. Not wanting to be overlooked, Goddard gave speeches and went to the press with his suggestion that rockets could carry radio transmitters. By 1926 expectations of an early Goddard space-flight were back in prominence, with the papers quoting astronomers and other scientists who hoped that he would launch soon.

The tremendous interest in rocketry was not limited to the plebeian press—such dignified outlets as *Review of Reviews, Literary Digest,* and *Scientific American* were on the bandwagon. Goddard's rocket was now gigantic, and bound for as far as Mars.[25]

His interest in rocketry, his obsession with patents, and the publicity that swirled around him made Goddard eccentric among his colleagues at Clark. That alone posed no difficulties. But oddballs could wear out their welcome if they failed to produce results—or if they failed to keep the research money coming in. Goddard had gotten years of mileage out of the expectation that his rocket would soon take to the sky. His time to send one up was running out.

HOPE THAT YOU WILL BE ABLE TO ACTU-ALLY SEND A ROCKET UP INTO THE AIR

To obtain support Goddard had to produce results, but he could not produce results without getting someone to pay for the work. He had a hidden asset, however, not immediately apparent even to himself: Charles G. Abbot, who became as much Dutch uncle as friendly patron. "We find these very tight times," Abbot told him in 1922. The next year he helped the rocketeer apply for a $3,000 grant from AAAS, which he hoped Goddard's speech at the annual meeting might earn him. That personal and jingoistic attack on Oberth backfired, and Abbot told him not to count on AAAS money.

Goddard's Clark grant was about gone, and in 1923 Riffolt had returned to working full-time for the university. Abbot decided to become more heavy-handed. When the rocketeer approached Watson Davis, the editor of the Science Service newsletter, early in 1924 with a plan to raise $4,000 by subscription, Abbot jumped on him. Generating another round of unseemly publicity would cause the Smithsonian to withdraw its support, he warned. Goddard backed off.[26]

His patron had been fishing for money on his own. The Smithsonian had access to the new Cottrell Fund, named for the chemist Dr. Frederick G. Cottrell, founder of the Research Corporation of New York, which administered the Hodgkins Fund. Abbot awaited some evidence of fiscal responsibility from the rocketeer, and received it in January 1924. Goddard told him that he had almost perfected a system for pumping fuels and that he had access to a supply of LOX, but without money he could not do much. Abbot surprised him by sending $500 from the Cottrell Fund, on the condition that Goddard cease generating moonship fantasies in the press and that he show results in place of promises.

The money was offered, Abbot said, "with the understanding that it is to be used for approaching a trial in the open in the most expeditious manner. In other words, the Institution feels that the chances of getting support will be excellent if a fairly successful trial in the open can be had, but that a further delay to try out some new scheme, even though it should promise to be a better scheme than that now in effect, would be apt to have a bad moral effect. If this grant makes satisfactory progress, it is not unlikely that we shall be able to give you some further support from the same source." Chastened by the realization that repeating the false hopes he had raised over the multiple-charge rocket could leave him penniless, Goddard accepted Abbot's terms.[27]

Abbot then had to contend with the enthusiasm of A. A. Hamerschlag, president of the Research Corporation, who wanted to publicize the grant. The Smithsonian, he told Hamerschlag, wanted no more publicity like that that had greeted the 1919 publication of "A Method of Reaching." Two months later, when the corporation president again wanted to crow to the press, Abbot squelched him: "I think the less publicity given to Dr. Goddard's work the better," he said, "as it results in a flood of correspondence and in newspaper interviews, which only give trouble and consume time."[28]

Goddard could not resist a little crowing of his own, telling Atwood about his grant from the Smithsonian Institution: "This is the strongest kind of endorsement, and recognition of the importance of the work that has been made possible by Clark University during the past two years."

When he received a surprise grant of $190 from AAAS, he passed the happy news on to Abbot, who, still dangling carrot and wielding stick, answered: "I am glad to know that you are getting more support. We also can give more on receipt of your satisfactory report of progress and analysis of just what remains to do to get a test, with estimate of needed funds." He never missed a chance to tell Goddard that, if he wanted more money, he would have to prove that he was worth the investment.[29]

There remained Hamerschlag's enthusiasm. In March 1924 he wanted to put more money into the project, and he asked the Smithsonian for a progress report. Walcott sent Abbot to Worcester to see how things were going and also agreed to release another $2,500 of Cottrell money, although not all at once.

Abbot's visit inspired the rocketeer to submit a "Supplementary Report on Ultimate Developments," yet another excursion among the planets. It outlined specifications for various vehicles to reach the Moon or the planets. Given that Goddard had not yet seen a liquid-fuel rocket get off the ground, it was a far-fetched essay, written as if he felt Oberth breathing down his neck.[30]

Such digressions kept Abbot uneasy about whether Goddard's promises would ever fly. When Abbot sent him another $500 in April 1924, Goddard replied that his engine was now working well with compressed air but not fast enough with LOX. He could move faster, he said, if he could get his hands on more than two liters per day of LOX. It was another excuse for lack of progress.

Two years later, in 1926, Walcott offered Goddard another $250, and asked: "Would this be sufficient to bring the apparatus to a test?" The rocketeer promised: "I believe this will be sufficient to bring the apparatus to a test, as I am at present preparing for the out-of-door experiments." For the first time, a Goddard promise of an early flight became reality, a consequence of which was continued funding from the Cottrell Fund for the rest of the decade.[31]

To get to that stage required following a long, hard, and meandering road. As Abbot knew, Goddard had difficulty focusing his mind on a problem; often he would drop an approach and chase another because it suddenly occurred to him. Equally often he would digress from the mechanics of rocketry to wander among the planets.

These shortcomings were especially evident during periods when money was no object. When circumstances forced him to concentrate, Goddard's focus improved, as when the army gave him a deadline in 1918, or through the 1920s when Abbot held his funding in a tight grip. Later in the 1930s, when he had money in his pocket and was left on his own, Goddard reverted to his haphazard way of doing things.

Navy rocketeer Robert C. Truax was inspired by Goddard to enter the field of rocketry, but became amazed at how the elder man practiced his art: "If you review the record of the tests he made, it is simply astonishing in light of present practice how he would take one component, which maybe had only one successful test on it; he would then combine it with a dozen other components, some of which had only one successful test and some of which had never been tested; and he would put them all together in a rocket and try to fly it. . . . And the most remarkable thing of all is that sometimes it worked."[32]

Goddard at least partly overcame the burden of his unsystematic practices through his persistence, which, as in his refusal to give up on the multicharge idea, could become stubbornness. This was interpreted by those who knew him as an amazing patience, in the spirit of Thomas Edison. Roope recalled: "We could go out on a trial and the rocket would explode. Some of us standing around would say, 'Gee, this is terrible,' and Goddard would say, 'We learned something today. We won't make this mistake again. We'll correct it.' "[33]

L. T. E. Thompson believed it was Goddard's talents as a scientist that overcame his inventive limitations and allowed him to leave his mark on the field. He was, after all, a pioneer: "I say also that the analysis which he produced while working alone in a new field for almost a generation, was fully up to the standard set by most of the pioneers in comparable fields who had to do the roughing-out work.... When placed alongside the work of a [Samuel P.] Langley or even a Benjamin Franklin, or of others who opened up major zones of scientific interest by following the experimental approaches, Goddard's analytic work looks good."[34]

Goddard embarked on the work that would make him a "father" of modern rocketry—liquid-fuel propulsion—in 1921, with Riffolt's help. There were two major technical problems. One was that the combustion chamber would have to receive, mix, and burn the fuel and oxidizer, and exhaust the combustion gases, in an efficient manner; a chamber for solid fuels was comparatively simple. The other challenge was to find a way to feed the fuel and oxidizer into the chamber; this required pressure, and for the moment Goddard settled on the obvious step of using pumps.

He also found a supply of LOX, which he obtained from the local outlet of Linde Air Products Company. On 13 September he conducted his first LOX experiment: "Lost part of it, and used up all rest without getting it to reach chamber. Planned new apparatus."[35]

He was now committed, and on 22 November he blew up his solid propellants at Aunt Effie's farm. He concentrated on testing various pump designs, but suffered one failure after another as the pumps blew out, seized up, or simply fizzled. By March he was optimistic enough to tell the Clark trustees that he had abandoned solid propellants and diverted their grant into liquid-fuel work. He said he had proven that he could mix LOX and ether and burn them without an uncontrollable explosion.

Pumps were another matter, along with other details, reflected in "shots" at Aunt Effie's in the spring of 1922: "April 7. Back pressure forced mixture into O[xygen] pipe and it exploded and all the fire came from this break. April 12. Chamber smoked and one rubber tube at O tank burned through. April 19. The rubber tube from O blew off, and showered drops of E[ther], and then E tank top blew off, broke cast-iron clamps on O tank and hit bottom of gauge, also chamber was pushed upward, cracking E pipe below tank."[36]

Goddard continued to wander from one thing to another. He tried cotton soaked in kerosene inserted into the chamber as an igniter. He labored over pumps, and they failed. Ether was both dangerous and inadequate, so he experimented with glycerine and with ammonium nitrite. In 1923 he claimed that he had succeeded in developing a "very light device ... for pumping the two liquids ... into a combustion chamber; and tests are in preparation for obtaining the proper mixture and combustion." He used a modified rotary pump, but began to believe that a piston pump might work better.[37]

"Whatever Dr. Goddard had at stake," Riffolt remembered, "he tried to keep it to himself. Most of our tests ended with something jamming, or sticking, or the chamber burning through. Yet I never saw him discouraged. These were not failures, he usually said, but what he called valuable negative information." What Goddard had at stake was a shortage of money that would send the younger man back to university work.

That fiscal shortage was aggravated by Abbot's shortening temper. "I am, however, consumed with impatience," his patron told him in November 1923, "and hope that you will be able to actually send a rocket up into the air some time soon. Interplanetary space will look much nearer to me after I had seen one of your rockets go up five or six miles in our own atmosphere."

That set off an exchange between them. Goddard vented his frustration, especially with the continuing expense of obtaining LOX from week to week, as it could not be stored and was produced only as a by-product of other processes on no certain schedule. He had talked Linde into giving him the stuff free until the end of the year, but thereafter things were uncertain. He had asked the National Geographic Society to buy him his own LOX plant, but without result. Abbot responded sympathetically.

The Cottrell money the following year put them both into a better mood, and in March Goddard claimed he was well on his way to a flight before the end of 1924. He now had a supply of LOX, and used it to oxidize a gasoline-ether mixture. His rocket included a combustion chamber and nozzle, an igniter to ensure continuous combustion, a pump to drive the LOX and the fuel into the chamber, a "means of diverting a portion of the oxygen gas under compression to drive the pump," and a means to regulate the flow of the LOX. When Abbot saw all this during a visit to Worcester early in 1924, he told Walcott: "On the whole, it seemed to me that there is every reason to believe that a successful trial of the rocket might be expected . . . within the present calendar year."

The year was consumed by more excuses from Goddard, however. Riffolt completed his master's degree and left for a job with the navy. Goddard had no luck replacing him, he could not get the pump to work right, Linde was failing to provide LOX, there was trouble finding suitable alloys for rocket parts, and so on. Abbot kept nagging him, and then AAAS got into the act, wondering what had become of its grant. He had not spent it, Goddard replied, repeating his list of laments.[38]

Goddard was a part-time rocketeer. His first obligation was as a professor, but he bent his academic responsibilities to serve his personal interests. As he admitted in 1927: "With a problem as fascinating as the navigation of interplanetary space as a hobby, it has been natural to assign masters of arts thesis topics in this field, when the student did not already have a problem upon which he desired to work. The results of these theses in affording a check upon the theoretical conclusions have been of interest, although unknown to the students carrying out the investigations." Many of the theses supported Goddard's interest in ion propulsion for interplanetary spacecraft, and often they provided data that he used, without the students' knowledge, to obtain patents in his own name.[39]

Whatever Goddard's ethics as a teacher, as a rocket builder he found things looking up early in 1925. On Abbot's recommendation, Henry Sachs, a gifted instrument maker, took temporary leave from the Bureau of Standards to join him. He stayed until 1931, returned to the government, and came out of retirement in 1944 to assist Goddard to the end. He was a quiet and eternally skeptical man, with the attitude and skills to make rocketry happen.

Goddard continued giving Abbot optimistic predictions throughout

1925. He claimed completion of a displacement pump driving LOX and gasoline, and predicted an early test. He had, however, another laundry list of reasons for not producing results. "Please let us know briefly how the prospects are for the flight," Abbot requested, "and whether you have means in sight for the purpose. If not, what further means would be absolutely necessary." Goddard asked for more money, and Abbot agreed "if that will assist you in bringing about a flight, which we are very anxious to have you accomplish."

That did not produce a flight, although in June Goddard claimed a successful static test with a rocket clamped down in his shed next to the Science Building. Because it jumped up and down with every stroke of the pump, he added another engine and pump. Sachs solved the problem by combining the pumps into a single, double-acting unit. This proved to be a significant advance.[40]

It made possible what happened on 6 December 1925. Goddard and Sachs, along with Esther, went to the Science Building, where they had previously mounted their twelve-pound rocket motor in the shed. Once fueled and ignited, the thing screeched for twenty seconds, and for ten of those seconds it raised itself against its metered restraints.

This was the first liquid-fueled rocket to lift its own weight. "This test was an important one," Goddard told Abbot, "in that it demonstrated that the problems of pumping, governing, and control of heating were solved, but it also showed that a rocket on so small a scale as this model would not lift itself sufficiently to give a flight, including, as it did, devices which would not be necessary in a larger rocket."

He was ecstatic, but neither Esther nor Abbot could work up any enthusiasm. They wanted a rocket to fly, and Goddard decided to give them what they wanted.[41]

I THINK I'LL GET THE HELL OUT OF HERE!

He set out to design something that at least would fly, even if it had no practical value for the future. The first objective was to reduce weight to a minimum. He and Sachs devised a two-inch combustion chamber with a nozzle, then worked on adjusting the flow rates of LOX and gasoline, now settled on as the fuel. When they discovered that too much combustion continued outside the chamber in the nozzle, they produced a new injection system with multiple orifices for the gasoline.

Goddard had wanted to wrap his rocket in a shiny metal shroud and add a parachute recovery system, but discarded both as unnecessary weight. A carbon-dioxide tank replaced the heavier pumps, but he scrapped it in the belief he could find a lighter way to apply pressure to the fuel and LOX. On 20 January 1926, a static test demonstrated that the thing might work, and Goddard bought Esther a movie camera. Sachs built a launch frame of gas pipe to hold the rocket before takeoff.

Assembling a flight model took up most of February, and what the two men ended up with would scarcely be recognizable as a rocket today. The motor was at the top, meaning it would drag the rest of the machine behind it. Goddard adopted this "tractor" configuration in the belief that it was the best way to ensure straight flight, although he eventually realized that as long as the motor was on the axis of the rocket, stability was the same.

The rocket was held together by its fuel lines. The LOX and fuel tanks hung beneath the motor on those lines, the top of the LOX tank protected from the exhaust flame by an asbestos-covered metal cone. Cotton in a cup of alcohol underneath the LOX tank proved a cheap way to provide pressure, by vaporizing the LOX and driving it upward to the motor and downward to the gasoline tank, whence the pressurized fuel flowed upward to the motor. Cork floats in the tanks kept the pressures even. The whole spindly thing was about ten feet long. They gave it a shot on 6 March, but the nozzle burned off and the rocket jammed into the top of the launch frame, bending the piping.

A revised model was ready on 16 March, a clear, cold, still day. Goddard and Sachs went out to Aunt Effie's in the morning, hauling the rocket, frame, a can of gasoline, and a Dewar's flask filled with LOX. Esther and Roope, carrying a theodolite to gauge altitude, arrived in early afternoon.

The scene was a picture of wintertime loveliness. Aunt Effie's fields wore a glittering blanket of snow; her barns overlooked the launch area from the top of a rise, above the rocketeers bundled bear-shaped against the cold, moving around at the bottom of a wide swale, surrounded by a scattering of naturally flocked trees. Esther photographed one and all, including Robert posing beside the rocket in its frame. As Sachs poured in the fuel, Esther grabbed her movie camera, a "Sept"—French for "seven," because it ran for seven seconds on each winding of its drive spring—and stepped back to record the proceedings.

Sachs lit a blowtorch attached to a long stick, and touched off the igniter, improvised from match heads, at the top of the motor. Then he lit the alcohol lamp under the LOX tank, and stepped behind a propped-up wooden door for shelter. Goddard turned a valve, which let pressurized oxygen from a tank enter the fuel system, donating a boost to the vapor pressure rising from the LOX tank, heated by the alcohol flame. The additional pressure forced both gasoline and oxygen into the combustion chamber, and a small, smokeless flame emitted from the nozzle with a growl.

Goddard recorded what happened next: "Even though the release was pulled, the rocket did not rise at first, but the flame came out, and there was a steady roar. After a number of seconds it rose, slowly until it cleared the frame, and then at express-train speed, curving over to the left, and striking the ice and snow, still going at a rapid rate. It looked almost magical as it rose, without any appreciably greater noise or flame, as if it said, 'I've been here long enough; I think I'll be going somewhere else, if you don't mind.' "

Esther admitted years later that he really said: "I think I'll get the hell out of here!" What had delayed takeoff was the fact that the 5¾-pound rocket weighed about 10¼ pounds with its fuel and oxygen load, while the motor produced under 9 pounds of thrust. It had to burn off excess fuel before it could lift itself. It caught Esther by surprise, as her movie camera ran down before the contraption rose, and by the time she had grabbed a still camera it was on the ground, leaving a faint vapor trail hanging in the frosty air.

The flight was not perfect. The motor fired about 20 seconds before liftoff, and burned off part of the lower end of the nozzle; the remaining nozzle was enough to provide thrust, but it was uneven. The rocket rose in a

tightening arc, topping out 41 feet above ground, and completing a semicir-
cle before hitting the ground 184 feet away. "The average speed," Goddard
told Abbot, "from the time of the flight as measured by a stopwatch, was 60
miles per hour. This flight was significant, as it was the first time that a
rocket operated by liquid propellants traveled under its own power."

A year later, he was still swollen with pride, comparing Aunt Effie's
frozen strawberry and cabbage patches to Kitty Hawk: "As a first flight it
compares favorably with the Wrights' first airplane flight, however, which
was 120 feet, with a height of 10 feet and a time of 12 seconds, and the event,
as demonstrating the first liquid-propelled rocket, was just as significant."[42]

Goddard must be given his due. The first flight of a liquid-propelled
rocket may not have looked like much; in fact, it did not greatly impress
Abbot. The rocket man himself knew that it was a stunt, with few of its technical aspects useful for high-altitude rocketry. But nothing like it had ever
happened on Earth before. Aunt Effie's truck farm was the Kitty Hawk of
modern rocketry, and Robert H. Goddard was its Wright Brothers.

Given his ego, yen for publicity, and hunger for recognition, he might
have been expected to shout the news from a rooftop, or at least in the direction of *The New York Times*. He did not. Reporting the achievement to Abbot,
he asked that the Smithsonian keep the matter quiet until he had completed
his next step. "My reason," he explained, "is that this rocket work is being
made almost a national issue in Germany."[43]

Just how news of American achievement of the first liquid-fueled flight
would abet German nationalism, Goddard did not explain. He was fully
aware that, lacking means to sustain fuel pressure, his little puddle jumper
was not the way to outer space, so it could invite German derision.

But no one else had ever achieved what he had just done—getting a
liquid-fuel rocket, however small and for however short a flight, off the
ground under its own power. While subsequent Goddard rocket launches
would generate their own headlines, however, the world would not know
for ten years that he, in this most important instance, could say without con-
tradiction: "I had it first!"

SEVEN

NELL

Granting that the question at present was simply that of sending a projectile up to the Moon, every one must see that that involved the commencement of a series of experiments.

—Jules Verne, *From the Earth to the Moon*

The August 1928 issue of *Amazing Stories* carried a story by Philip Francis Nowlan called "Armageddon—2419 A.D." It followed the adventures of a pilot who becomes a war hero in the future. A sequel, "The Airlords of Han," appeared in the March 1929 issue. The illustrated tales continued in a newspaper comic strip that ran from 1929 to 1967. That in turn engendered movies, radio and television series, legions of imitators, and innumerable toys, school lunchboxes, and other souvenirs. It was called *Buck Rogers in the 25th Century*.

The hero did not fight evil alone. Dr. Huer, a bald, mustached, and eternally optimistic scientific genius, invented Buck's rocket ship and other wonders. Thanks to his inventiveness, "Buck Rogers" became a metaphor and adjective for anything related to futuristic technology. Dr. Huer was modeled on a man whose face and ambitions to send rockets into space were familiar to newspaper readers everywhere—Robert H. Goddard.

In 1956, Esther Goddard told Milton Lehman that her husband had been proud of the association—once they even went out of their way to see a Buck Rogers exhibit in Chicago. Whenever there was a problem with Robert's rocket experiments, she said, she quoted Dr. Huer's refrain to him: "Now don't you worry, the old doctor will take care of that." Four years later, she told another interviewer: "I think this rather hurt my husband's feelings, though I doubt if it damaged his scientific reputation."[1]

IT IS PHYSICALLY POSSIBLE TO SEND A
ROCKET TO THE MOON

With or without Dr. Huer, Goddard was a constant presence in the newspapers. After he had been enjoined by Abbot to cease generating moon-shot fantasies, he tried, but he could not keep his mouth shut altogether. He emphasized that his target was high altitude, rather than the Moon, but always pointed out that the former was a step toward the latter.

In February 1927, he told the *Boston Herald* that he was near completion of a new rocket. In May, newspapers all over the East Coast carried an article, illustrated with pictures of Goddard and his inventions, under the headline WANT TO BE FIRST TO VISIT THE MOON? APPLY TO ROBERT GODDARD, CLARK UNIVERSITY. After Charles Lindbergh flew to Paris, Goddard coyly declined to say whether the Lone Eagle would pilot his moonship.

When the German Max Valier announced in 1927 plans to build a transatlantic passenger rocket, Goddard called the papers to scream patent infringement. Pressed by *The New York Times* for more information, he said: "This is no idle dream, but an actual scientific possibility. The idea of combining rocket and airplane is an offshoot of the space rocket on which I have been working for the past eleven years and whose possibilities I saw clearly as early as 1912."[2]

As this incident reflected, by this time Goddard had come fiercely to hold rocketry as his private domain—one that he had created exclusively during his boyhood, and to which he held sole title. Accusing someone of patent infringement in the case of an "offshoot" of an old idea of his stretched the bounds of reason. Yet he got away with it, thanks to a national press that seemed to ratify his belief that he was the world's first, foremost, and only legitimate rocket scientist. The rocket clubs and individual rocketeers popping up everywhere were, in his view, neither colleagues nor even competitors. They had become besiegers.

Widespread speculation about rocketry's possibilities attracted the notice of the science writer George W. Gray, who corresponded with and visited Goddard, producing an essay called "Speed" in the March 1930 *Atlantic Monthly*. It followed a front-page piece in *The New York Times* under the headline: A GREAT ROCKET FOR EXPLORING OUTER SPACE; PRO-FESSOR GODDARD'S MISSILE, TO BE FIRED THIS WINTER, IS TO DO PATHFINDING FOR SCIENCE AND IS EXPECTED TO REVEAL MANY SECRETS OF THE ETHER. That also reflected Goddard's input, including a visit to the *Times* office.[3]

When the editor of *Science and Invention* asked whether "any scientists of standing believe that interplanetary travel will ever be achieved," Goddard answered that his years of research showed that space travel was possible with present technology. He made only feeble efforts to restrain the wildest speculation. "The impression seems to prevail," he told an editor, "that I am building a rocket intended to reach the moon. Such is not the case, as my attempts are merely toward exploring the atmosphere." That merely set up his punch line: "It is physically possible to send a rocket to the moon, how-ever, and the exploration of the earth's atmosphere is naturally the first step."[4]

Goddard had become a publicity hound in the early 1920s as a fund-raising ploy, and he continued to court the press to ensure that he remained the world's foremost rocket scientist. Late in the decade, he wrapped his ego in the flag. When David Lasser founded the American Interplanetary Soci-ety in 1930, he granted Goddard honorary membership and asked him to become a member of his advisory committee and a guest lecturer. The rocket man accepted the first but declined the two other offices. "I neverthe-less feel," he concluded, "that America can stand a stronger popular interest in the subject, and I wish you success, and feel honored by your invitation."[5]

Goddard wanted his countrymen to share the attention to his work shown by others around the world. But he viewed foreign rocketeers with growing hostility. When the Weather Bureau sent him a copy of a 1928 paper by Robert Esnault-Pelterie, "Exploration by Rocket of the Very High Atmosphere and the Possibility of Interplanetary Voyages," he was incensed that REP claimed to have invented the space-going rocket idea in 1912. He refuted that by pointing to his student theme of 1907. When REP sent God-dard a copy of his latest book and again urged him to compete for the Hirsch Prize, the American's response was polite but cold.[6]

He was more gratified by the uniformly favorable publicity he received in the USSR. Even in the workers' paradise, however, the American attracted crackpots, including the "All-Inventors' Vegetarian Club of Interplanetary Cosmopolitans." They planned to travel to the Moon and live there, claiming that Goddard had inspired them.

But Russia lost its luster for him after a 1927 Moscow exhibition on "Interplanetary Apparatus and Devices." It featured Goddard's writings and pictures, but Esther wrote all over a souvenir scrapbook of the exhibit that it gave him insufficient recognition. The Soviets had unearthed Tsiolkovsky, and he was now the Russian "father" of rocketry.

Following his own inclinations and prodded by Esther's defensiveness, Goddard was not pleased at this turn of affairs. Whatever his sentiments, however, a Soviet rocket program developed in the 1930s mostly without reference to him, tracing its origins to amateur rocket clubs arising in the 1920s, and drawing their inspiration from the bearded Russian "father."[7]

The Germans vexed him the most, as he told Abbot in 1927: "I have read carefully the books that have been written in Germany recently on the application of the rocket method to the problem of interplanetary flight . . . and in every book disparagement is made of America's contribution to the subject. I believe that, unless I can present the case in the proper way, when the time comes, my own and the Institution's part in the problem will be put in an unfavorable light."[8]

When Robert Lademann of Berlin asked for permission to translate and publish his 1919 "Method of Reaching," Goddard jumped at the chance, and Abbot gave the Smithsonian's blessing. However, the German publisher backed out of the undertaking, then passed Lademann's translation on to Max Valier without permission. The ensuing contretemps caused German science writer Willy Ley to contact Goddard. "There is a confusion of contradictory and irresponsible views prevailing here in Germany about you," Ley told him.[9]

Germany was a hotbed of rocket activity. Herman Potocnik, writing under the pseudonym Hermann Noordung, explored rocket efficiency for space travel, and described a manned space station. More down to earth was

Valier, who not only made Oberth's ideas fascinating, but was a persistent experimenter—an area that Goddard claimed to be alone in. With backing from the Opel automobile company, he devised a rocket-powered automobile, which Goddard dismissed as a stunt. It involved solid fuels, he said, whereas only he was working with liquid propellants. Valier ceased to worry the American rocketeer in May 1930, when the German's rocket blew up and killed him.[10]

Other Germans remained a threat to his ego. In October 1929, Goddard wrote comments on articles appearing in publications ranging from the *Science News Letter* to *The New York Times*, and sent them to the editors. He corrected what he said were errors appearing in print, but mostly, he decried what he perceived as a conspiracy at work. "It looks to me from this," he said, "together with the recent flood of publicity, like a concerted effort to bring the entire rocket problem to Germany."[11]

Goddard's patent on a "Vaporizer for Use with Solar Energy," a solar-powered boiler, generated more news articles. Most of them quoted him, and he digressed from the boiler to rocketry to Oberth. The German's work was old hat, he growled. When the Chicago Museum of Natural History planned to publish a paper called "Rocketing Through the Solar System," Goddard erupted over its equating him with Oberth. Only he had undertaken successful trials, he cried, while Oberth was little more than a thief.[12]

When he wrote that, another storm of publicity swirled around him. By the happiest of accidents, it gave him something that his earlier press assaults had failed to achieve—access to real money.

UNTIL YOU GET A GOOD HIGH FLIGHT I DO NOT CONSIDER YOUR ROCKET DEVELOPED

Goddard did not immediately tell Abbot about his success on 16 March 1926, because he wanted to replicate his experiment. The first chance came on 3 April: "It rose after some time and landed about 50 ft. away, occupying 4⅕ sec in the air," he recorded in his diary. Ten days later: "Lifted a foot, fell back, and had a large gasoline flame. Nozzle half burned through near the top. Chamber OK." On the twenty-second: "No lift. Chamber burned out near bottom after about 1 minute. . . . Planned model with chamber at bottom."

He realized by this time that putting his motor at the nose end was unnecessary. Thereafter, Goddard's rockets generated thrust from motors at the tail end, and by May he had a shorter, but still spindly, contraption about six feet long. He tried and failed to get it off the ground on 4 May, then again the next day.

He next completed a long report to Abbot. It described the December static tests, subsequent modifications, the flight of 16 March, and a disingenuous summary of his tests since, implying that they had replicated the 16 March results. He had proven his case, he avowed. He could do more of the same with another $500, but that would be a waste; reaching high altitudes would require a bigger rocket with pumps, and far more money.

While he wanted the news withheld from the Germans for the time being, he was bent on receiving publicity via the Smithsonian: "Another point," he closed, "is that Science Service is very anxious to secure statements from a number of authorities regarding the importance of this rocket work in time for publication when any announcement is made by the Smithsonian Institution."[13]

Goddard faced two challenges in 1926. One was technical—to increase exhaust velocity by feeding propellants under pressure. To reach high altitude, his rocket also must carry a greater supply of fuel and LOX, meaning that the whole thing had to grow bigger.

His other problem was political, and of his own making. He had drilled the Smithsonian along for over a decade with promises that his rocket was about to leap toward space. Abbot and Walcott had as a result become highly skeptical; there would be no official announcement of the 16 March flight of a little rocket with no high-altitude potential.

Focused on getting to the upper atmosphere, Abbot concurred with Goddard's proposal to develop a larger rocket, and asked: "May I inquire what funds would be required to make immediately a large enough rocket to surely fly to a high altitude, so that a conclusive spectacular demonstration could be made with it?" It was possible the Smithsonian could find the money, if "we could depend upon a financial response to appeals for commercial exploitation." Goddard answered that the high-altitude missile would cost $2,500, but he saw no immediate commercial application. He pressed Abbot for the money.[14]

The correspondence revealed Abbot's skepticism and Goddard's tendency to pelt his patron with promises. Hamerschlag of the Research Corporation agreed to release $2,500, but Abbot resisted simply passing it on. To convince the public of the value of rocketry, he said, "We should have a really high flight to clinch the matter," and not a crash. "Before we definitely authorize you to undertake the new construction, I wish you would write exactly what you would expect to do if the Institution should put $2,500 more at your disposal to bring about a really spectacular flight."

Goddard's response was testy. He was not about to put up a rocket that could not be salvaged and reused, he said, as most of the money went into testing parts. The rocket itself would be cheap. With $2,500 he would "carry through the development of a rocket large enough to give a spectacular flight; to have more than a single rocket on hand; and to take all precautions possible to avoid damage to the rocket on landing."

That did not satisfy Abbot. It appeared to him that Goddard would spend the money on "further developmental work . . . without the spectacular flight to show for it." He required "something a little more specific and definite, almost in the nature of a proposal."

Then he dropped a hammer: "Hitherto," he said, "for the past two or three years, we have been supplying additional sums of $500 at a time with the expectation that each in succession would bring the matter to a climax. That is just what we do not wish to do any longer, but to go about the matter with a well-grounded assurance that it is going to bring it to a fruition, or else drop it right here."

Stung by that, Goddard explained the problems involved in scaling his rocket up to twenty times as large as before. He included an itemized cost estimate for every stage of the work. Abbot and Walcott, satisfied that he had a clear plan, sent him $500.[15]

Goddard had already begun designing a big rocket, one with a combustion chamber eight inches in diameter and a length of over ten feet. He sketched the ideas, and Sachs made them real, but jumping to such an increase over the earlier dimensions presented a daunting challenge. They worked on new valves for the coupled pumps, venturi tubes for flow regulation in the fuel system, new designs for the chamber and nozzles, larger gasoline and LOX tanks, and new ways to maintain propellant pressures. Every step required experiments, failures, and modifications.

Goddard continued to be optimistic. "I have, without going into detail," he told Abbot in October, "made more rapid progress than I anticipated, and there remains not much more than the completing of the chamber—the piping of which requires considerable time—the completing of the tanks, and the assembling in position, before the testing of the rocket in a frame to measure the lift and make any necessary adjustments, is begun." He had turned to "multiple feeding" of fuel and LOX, via manifolds that distributed them across the top of the chamber.

He acquired an old windmill tower that he adapted as a launch tower at Aunt Effie's. When Abbot visited before Thanksgiving, Goddard gave him a tour of his lab and the launch site, and his patron left suitably impressed. Abbot expressed hope that he could report a successful launch at a special Smithsonian meeting, to be attended by dignitaries including President Calvin Coolidge, on 11 February 1927. By this time, Goddard and Sachs were hauling their stuff on a trailer behind a Model A Ford back and forth to Aunt Effie's, rigging the tower, tinkering with the rocket, and getting stuck in the snow. He told Abbot not to count on a launch before the meeting.

The rocket stood in the tower by April. Like its predecessor, it was a skeletal arrangement held together by its fuel piping, but now with chamber and nozzle below, tanks and pumps above, and a conical cap housing a parachute on top. Goddard tried to launch on 16 and 20 April; the igniter failed the first time, and he could get no fuel pressure the second. He told Abbot his rocket was ready notwithstanding, explaining that its launch had been delayed by winds, a high fire danger in nearby woods, and various "inconveniences."

Weighing about 150 pounds fueled, the rocket was ready for a test on 3 May, when it "jerked upward," lost its oxygen, and the gasoline tank exploded in a ball of fire. Goddard and his crew picked up the pieces, and Sachs rebuilt the thing. It was back in the tower by the middle of July; the most important change was a manual fuel cutoff to avoid a repetition of the accident. It was no use. On 20 July: "Engine did not seem to generate pressure, control lines became entangled, and chamber burned through at middle." It was ready again on 31 August, when a similar failure occurred.

In September Goddard gave up, and began working on a smaller missile. He had taken on more than he could handle trying to build such a big rocket.[16]

Goddard was justified in interpreting his failed experiments as providing "valuable negative information." He was the pioneer in his field, trying to do things that no one had ever attempted, and even in its primitive stages, liquid-fuel rocketry was a complicated line of work. Decades later, when billion-dollar rockets created by the best brains in the business blew up or went astray, such failures were viewed as steps on the road to progress. Goddard's experience demonstrated that, in rocketry, it was ever thus.

He remained easily distracted, as in July 1927 when the Carnegie Institution sent two experts to visit him. They asked for a cost estimate for a high-altitude research flight, and Goddard told Abbot that he had proclaimed his loyalty to the Smithsonian, but his money was about to run out again. He asked whether he should give Carnegie an estimate or defer to the Smithsonian.

Abbot told him to go for the money, but "until you get a good high flight I do not consider your rocket developed, and I do not see how others could regard it otherwise. I hope you will concentrate on this preliminary flight and not be drawn away to spend any time on applications until you attain it. There can be no question that the support for the program you propose will come forward when the high flight is once made. Concentrate upon it!"[17]

Goddard and Sachs spent the next year devising a rocket one-fifth the size of the one they had abandoned. Now they used choking tubes to control propellant flow, reverted to an alcohol burner to induce pressure, and redesigned the tanks and the chamber with conical ends to avoid the bursting experienced earlier. If it did fly, a parachute would bring it down.

They were back at Aunt Effie's by October 1927, with Esther, Roope, and a student. They watched failures, often explosive, on four successive tries. On the fifth, on 6 December, it all worked better, but produced insufficient pressure to rise. More explosions followed in January and February.

He soldiered onward. By May 1928 he had built a new model weighing seventy pounds, which on the third generated two hundred pounds of lift for a few moments, but did not fly. Another series of fiery explosions followed during the year, while he hired his machinist brother-in-law, Albert Kisk, and Clark graduate student Lawrence Mansur to help with the work. Meanwhile, his father, Nahum, who retired from Hardy Company in August 1928, died of throat cancer a month later, at the age of sixty-nine, a cigar smoker to the end.

The flight attempts resumed, with further digressions. In December Goddard achieved a short hop with a rocket, but rather than leave well enough alone he began to tinker with gyroscopes to stabilize higher flights. But he soon dropped that, because his chambers, thanks to improvements in fuel feed, started to burn out. Injecting excess gasoline to cool the chamber seemed to work, so Goddard resumed testing gyroscopes. When, at the end of December, he watched his rocket roar 204 feet above the tower—a signal achievement after so many failures—he was a happy man posing for Esther's camera beside the vehicle's wreckage.

Abbot continued prodding him, so Goddard again promised an early flight. He also suggested that his experiments might be safer in a desert environment. Abbot consulted with his friends at Mount Wilson, and citing Germany's potential military development he asked Congress for $10,000 for the work. He failed in that, but promised Goddard he could scrape up half that amount from Smithsonian funds if, once again, the rocketeer could promise a successful flight. Goddard responded with a proposal for $5,000.

Robert Goddard at age 10, with his parents in
the backyard at Maple Hill.
(Source: Clark University Archives)

"Gram," Mary Pease Goddard,
Robert's grandmother, about 1910.
(Source: Author's Collection)

Worcester Polytechnic Institute in 1906, while Goddard was a student. The Magnetic
Building is at the lower left. *(Source: WPI Archives and Special Collections)*

A. Wilmer Duff, Goddard's first patron, WPI physicist.
(Source: WPI Archives and Special Collections)

Konstantin Tsiolkovsky, late in his life as a Soviet "Hero" and the Russian "father" of modern rocketry and space flight.
(Source: Author's Collection)

Arthur Gordon Webster, Goddard's second patron, Clark University physicist.
(Source: Clark University Archives)

Charles T. Hawley, patent attorney, who served the Goddards from 1913 to 1957; photo from early in his career, around 1920. *(Courtesy Clark University Archives)*

Goddard's first proof that a rocket would provide thrust in a vacuum, about 1914. The muzzle blast from a blank cartridge would cause the pistol to twirl around the spindle. This simple demonstration was popular during "Sub-Freshman Days" in the 1920s.

(Source: Clark University Archives)

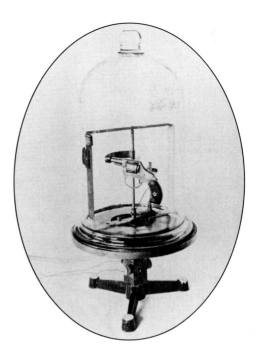

Circular tube used in proving further that a rocket would provide thrust in a vacuum, 1915.

(Source: NASA)

The best-dressed man on campus: Goddard posing with a rocket test stand on the Clark University campus, about 1914–16. *(Source: Clark University Archives)*

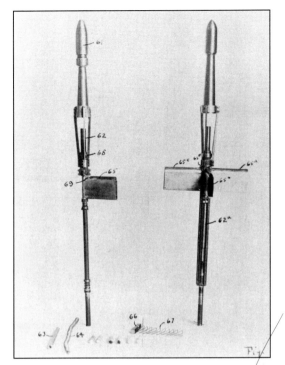

One of several versions of a self-reloading, multiple-charge, solid-fuel rocket, the "machine gun firing blanks," from an illustration for a report to the Smithsonian. This, like all other versions, could not be made to work.

(Source: Clark University Archives)

Goddard loading one of his tube-launched infantry rockets, near Mt. Wilson, California, 1918. *(Source: NASA)*

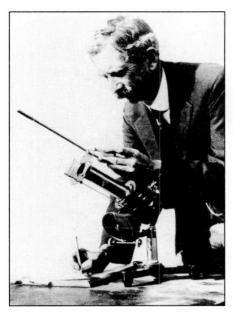

Charles G. Abbot of the Smithsonian Institution, Goddard's third patron, observing the sun with one of his many inventions, around 1920.

(Source: Smithsonian Institution Archives)

Esther Christine Kisk, age 17, as a senior in high school, 1918.

(Source: Author's Collection)

Wallace W. Atwood, president of Clark University.

(Source: Clark University Archives)

Hermann Oberth, soon after the publication of his 1923 book, which made him the German "father" of modern rocketry and space flight. *(Source: Author's Collection)*

Goddard in the laboratory at Clark University, late 1920s.
(Source: NASA)

Diagram of the first liquid-fuel rocket to fly, 16 March 1926.
(Source: NASA)

Goddard posing with the rocket in its launch frame, shortly before the first liquid-fuel rocket flight, "Aunt Effie" Ward's farm, Auburn, Massachusetts, 16 March 1926. This is the most familiar photo of Goddard ever taken, distributed by the photographer, his wife Esther, in the decades after his death. *(Source: NASA)*

Goddard and crew posing for Esther's camera with the wrecked rocket after the attention-grabbing flight of 17 July 1929, just before the police, ambulances, and reporters arrived, near Auburn, Massachusetts. *Left to right*: Larry Mansur, Goddard, Henry Sachs, Al Kisk, Percy M. Roope. *(Source: Clark University Archives)*

Daniel Guggenheim, about 1920. "Mr. Dan" became Goddard's benefactor, pledging $100,000 from his own pocket on the advice of Charles Lindbergh, enabling Goddard to begin his experiments in New Mexico in 1930.

(Courtesy Clark University Archives)

Main Street, Roswell, New Mexico, 1930s. *(Source: Historical Society of Southeastern New Mexico)*

Rear (north) side of the house at Mescalero Ranch, 1930, shortly before the Goddards leased the place from "Miss Effie" Olds and Esther started remodeling. Among her changes was the screening of the porch. The front (south) façade of the house was, in keeping with the New Mexico Territorial Style, nearly blank. *(Source: Clark University Archives)*

The first truck and trailer used to haul rockets from Mescalero Ranch to Eden Valley, 1930–32. *(Source: NASA)*

Loading an early rocket into the launch tower at Eden Valley, 17 October 1931. *Left to right*: Larry Mansur, Al Kisk, Charles Mansur, unidentified, Goddard, Ole Ljungquist.

(Source: Clark University Archives)

Maple Hill (1 Tallawanda Drive), Worcester, Goddard's birthplace and lifelong home, as it looked on his return from New Mexico, 1932. The porch was closed in during the 1960s and utilities have been upgraded, but otherwise the place looks the same now as it did following Esther's remodeling in the 1920s. *(Source: Clark University Archives)*

Group photo taken by Esther during the visit to Roswell by Goddard's last two patrons, Charles Lindbergh and Harry Guggenheim, 25 September 1935. *Left to right*: Al Kisk, Guggenheim, Goddard, Lindbergh, Ole Ljungquist, Charles Mansur.
(Source: Clark University Archives)

Goddard at his launch command post, Eden Valley, late 1930s. The telescope focused on gauges connected to pressure lines on the rocket in the launch tower. When readings were right, he could operate fire, release, or stop keys on the panel in front of him. *(Source: NASA)*

Goddard leaving the ranch house for the shop one morning in 1937. Notice the shabby clothing; one of his crew described him as "very sloppy."

(Source: Clark University Archives)

Robert and Esther Goddard at home in Roswell, 1937. Photographs of the two of them together are very rare. *(Source: Clark University Archives)*

The official NAA record flight, 9 August 1938. Esther made this reverse print from her movie of the flight.

(Source: Clark University Archives)

Howard and Marjorie Alden, NAA observers, checking the barograph after the record flight of 9 August 1938.

(Source: Clark University Archives)

A P-series pump rocket explodes at ignition, 14 June 1939.

(Source: Clark University Archives)

Nell at her height: A P-series pump-turbine rocket in the shop at Mescalero Ranch, 1941. After the German V-2 campaign began, Goddard gave Harry Guggenheim a copy of this photograph bearing an inscription claiming that it was "practically identical with the German V-2 rocket." The Guggenheim Foundation later distributed thousands of copies of this photo and the inscription as part of the campaign to prove that the V-2 was a theft of Goddard's work, as was use of the V-2 by the United States government. *Left to right:* Ljungquist, Kisk, Charles Mansur. *(Source: Author's Collection)*

Nell P-23 in the tower, being prepared for launch, March 1940. The flight was not successful. *(Source: NASA)*

Frank Malina (in suit) and Homer Boushey (in flight jacket), with Army and GALCIT technicians, just before Boushey became the first American to pilot a rocket-propelled aircraft, the JATO-equipped Ercoupe behind them, August 1941. *(Source: NASA)*

German technicians preparing an A-4 (V-2) for a flight test, 1944. Compare its size with Goddard's largest rockets, the "P" series.

(Source: NASA)

Goddard examining the combustion chamber of a captured V-2, Annapolis, April 1945. The rocketeer's physical deterioration is evident four months before his death.

(U.S. Navy photo courtesy Clark University Archives)

Theodore von Kármán, co-founder with Frank Malina of the Jet Propulsion Laboratory out of the former GALCIT. *(Source: NASA)*

Wernher von Braun, Esther Goddard, and former NMMI Superintendent Gen. Hugh Milton, posing in front of the launch tower on the grounds of the Roswell Museum, during the dedication of the museum's Goddard Wing, 1959. It was the first of a series of dedications at which von Braun would declare Goddard to be his "boyhood hero."

(Source: Roswell Museum and Art Center)

Wernher von Braun (right) with his other "boyhood hero," Hermann Oberth, in 1961.
(Source: NASA)

Goddard Commemorative Airmail Stamp, first-day covers, 1964.
(Source: Author's Collection)

By May 1929 his latest eleven-foot, thirty-five-pound machine was ready. It was equipped with a barometer, a thermometer, and a camera to photograph the instruments when the parachute released at the high point of its flight. When he tried a launch on the seventeenth, however, a fuel pipe blew off and the nozzle burned through. He expected to be ready for another try in July.[18]

MAN IN MOON SCARED GREEN

Everything was set by 17 July. From the top the rocket featured a nose cone carrying instruments and parachute, below that the propellant tanks, and at the tail end the chamber and nozzle surrounded by four stabilizing vanes. Not far away were a wooden shelter and the crew. Roope stood to the rear of the shelter with his theodolite and stopwatch, while Esther was at the right end, armed with a Kodak movie camera. Beside her was Al Kisk, watching the cotter pin that held the rocket down until liftoff. Goddard was in the middle, ready to pull a cord on the pin when the rocket raised itself three inches. Sachs crouched to his left, operating a pressure-generating tank and two other control cords. Larry Mansur stood at the far left, his only job being to watch.

The object of all this attention came to life at two in the afternoon, when Sachs lit the alcohol burner under the LOX tank, waited 30 seconds, touched off the igniter, and gave the system 125 pounds of pressure from his tank. Goddard pulled the release 13 seconds after ignition. He heard no change in the sound of the motor, assumed that the release had jammed, kept pulling the cord, and missed everything that happened next.

Emitting a tremendous roar from a 20-foot exhaust flame, the rocket rose out of the tower at 17 seconds after ignition. It climbed up 20 feet farther, tipped to the right, ran slightly over horizontal, then dived to Earth. After topping out at about 80 feet aboveground, it landed 171 feet away 2 seconds after clearing the tower, a smoke trail marking its thunderous passage through the air.

The gasoline tank exploded when it hit, but otherwise the rocket was in good shape. Esther caught the flight on movie film, then took stills of the men standing around the wreckage, toasting their triumph with ginger ale. The parachute had not deployed, the instruments were smashed, and the gasoline cap was missing.

Finding the cap delayed the group's departure for home. They had finished packing when about a dozen automobiles, two ambulances among them, came over a nearby hill looking for a crashed airplane—the racket was heard two miles away and caused neighbors to report such a calamity. Goddard asked two policemen present to keep the facts quiet, but the officers pointed to the backs of reporters heading for their news desks. He visited the editors of the two local papers to squelch the story, but they already had extra editions out on the street.[19]

GODDARD EXPERIMENTAL ROCKET EXPLODES IN AIR, CLARK PROFESSOR MAKING TESTS ON AUBURN FARM, roared the *Worcester Evening Gazette*, with subheads saying NOISE OF BLAST HEARD FOR MILES—FALLING EMBERS LED TO BELIEF THAT MAJOR AIR CATASTROPHE HAD TAKEN PLACE—TWO AMBULANCES SUMMONED BUT NO INJURIES WERE SUFFERED. The *Evening Post* was not to be outdone: TERRIFIC EXPLOSION AS PROF. GODDARD OF CLARK SHOOTS HIS "MOON ROCKET," WOMAN THOUGHT ROCKET WAS WRECKED AIRPLANE; AMBULANCE RUSHED TO AUBURN TO CARE FOR "VICTIMS OF CRASH," FIND CLARK PROFESSOR AND ASSISTANTS MAKING EXPERIMENTS WITH ROCKET—PLANE FROM LOCAL AIRPORT ALSO MADE SEARCH FROM AIR FOR REPUTED "VICTIMS."[20]

Reporters were on Goddard's doorstep that evening and the next morning. On the eighteenth only the *Worcester Evening Gazette* paid much attention to his emphasis on the technical aspects of his work: PROPEL GODDARD ROCKET WITH LIQUID EXPLOSIVE, CLARK UNIVERSITY PROFESSOR ABANDONS ORIGINAL DESIGN; ONE EXPLOSION, LONG CONTINUED, TAKES PLACE OF SERIES OF EXPLOSIONS.

It was the 1920 moon-rocket uproar all over again. Said the *Boston Globe* on 18 July: "MOON ROCKET" MAN'S TEST ALARMS WHOLE

COUNTRYSIDE, BLAST AS METAL PROJECTILE IS FIRED THROUGH AUBURN TOWER ECHOES FOR MILES AROUND, STARTS HUNT FOR FALLEN PLANE, AND FINALLY REVEALS GODDARD EXPERIMENT STATION.

Other headlines echoed the same day: ROCKET STARTS FOR THE MOON BUT BLOWS UP ON WAY (*St. Louis Post-Dispatch*). ROAR HEARD TWO MILES IN MOON ROCKET TEST (*Washington Post*). MAN IN MOON SCARED GREEN (*Los Angeles Times*). METEOR-LIKE ROCKET STARTLES WORCESTER, CLARK PROFESSOR'S TEST OF NEW PROPELLANT TO EXPLORE AIR STRATA BRINGS POLICE TO SCENE (*The New York Times*). The story made *Time* magazine on the twenty-ninth.[21]

Goddard had planned to make no statements, but late on the seventeenth he issued the following: "The test this afternoon was one of a long series of experiments with rockets using entirely new propellants. There was no attempt to reach the moon, or anything of such a spectacular nature. The rocket is normally noisy, possibly enough to attract considerable attention. The test was thoroughly satisfactory; nothing exploded in the air, and there was no damage except incident to landing." The next day he reissued the statement, replacing "entirely new propellants" with "liquid propellants."[22]

His words had some effect. By the nineteenth they had inspired editorials favoring a continuation of his experiments, including in major Washington papers. Then the cartoons and jokes appeared, despite Abbot's issuing a statement in Goddard's support. The news was all over Germany by the end of the month.

Goddard heard from friends baffled by the uproar. The Science Service was miffed that it had not been alerted in advance, while George W. Gray was simply confused by conflicting accounts. Goddard issued another statement, directed to Howard Blakeslee, the science editor for AP. He denied that his rocket caught fire or exploded, and emphasized that it carried instruments, to realize rocketry's potential for research. He closed by dismissing public speculation about "interplanetary communication," saying "real progress toward such an end can only be made a step at a time, as in any research problem. The recent successful test of the liquid-propellant rocket is the first step."

It was another futile attempt to replace speculation with dull academic restraint. Goddard was indelibly known as the author of the "moon rocket" about to take off, and the papers followed his every move.[23]

The uproar produced one important compensation for both of the Goddards, by providing something that had been missing from their lives. They were a couple without children or pets, and prospects for neither. Larry Mansur inadvertently answered their need when the press agitation was at its loudest, saying, "They ain't done right by our Nell." He alluded to Edward B. Sheldon's *Salvation Nell*, a maudlin melodrama about a downtrodden girl that premiered on Broadway in 1908 and remained popular for years.[24]

Goddard's rockets were called "Nell" from then on. To Robert and Esther this personification of his inventions held a special, intimate meaning. As she recalled in 1950: "The name stuck. All during the New Mexico adventure, and afterwards, the rocket was always 'Nell,' a member of the family. Always, too, when there were drinks at a party, no matter whose name was mentioned as being toasted, my husband's eyes sought mine, even across the room, and our drinks were silently for 'Nell.' "[25]

Nell was about to be evicted from Aunt Effie's. At the instructions of State Fire Marshal George C. Neal, District Fire Inspector Robert E. Molt visited Goddard at the launch site on 25 July. He reported that Goddard's tests were perfectly safe, in a clear area, and threatened no one. However, some of the press accounts had mentioned the burned area at the base of the tower, and Goddard knew what to expect even before he went to Boston to see Neal on 1 August. His flight tests were forbidden in the state.

He had already decided to move, but for different reasons. Aunt Effie's was too visible, and he worried that people would mistake a static test for a failed flight attempt. He found a new proving ground at the Hell Pond area of Camp Devens, an army post northeast of Worcester; it was federal land, beyond Neal's reach. He asked Abbot to secure permission from the War Department, emphasizing the antiaircraft potential of his rockets.

The secretary of war acceded to the Smithsonian's request, and issued a license to use Hell Pond in late October. Goddard had to confine his tests to the aftermath of rain or snow, install fire extinguishers and no-trespassing signs, subject himself to rules imposed by the commanding officer, assume responsibility for damages, and restore the site to its original condition when he was done.

It was a miserable place. It was inconveniently located, the roads were bad, and the weather could be wretched. The return of mosquito season in the spring would only make conditions worse.

The rockets were demonstrating their stuff at Camp Devens by early December. The first two tries failed, but the third gave a 25-pound lift before "chamber somewhat bulged, but apparently undamaged inside." By Christmas, with Goddard testing a wider chamber and the gasoline spray he now called "curtain cooling," his engine achieved 150 pounds of thrust before it blew apart in a ball of flame and metal fragments. So it continued, fitfully and with mixed results, into the summer of 1930.[26]

The delay in gaining access to Hell Pond had given Goddard time to fill, which he used to gather and update all his notes concerning rockets and space travel. He sent Abbot a "Report on Conditions for Minimum Mass of Propellant," to be filed with his earlier excursions into space, dating the inception of his many ideas. He also assembled two volumes of experimental results and photographs, in an effort to certify his place in history.[27]

That did not discourage Abbot from sending him another $500 for his research, along with the customary prod. Clearance to use Hell Pond allowed flight tests, but imposed conditions that made them unlikely. Goddard used some of the money to hire Larry Mansur's brother Charles, who was working his way through Clark as a janitor. "I asked Goddard for work when he had it," Charles remembered, "and sometime in the late summer or fall of 1929 he told me that he had a hundred dollars and put me to work. He said, 'When you think you've earned a hundred dollars, you quit.'" He stayed until Goddard's death, and became a distinguished rocketeer in his own right.[28]

Goddard's casual approach to hiring Charles Mansur reflected an increasingly stoic patience that emerged over the following winter. A photo-graph taken at Camp Devens in the spring of 1930 shows him relaxing with his men in an old chicken coop they used as a workshop. Charles Mansur, Henry Sachs, and Al Kisk are wearing work clothes, Goddard a white shirt, necktie, and slacks. He holds a coffee cup in one hand and a cigar in the other, and his face bears the placid expression of someone who expects good news to arrive at any minute.[29]

IN THE FIELD OF AVIATION THEY WERE WAY BEYOND THEIR TIME

The first hint that better times might be in the offing had arrived the previous November, when Colonel Charles A. Lindbergh, USAR, appeared on Goddard's doorstep to inquire about his rocket experiments. Tall and lanky, with boyish good looks and an engaging shyness, he and Goddard had several things in common, including introverted manners and lives in the glare of publicity.

The attention drawn by the man who had flown solo from New York to Paris in 1927 absorbed Goddard's own place in the public arena, and the rocketeer was thereafter in the aviator's orbit. LINDY GUEST OF GOD-DARD ON VISIT HERE the *Worcester Sunday Telegram* shouted. Within a few days officials and scientists in Washington were abuzz about what the two were up to. Before long the moon shot was current again, but there were no jokes. The fact that Lindbergh was interested in his work gave God-dard increased respectability.[30]

The events that carried Lindbergh to Goddard's door had actually begun before his 1927 flight, when he met Harry Guggenheim, the president of his father's Daniel Guggenheim Fund for the Promotion of Aeronautics. When "Slim" Lindbergh made it to Paris, Harry recognized that the Lone Eagle was the best thing that could have happened to aviation's public image.

Harry took the flyer under his wing, offering a haven at his Long Island estate, Falaise. There Slim wrote his memoir of the flight, *We*, and Harry trained him in social graces to shed his aw-shucks bashfulness in public. He also sent the hero to his tailor and bought him a complete wardrobe. Knowing that fortune might accompany fame, he turned Lindbergh over to his legal majordomo, Colonel Henry C. Breckinridge, who became Slim's legal adviser. Breckinridge set up an account with the House of Morgan, and by the time Lindbergh met Goddard he was a millionaire.

His money was nothing compared to that of the Guggenheims. The American branch of the family had been founded by Harry's grandfather, Meyer, a Jewish peddler who emigrated from Switzerland in the 1850s. He invented a profitable stove-cleaning solvent, developed and then sold a lace-manufacturing industry, and parlayed the proceeds from that into a copper smelting and refining empire spread across the Americas. He divided his fortune among his seven sons, commanding them to grow their wealth and plow their profits into good works.

One of the sons, Daniel, developed an interest in aviation. Through the 1920s his Daniel Guggenheim Fund spent nearly $3 million to promote education and research to make the skies safe for people and commerce. It shut down in 1929, in the belief that it had provided "seed money" for progress.

Daniel's son Harry F. Guggenheim, born in 1890, studied for a while at Yale, worked in the family's smelting operations in Mexico, then earned a bachelor's degree at Cambridge University in England in 1913, following it with a master's five years later. In the interim, he became a pioneer naval aviator, served in Europe during World War I, and retained his reserve commission thereafter.

Tall and athletic, with a Roman nose and electric-blue eyes, Harry exhibited every talent he inherited from his father and grandfather, but not their personal warmth. He joined the family firm, Guggenheim Brothers, on his return from England in 1913, but left ten years later in a business dispute with his uncles.

Thereafter he managed his own portfolio and those of his parents, his father's charitable foundations, and a succession of new enterprises that he started on his own, until he returned to take charge of Guggenheim Brothers after World War II. He was married three times, became the founder and publisher of the newspaper *Newsday*, and was a famous breeder of race-horses. Ceaselessly promoting aviation, Harry was a member of several international conferences and of the National Advisory Committee for Aeronautics (NACA), was the United States ambassador to Cuba from 1929 to 1933, and wrote dozens of articles and two books.

When the Lone Eagle brought acclaim to aviation he and Harry became inseparable, and were closer to each other than either was to his family. The entry of Goddard into their relationship created a triangle that would hold together long after the rocketeer's death.

It was one of the more remarkable friendships of the twentieth century, involving three highly intelligent, energetic, and ambitious men. Each guarded his private self behind a wall of taciturnity, yet all three were eminently famous. Their ages spanned the last generation of the Victorian era—Goddard was born in 1882, Guggenheim in 1890, Lindbergh in 1902—but all devoted themselves to the future of human flight. They grew up in an age of formal manners—they addressed each other in letters as "Dr. Goddard," "Colonel Lindbergh," and "Mr. Guggenheim," although face-to-face they were "Bob," "Slim," and "Harry." Each evinced the profoundest respect and admiration for the others, but none would ever reveal his inner sentiments, no more to one another than to the world at large.

Despite their commonalities, they were different people from different worlds. Harry was a fabulously wealthy capitalist and a modern Renaissance prince with wide-ranging interests and the wherewithal to pursue them. He had a gift for making money, and a greater one for giving it away in a quest to improve the world. Yet he was the sort of man who loved humanity in the abstract, but could not get along well with individual members of the species—with the notable exception of Slim.

The Lone Eagle started out focused on a solitary ambition to fly, became at an early age the world's foremost celebrity, then spent the rest of his life seeking a personal sense of himself by widening his interests while ferociously safeguarding his privacy and that of his family. Through it all, Goddard was his beacon to the future, while Guggenheim was his only close friend.

As for the rocketeer, a shy man with limited social skills, rocketry was his life, and everybody and everything else was subordinate to his goal of reaching "extreme altitudes," in his own way and on his own schedule.

The hypotenuse of this upright triangle was Slim Lindbergh, whose life-long attitude toward Goddard was one of abject admiration for his intelligence, his imagination, his intellectual accomplishment, and his potential to raise men to the stars. Until near the end of Goddard's life, Guggenheim kept a certain distance, observing the rocketeer through Lindbergh's eyes. He was so enthralled by the Lone Eagle's grasp of aviation technology that Slim's judgment about Goddard's work was perforce Harry's judgment as well. To Goddard, always slow to form friendships, Lindbergh and Guggenheim were objects of gratitude, with gradually emerging signs of reciprocal regard for Slim, and near the end of his life an almost helpless clinging to Harry.

Lindbergh, accompanied by his wife and copilot, Anne, spent much of his time in the air after 1927. He promoted aviation and searched out potential airport sites, set flight records, and thought about how flying might improve further. His mind soared beyond his aircraft on long flights, and by 1929 he concluded that propeller-driven airplanes had realized their potential. What, he wondered, might go higher and faster?

It was just such a problem that he and Harry were discussing at Falaise one day in the fall of 1929 when Harry's second wife, Caroline, made a discovery that caused Harry to mount a brass plaque on the wall next to where she sat. She was bored to tears—often literally, as Slim's frequent presence at Falaise further strained an already tense marriage—with ceaseless boy talk about flying machines. Leafing through a pile of magazines, Carol, as she liked to be called, came across an account of the excitement in Massachusetts over the "explosion" of Goddard's rocket. She showed it to the men, and Harry suggested that Slim check the story out.

He consulted first with C. Fayette Taylor of the Massachusetts Institute of Technology (MIT), who verified that Goddard was a reputable physicist, and not a crackpot. Lindbergh called Worcester, and he and Goddard arranged a visit for 23 November.

The rocketeer gave the aviator a tour of his laboratory and showed him stills and movies of his experiments. Then he took him home, where Esther had milk and chocolate cake waiting. When she saw a sling on Lindbergh's arm, she asked if his plane had crashed on the way to Worcester. He had driven there, he answered, and had dislocated his arm trying to retrieve a puppy hiding under his bed.

The thin, young pilot and the thin, middle-aged professor spent hours talking about aviation and the future of rocketry. Lindbergh asked whether it would be possible to rocket to the Moon. Goddard answered that it could be done with a multi-stage missile, but it would cost at least a million dollars, and that kind of money was out of the question. Goddard said he wanted $25,000 per year for four years, and with that he could perfect high-altitude rocketry.

Lindbergh could not come down to Earth on the drive home. "The thought of sending a rocket to the moon set my mind spinning," he said in his memoirs. "If man could reach to the moon, then why not thence to the planets, even to the stars? My interest in Goddard's accomplishments and ideas began a collaboration and friendship that continued until his death."[31]

"Colonel Lindbergh, whom you doubtless know," Goddard told Abbot, "impressed me very favorably throughout these events, and I believe he is a keen-minded young man." The aviator set up a meeting with DuPont Chemical Company officials in Delaware for 26 and 27 November. They had previously told Lindbergh that liquid-fuel rocketry was impractical because the combustion chamber required a heavy ceramic lining. At the meeting they questioned Goddard so closely that he believed they were trying to steal his technology. As Lindbergh flew the rocketeer home—it was his first trip in an airplane, and he tried to keep it his last—they agreed that DuPont was hopeless.[32]

While they were in Delaware, President Atwood was in Washington meeting with John C. Merriam of the Carnegie Institution. Lindbergh had already told Merriam about his visit to Worcester, and through Atwood the Carnegie president invited the pilot and the rocketeer to visit him. Goddard began feeding cost estimates to Lindbergh, while Merriam visited Abbot to obtain the Smithsonian's clearance to deal with the rocket scientist. Abbot told Goddard to expect an invitation, and to get his exhibits and movies together. Merriam had told him that there should be $100,000 available and that Lindbergh was just the man to find it.

Merriam set up a meeting at his home for 10 December. Besides Goddard and Lindbergh, he invited Abbot, a delegate from the Weather Bureau, and three Carnegie scientists. After lunch, Merriam kicked it off by thanking Lindbergh for bringing the subject to his attention. Lindbergh replied by saying that airplanes had reached their limits and rockets were the way to the future. Goddard summarized his work and showed movies, then Abbot gave his own summary. He ended by asserting the justice of allowing Goddard to patent any future developments; Lindbergh seconded that, while the others fairly slavered over the prospect of getting their instruments high into the sky.

The meeting ended when Goddard agreed to present a budget. He sent Merriam proposals for a two-year program costing $50,000 and a five-year effort costing $100,000. Location of the work, he said, would be determined by proximity to LOX plants; he favored a return to the Mojave Desert near Mount Wilson. In response, Merriam gave him $5,000 in Carnegie funds. Goddard was grateful, but feared that it might be deducted from the larger budget, until Merriam reassured him that the grant was good regardless of any other funding.[33]

Lindbergh was pleased, but not entirely. Goddard's work was the future of aviation, he had decided, and it needed more support than the nickel-and-dime contributions that places like the Smithsonian and Carnegie offered. He knew where the real money was, but he was reluctant to impose on his benefactors. Harry was in Cuba, but "Mr. Dan" was around. "Dan Guggenheim," he said later, "had vision and courage. He was short and heavy-set and had penetrating eyes. He was a man of great decision.... Dan Guggenheim listened and understood. Harry inherited his father's vision. In the field of aviation they were way beyond their time."

Daniel was also famously generous with his wealth, fully aware that a small grant from a millionaire could make all the difference in the world to someone less fortunate. "A million dollars isn't much," he reputedly observed, "unless you don't have it." Lindbergh feared that Mr. Dan's generosity could make him an easy touch for a friend. Hesitating to presume on his relationship with Daniel, he flew out West to scout locations for airfields.[34]

As winter passed into spring in 1930, Goddard thought that Slim had lost interest, and bombarded Abbot and Merriam with inquiries. In May he asked Merriam "if the large-scale plan is likely to go through, or will probably not be realized." A week later, he received a call from Atwood, who told him that Merriam had talked with Lindbergh. The next day, according to his diary, "Lindbergh called up, saying he was working on the project, and expected to get results." On the twenty-ninth: "Lindy called about Daniel Guggenheim." What Slim said caused him to take Esther to a Chinese restaurant to celebrate.[35]

Lindbergh had steeled himself to talk to Mr. Dan about rocketry. His word was good enough for the old man, who wanted only his assurance that the work had potential and that Goddard was the one to see it through. He pledged $100,000 out of his own pocket, with half guaranteed for the first two years at $25,000 each. The balance would follow when Lindbergh and an advisory committee certified Goddard's satisfactory progress.

By the end of May, Abbot had Lindbergh, Merriam, and Breckinridge in his office working out details. On 5 June Lindbergh summoned Goddard and Atwood to Breckinridge's New York office, then they went on to Washington to see Merriam and Abbot, after which they visited Breckinridge again.

The lawyer wanted to ensure that any liabilities arising from the rocket work would not fall upon Guggenheim. He insisted that Clark establish a Clark University Research Corporation to manage the grant and bear liability. Goddard would retain half interest in any profits from patents arising from the work; Clark and Guggenheim would share the balance. Atwood agreed, but it took a year of badgering him by Breckinridge and Goddard to get the corporation established.

The advisory committee had nine members to oversee the work and report to Guggenheim. They were Merriam, Abbot, Walter S. Adams of Mount Wilson, Atwood, Breckinridge, John A. Fleming of Carnegie, Lindbergh, chief of the U.S. Weather Bureau C. F. Marvin, and Robert A. Millikan, director of physical laboratories at CalTech. The committee was window dressing, because Daniel and later Harry Guggenheim heeded only Slim Lindbergh.

Breckinridge sent Atwood the first check for $25,000 on 23 June, by which time Goddard and his crew were packing for a move. A month later, Daniel Guggenheim, in Carlsbad, Czechoslovakia, announced that he was giving $100,000 to Goddard. "After investigating the matter and securing information from scientists well qualified to pass on the possibilities of a study of the atmosphere by means of rockets," his statement droned, "I am convinced the experiments deserve support."[36]

In the Smithsonian's annual report for 1930, and in a separate statement to the press at the end of the year, Abbot described his institution's years of support for Goddard's work. It would go forward thanks to the generosity of Guggenheim and the interest of Lindbergh, but: "It is a pleasure to record here that the Smithsonian has again been able to support during its more or less uncertain pioneering stages an investigation of great promise for the increase of knowledge."[37]

As for Goddard, his pockets were full, Clark had given him leave for two years, and he would have to endure no more nagging from Abbot. He was bound for New Mexico.

EIGHT

THE GREAT ADVENTURE IN NEW MEXICO

O for a Muse of fire, that would ascend
The brightest heaven of invention!
—William Shakespeare, *King Henry V*

At Lindbergh's suggestion, publicity on the Guggenheim grant was arranged by public-relations pioneer Ivy Lee, whose staff prepared a long statement released by Clark University on 9 July. It summarized the history of Goddard's work and emphasized the scientific value of being able to gather data high in the atmosphere. It also said that the work would take years to complete, but it did not mention space travel.

The announcement was widely reported, with attention to Lindbergh's "faith" in Goddard. The news accounts also emphasized trips into space. On this point, Goddard told Lee's assistant: "The problem is an especially difficult one to handle, for the reason that the interplanetary application is really physically possible. This is, in fact, rather widely recognized abroad, where an effort is being made to raise funds by popular subscription for the development of a 'spaceship' (*Raumschiff*)."[1]

"So annoyed by bizarre moon publicity is Clark University's Robert Hutchings Goddard, father of the rocket exploration idea," said *Time* magazine, "that he has shrouded his research with secrecy, and has refused all interviews." With that fanfare to his primacy, he need not blow his own horn. He told *Time* that, with Guggenheim's money, he would build a "new, giant test rocket."[2]

LEAVE THE GATES AS YOU FIND THEM

When the grant came through, Goddard asked Lindbergh for advice on where to move. Slim, who knew the aviation weather of North America better than anyone else, suggested the southwestern Great Plains. The rocketeer then consulted Charles F. Brooks, a meteorology professor. As Goddard said later: "The ideal location for this work should combine an even climate—year-round outdoor working conditions, few cloudy or hazy days, low average wind velocity, small annual rainfall—with level terrain, low vegetation, adequate transportation facilities, and suitable living conditions for the men employed in the work." Southeastern New Mexico, around Roswell, seemed the best choice.

As Goddard and his crew dismantled the operations at Devens, Abbot, Merriam, and Lindbergh all asked him to prepare a plan so they would know what he would do with the grant. He declined, pleading the demands of the move, and promised a detailed outline later. Meanwhile, he talked Clark and the Smithsonian out of machinery, instruments, and tools to take with him.[3]

Goddard was being disingenuous again, because he was eager to be gone before anyone quibbled about his work plan, hoping to avoid alienating his sponsors by promising a plan later. In fact, systematic planning was never his long suit, as he made up his work as he went along. But if he was too busy to say what he was going to do with the money, he was not too busy to look after his own interests. He negotiated with Bausch and Lomb Optical Company and the Automatic Electric Heater Company over his patent on a solar water-heating device, and sold them licenses to it before he left town.[4]

Robert and Esther, a freight car full of machinery and household goods behind them, arrived in Roswell on 25 July 1930. "Grazing land and desert for 75 miles from the town," he observed. "Trees begin about 6 miles from town. Clouds pretty, purple beneath. Blue to horizon on clear part of sky, and green at horizon near showers." The next day he picked up his mail at the Weather Bureau. So began what Esther called "the great adventure in New Mexico."[5]

"I am deeply interested in hearing how you are getting along with your work," Atwood wrote. "Remember that the big thing before you is success in this wonderful adventure. I want you to keep that in mind above everything else; keep your courage up and put your very best metal into that task, and I think everything will come out nicely for you in the end. I shall do everything I can to help you."

That encouragement was warranted, because a shadow had fallen over the "adventure." Daniel Guggenheim, age seventy-four, died of heart failure on 28 September. Goddard had never met him, but the obituaries mentioned his grant along with Guggenheim's other benefactions. "The whole of science, particularly aviation," he wrote to Daniel's family, "suffers a great loss in the passing of this noble character." Just how great this loss would be was not yet apparent. The task at hand was to vindicate Guggenheim's faith in Lindbergh, and Slim's in Goddard, by firing rockets over Roswell.[6]

The city of wide, paved streets shaded by willows and cottonwoods sat just west of the Pecos River in the center of Chaves County. The environment was semi-arid high-plains grassland, in the transition zone between the Great Plains and the Chihuahuan Desert. As Brooks should have told Goddard, the area received about a foot of precipitation a year, much of it in the "monsoon" season of July and August; violent weather was rare. At an elevation of 3,600 feet, Roswell's climate was moderate, with January and July mean temperatures of 40 and 79 degrees; prevailing winds were southerly.

Brooks neglected to mention the fierce winds of early spring, or the fact that the ambient dust level was high. He could not have predicted the abnormally hot summers, hard winters, persistent drought, and periodic dust storms that would make the ensuing decade the "Dust Bowl" years.

Nature's surprises notwithstanding, Roswell shimmered under a brilliant sun in a deep blue sky almost every day of the year. At night, the heavens blazed with the stunning light of more stars than the city-bred Goddards of New England had ever realized existed. Not least, the air was fresh and dry, and Robert breathed easier than he had in nearly two decades.

Roswell's population of 11,173 people in 1930 would grow to 13,482 by 1940; 90 percent of them were Anglo, the rest Hispano. Their community was founded in 1869, and acquired its name with a post office in 1873. It was part of Lincoln County until 1889, but played only a small part in the Lincoln County War of 1877–79. In the 1890s the railroad arrived and developers discovered its artesian groundwater.

By the turn of the century the town was a center of agriculture, and host to the New Mexico Military Institute (NMMI). Thanks to a series of land booms devoted to apple orchards and cotton and hay production, Roswell became the principal center of southeastern New Mexico, surrounded by irrigated fields, oil and gas wells, and sheep and cattle ranches. The town hosted cotton gins, creameries, oil refineries, and meat-packing industries. Early in the century it had also attracted "lungers," as people with tuberculosis were called locally.[7]

Considerable intellectual horsepower was at work in Roswell. Besides the capable NMMI faculty, there was the Institute's librarian, Paul Horgan. In 1930 he was already producing the scores of books in history, biography, fiction, and criticism that would earn him two Pulitzer Prizes. In charge of the local weather office was Cleve Hallenbeck, prodigious author and expert on the Spanish Colonial Southwest. And about sixty miles west at San Patricio, Peter Hurd and his bride, Henriette Wyeth, had begun their celebrated careers as painters.[8]

The Goddards did not associate with that brainpower for several years. They were equally aloof from other local celebrities, including the singer Roy Rogers and the rodeo champion Bob Crosby. They seemed oblivious also to the frequent visits of the humorist Will Rogers, whose son attended NMMI and who sponsored the Institute's polo team. Rogers was just as unaware that the rocket scientist was at Roswell. Despite their mutual friendship with Lindbergh and the moon-rocket publicity, he never mentioned Goddard in his columns or performances.[9]

Goddard's first order of business was to find places to live and work. As his crew trickled into town, Robert and Esther alternated exploratory trips in all directions, looking for a launch site, with shorter tours in search of a home. This they found in the Berrendo District of Chaves County on Mescalero Road about two miles northeast of the town center. Mescalero

Ranch, a rambling adobe house and outbuildings on about fifteen acres, had been built in 1908 by "Miss Effie" Olds, who moved to Roswell from Columbus, Ohio, during an orchard boom. She raised the three orphaned children of her best friend from Ohio there, and succeeded in the apple business.

Miss Effie no longer needed the large house, and recent harsh winters had ruined her orchard, so she rented the property to the Goddards for $115 per month for two years. Goddard hired attorney Herman Crile to write the lease on the property to protect himself in case he had to remove any improvements, such as the shop he planned to build. Esther overhauled and cleaned up the house. The freight car full of personal and rocket goods was on a siding by 4 August, and Kisk began unloading it; the rest of the crew arrived in the next few days and pitched in. Sachs started designing a workshop on the eleventh.

Goddard's entire crew followed him from Worcester to Roswell. Although jobs were scarce during the Depression, for most the decision to pull up stakes and move two thousand miles—to what to New Englanders must have seemed a biblical wilderness—was based on their loyalty to the rocketeer. They chose housing according to seniority. Mr. and Mrs. Sachs opted to live at the ranch, as did Al Kisk and his wife and the bachelor Charles Mansur. Larry Mansur and his wife lived in town because she planned to look for work.

They were in the ranch house by the middle of August, and Goddard let a construction contract for the shop building. Meanwhile, he had introduced himself around town, and established a business relationship with hardware merchant Dan Wilmot, from whom he bought supplies and a Ford truck. Much of September was spent completing the shop and installing machinery, along with erecting a static-test tower. Goddard next looked for a place to launch his rockets.[10]

He found it on 25 September, with the help of Cort Marley, a cattle rancher at the end of Pine Lodge Road, northwest of Roswell. It was called Eden Valley, a shallow depression on Oscar White's ranch just beyond Marley's, about ten miles from Mescalero Ranch. "He just told us that it was to be kept very confidential," Marley's widow remembered, "and he'd been told that we could be trusted, and he wanted a place that would be open spaces where he could fix his paraphernalia." Construction soon began on a launch tower and control shelter.

To get to the launch site, the rocketeers traveled over dirt roads across the rolling prairie landscape of grass, scrub, and the occasional cholla, until they entered Marley's private ranch roads, then went on to the shelter on Marley's land and the launch tower on White's. All Marley asked in return was "that they just leave the gates as you find them."[11]

Expeditions to Eden Valley were only occasional, and the tower was exposed to the elements and to animal nuisances. An unnoticed hawk's nest, which incorporated several pounds of wire, once ruined a launch, and Goddard and his crew learned to inspect the sixty-foot modified windmill before using it. They found everything from rodents and black widow spiders to a horse that had hung itself up in the tower base and starved to death.[12]

Most of the time, the "Guggenheim Rocket Research Project," as Goddard styled it on his letterhead, kept close to the ranch. There were a few other residences in the vicinity, close enough to hear the roar of static tests, and a nearby dairy farm complained that the noise cut into milk production. Jose and Maria Gonzales let Goddard store tubing and sheet metal in their barn. Mostly, the rocketeer and his crew worked in mysterious isolation reminiscent of his time in WPI's Magnetic Building years earlier. Questions buzzed among people in town about what he was up to, exciting NMMI cadets and local teenagers especially.[13]

There was only one personnel change during Goddard's first tour in Roswell. Henry Sachs accepted an offer to join the Coast and Geodetic Survey as an instrument maker, because his wife wanted to move back east, and Goddard offered the job to Nils Thure "Ole" Ljungquist, Kisk's old Worcester acquaintance who was then working as a mechanic in California. He arrived on 2 May 1931. Thus began a fourteen-year association that Ljungquist called "the best years of my life." He became Goddard's right-hand man, and went on to his own career in rocketry.[14]

The population at the ranch was a combination of New Englander and Swede, with a native reserve that complemented the phlegmatic personality of the "Boss." "This was a peculiar bunch of men," Ljungquist's wife recalled, "and never showed much emotion at times of failure and not much on success, either. Regardless of how they felt, they didn't show it." Roope—who filled in as department chairman during Goddard's absence from Clark—remembered them as "certainly a wonderful group of people—and they stuck right with him."[15]

Esther presided over Mescalero Ranch as its queen bee. Having overhauled the house, she continued to manage Robert, trying to enlarge his social life. In October they decorated their car with ribbons and drove it in the Cotton Carnival parade as the Massachusetts "float." She dragged him to a church supper in December, and although he met a ranch couple "who used to hire 'Billy the Kid,'" he would not go to another. Occasionally Robert and Esther went to dinner or a movie together, but that was as far as she could drive him from the ranch. She made him take time off anyway, and spent considerable sums at Cobean's Stationery buying art supplies. He soon was slinging paint at canvas on many a Sunday.

Esther participated in Robert's work as his typist, bookkeeper, business manager, and majordomo. Her most important contribution was photography, an area where she came into her own during the 1930s. She installed a darkroom in the house, and did her own developing and lab work. She experimented with technique, and became a pioneer in photographic documentation of technology, taking pictures of tests and flights—including deriving stills from movies—and illustrating Robert's notes with detailed photographs of everything manufactured in the shop and assembled into rockets.

Robert, Esther, and the others socialized mostly among themselves, with occasional potluck suppers and card games. On or off the job Esther put her stamp on everything. She had a habit, annoying to the men, of bestowing nicknames. Goddard was "Doctor" or "Doctor G" to her, although the crew more often called him "Boss." She and Robert called the rockets "Nell" or "Little Nell." Out of her hearing the other men called their machines "the rocket," dismissing Esther's "pet name." She could be too cute. Charles Mansur was "Bonnie Prince Charlie," and their truck and trailer together were "Good Queen Bess."[16]

It was not Esther who held the men's allegiance, but Robert. He amused them by figuring their pay on a slide rule, by maintaining his desk in the shop rather than in a separate office, and by becoming a slob. "Very sloppy," Ljungquist remembered, "old stuff. An old suede coat, old comfortable baggy pants, old comfortable shoes." The former best-dressed man on campus even worked in his undershirt, and sat around in the evenings scantily clad. He joined them in morning salutes to "Langley," the old Smithsonian lathe. He taught them to make do with what they had, fashioning fine instruments, according to Charles Mansur, "out of whatever scraps we could find."

They wasted nothing. Mansur said that a major crash was when "there was nothing left but scrap metal, and then we'd have to turn around and rebuild it one hundred percent." A minor crash was when only part of a rocket was destroyed. In between was the moderate crash, where damage extended to instruments or other rare equipment.

Goddard's most memorable traits were his patience and his humor. "At one period in Roswell we had a series of test troubles," Ljungquist recalled. "After coming back to the laboratory from the test area we were all dejected and in poor spirits. I said, 'Doctor, you sure have the patience of Job.' With that he replied, 'Job only had boils.'"[17]

Goddard may have been the boss, but he was anything but domineering. Paul Horgan described him best:

If he stood up straight he would have been quite a tall man, but he was stooped. You had a sense of a large and thin body with this rather hollow-chested effect that TB gives even arrested, and a forward thrust to his shoulders and his neck, and his head out beyond his breast. Completely bald with a fringe of hair around the back and sides. A great, tall brow, with eyes under well-modeled, very definitely modeled brows. Mustached. A heavy beard which always showed when clean shaven. The most marvelous features he had were the eyes. They were wonderfully expressive, extremely kind, extremely gentle, very intelligent. They looked like the eyes of someone who could be easily hurt and yet who had a dancing humor somewhere. They were

*very dark, almost black, a tremendously expressive feature. You couldn't help
liking a man with eyes like that.*[18]

COME AND SEE GODDARD THE MAN
WHENEVER YOU FEEL LIKE IT

This gentle, unassuming man later acquired a reputation for almost para-
noid secretiveness, but it was an image no more warranted by the facts in the
1930s than during the 1920s. To a great extent Goddard's reputation was
developed in the 1950s and thrust upon him retroactively. His authorized
biographer belabored interviewees to emphasize his closed mouth, and they
obliged, but only in part. Lindbergh said his secretive nature was typical of
inventors. Harry Guggenheim called him a "lone wolf," who "became obviously disturbed if his scientific findings were attacked." Charles Mansur
remembered that his men were expected to keep their work to themselves.

Robert Truax alone went further, describing Goddard as "somewhat
abnormal," because most rocketeers could not keep their mouths shut about
their work. "I think in general," he said, "most of the other people in the field
were somewhat resentful of Goddard because in the early days he was the
only one who had any money and the only one who had done any appreciable amount of work on it, but wouldn't tell anyone else what he was
doing. They felt that he was sort of cheating everyone else by keeping it all
to himself."[19]

Goddard's playing close to his vest should be viewed in context. He had
long encouraged public belief that his rockets were about to strike out for
space. He felt no compulsion to reveal technical details, because he was not
as far along as he wanted people to assume; his greatest secret was that he
had no real secrets to unveil. This circumstance became aggravated in the
1930s because Goddard faced competition everywhere—not just in Germany, but also in the United States. When CalTech's Robert Millikan offered
to send graduate students to assist Goddard, he said thanks but no thanks.
He wanted no help, and would give none.

He knew that in Germany and the Soviet Union rocket clubs and government projects were trying to beat him into the sky. Urged on by Esther as well as by his own inclinations, he reacted fiercely if an account gave anyone else credit in rocketry. When the Interplanetary Society's journal erroneously said that his 1929 rocket had used solid fuel, he bombarded the group's leader, G. Edward Pendray, with the news that only Goddard had successfully used liquid fuels. As for "my reticence in giving my results to the public," he said that "so many of my ideas and suggestions have been copied abroad without the acknowledgment usual in scientific circles that I have been forced to take this attitude."[20]

Generating publicity as a way to seek financial support was no longer necessary, since he had found the pot at the end of the Guggenheim rainbow. Nevertheless, he remained hungry for public acclaim, and took advantage of every opportunity to grab his share of glory. With his considerable help, *Scientific American* carried a long article on the state of the art of rocketry, which emphasized his contributions above others. It was the first in a series of features that he engineered in 1931.

The New York Times carried laudatory articles about him every few months, and sometimes Goddard promoted himself over his own byline, as in *The New York Times* in September 1931 and in the March 1932 *Scientific American*. He did not need to reveal technical details as long as the press sang his praises, which it persisted in doing. He was still a publicity hound, only somewhat more restrained than he had been a decade earlier.[21]

There was reason for restraint, as Goddard told the Roswell and regional press when he first met with them in August 1930. The editor of the *Roswell Morning Dispatch* recalled after Goddard's death: "When he first came, he called us in and said: 'Boys, my work is made possible by the money of Daniel Guggenheim. He wishes that fact to remain out of the papers and also the details of such progress as I may make. When it is all perfected I will tell you about it. Please just forget that Goddard the investigator is here, and come and see Goddard the man whenever you feel like it.'" The legendarily publicity-shy scientist played the press like a piano.[22]

His arrival in Roswell attracted considerable attention. On 14 August

1930, he addressed the Rotary Club to explain his work, which was aimed at high-altitude research, he said, not outer space. The Wright brothers, he reminded his audience, did not try to cross the Atlantic the first time up. Through the fall the local papers occasionally printed his picture or reminded readers that he was in the vicinity.

On 1 December he distributed a statement repeating many of his Rotary remarks, and saying that he expected to begin flights within the next few months. "Curiosity of the public in other places where Dr. Goddard has made previous experiments," said the *Morning Dispatch*, "has made it necessary to exclude visitors from his workshop. This was done in order to permit unhampered work on the part of the doctor and his assistants."

On 12 December, when an AP story covered the Smithsonian's promotion of his work, Goddard talked to reporters again, telling them that he had "perfected the long distance rocket to his own satisfaction." AP picked that up and it was all over the country the next day. Thereafter, the Roswell papers contented themselves with daily stories about the Lindberghs and less frequent accounts of Will Rogers's visits. Goddard remained national news, but he nearly vanished from the local press until he left town in 1932.[23]

"The people of Roswell and the surrounding country have been most considerate of us and our problems in carrying on our work in their community," Goddard wrote in 1939. "My assistants and I are permitted to follow our own rather unusual devices with no more ado than if we were farmers, cattlemen, or oil scouts." Roswellians minded their own business, and it was not they who established the legend of Goddard's secretiveness.

If any fact lay behind the story, it was that the rocketeer kept his inventions under tarps. Some visitors during the 1930s noticed that, and Goddard's biographer emphasized it. In 1939 he offered an explanation that would be familiar to anyone who had ever lived in the Southwest: "Then there are the figurative thorns in our Eden. Dust is the greatest of these. It is the price we pay no doubt for the sunny skies and dry air. We must fight against it continually, for a few dust grains in the working parts of delicate mechanisms could be disastrous to a flight. When a rocket is to be hauled from the shop over ten miles of country roads to the launching field it must be wrapped up very carefully."[24]

SEND ME A REPORT OF PROGRESS SO THAT I COULD HAVE A LITTLE NEWS TO GIVE HER

Two months after Goddard arrived in Roswell, Merriam wrote with some technical questions, and suggested that he go east for an advisory committee meeting. Atwood also wrote, after hearing from both Merriam and Lindbergh, and told him that all three thought a meeting would be a good idea. Despite his promises, the rocketeer had not provided his patrons with a work plan, and he turned the meeting request aside. "I imagine there is a good deal of general interest in the undertaking," he told Atwood, "but, after all, the actual work is of the usual hard, and rather prosaic, kind."[25]

Goddard was working on rockets, but he remained easily distracted. Lindbergh thought that was one of his more endearing qualities, but it interfered with the work. In the fall of 1930 he developed an elaborate proposal to use thermocouples to study changes in light intensity and another on methods of proton acceleration, and sent them to other scientists. Both ideas, he learned, had already been tried and proven invalid. He also sent fan letters to such writers as H. G. Wells and Rudyard Kipling, and spent a fair amount of time helping Gordon Garbedian write a book about scientific mysteries.[26]

Some distractions were unavoidable. The greatest was a ceaseless hunt to find a dependable supply of LOX; the Linde Air Products Company outlet in Amarillo, Texas, did its best, but that was not good enough. Abbot tried to locate a LOX plant for him, but such equipment was expensive and inefficient, and required continuous operation. So Goddard turned to liquid nitrogen to raise oxygen pressures in his rockets, thus reducing the requisite quantity of LOX. He never solved the supply problem, which was aggravated by the Depression, with plants operating part-time.[27]

There was other scrounging as well. John Fleming of Carnegie assembled instrument packages to take aloft whenever the rockets were ready, and Lindbergh tracked down rare metals. Atwood was startled to discover that Goddard had used him as a credit reference for a wide range of suppliers.

Goddard also had a liability problem. He was turned down for workmen's compensation insurance by one company after another, because they had no experience with his line of work. Establishment of the Clark University Research Corporation relieved him of personal liability in the summer of 1931.[28]

Finally, among Goddard's distractions were his patents. He filed five more successful applications during his first two years in Roswell. More worrisome to him were his two 1914 patents on rockets, which were scheduled to expire in 1931. He prepared a statement justifying renewal of them on national defense grounds, which would require an act of Congress. Fleming at Carnegie was in favor, but Abbot was opposed to patent extensions on principle, and he rounded up people in the War Department who held the same view. Without support from the Smithsonian or the army, Goddard's proposal died at birth.[29]

He forged ahead on rockets. The actual work was done by the men, though Goddard supplied the ideas and he was always there. The place was kept spotless, befitting the precision of the work, which in Larry Mansur's words was making "rockets from scratch."

On 20 October Sachs began assembling the first model, and nine days later came the first of many static tests at Roswell. "Nozzle closed up by the pressure," Goddard wrote, "probably making liquid irregular, and perhaps causing the gas tank to explode, although the cork float may have stuck." It was not the last failure, fiery or otherwise. Like others it was followed by a "post-mortem" examination to find the cause and correct it, after which Goddard would light a cigar. When a test succeeded, he lit up immediately.[30]

Goddard kept everyone busy with separate projects for gyroscopes, parachutes, and instruments, while he and Sachs concentrated on the rocket itself. Seeking the higher pressure that would provide greater acceleration of exhaust gases, and giving up on pumps, they pressurized the system by installing a large tank of nitrogen gas between the gasoline tank above and the oxygen tank below. The three tanks and the combustion chamber took the form of cylinders capped at the ends with sixty-degree cones, shapes easily formed and welded from steel rectangles and semicircles.

What they ended up with was the last of their open-sided rockets; all later models would have skins to improve aerodynamics. At the top was a conical nose cone with a parachute that would release after the rocket stopped climbing. Below were the three tanks, and at the bottom was a four-inch combustion chamber with nozzle. Four large duralumin vanes provided stability in flight; gyroscopes were not yet ready to test. The device was about eleven feet tall, and on it were mounted four launch shoes, each with three rollers, to engage the two guide rails inside the tower.

Static tests went well, and soon the crew was ready to fly. They hauled it to Eden Valley on 22 December, but when they tugged ropes to open valves and start the system, they pulled the rocket out of the tower. They redesigned the launch system, a gimcrack arrangement of ten ropes that must be pulled in sequence to close the oxygen tank, open the nitrogen tank, start fuel flow, ignite the motor, and finally release the rocket.

All was ready on 30 December, when Nell erupted into life with a thunderous roar atop a column of fire, leaving a feathery tower of soot in her wake. "It went about 2000 ft up and 1000 ft away," Goddard recorded in his diary, "time of ascent 7 sec, no explosion, and little damage. Parachute opened partly, and was not much damaged."

Despite that quiet report, the launch was a triumph, the highest liquid-fuel rocket flight to date. It was still ahead of everyone else; unbeknownst to Goddard, the German Johannes Winkler would become the second person to launch a liquid rocket, the following March.

The Roswell rocket needed work, however. The wind had caught the vanes after launch, causing an arcing rather than a vertical flight, and the resulting aerodynamics had not opened the parachute nose cone properly. Nevertheless, on 3 January Goddard crowed about his achievement in a letter to Merriam, and agreed to attend a committee meeting.[31]

Merriam arranged the meeting in his Washington office for 20 February.

Goddard showed movies and summarized his work, but he was most interested in looking after his financial interests. Since half the profits from his patents would go to science, via Guggenheim and Clark, he persuaded the group that the Guggenheim grant should pay half the cost of patent applications, subject to approval by Abbot and Breckinridge.[32]

The rocketeer then became uncommunicative toward his sponsors. He did not tell them that he had been contacted by the Army Ordnance Department, which sent delegations to see him twice in 1931 and 1932. He declined to show the officers anything, referring them to Breckinridge to secure copies of his reports. He told the committee nothing else either, until October when Atwood said that he had a meeting with Mrs. Daniel Guggenheim coming up. "Could you not send me a report of progress so that I could have a little news to give her?" he asked. Goddard declined, saying that anything he could say would be too technical for her.[33]

Merriam and Atwood together demanded a progress report, and Goddard sent one in December. It described a series of static tests to increase exhaust velocities, with limited results. Since then he had been developing a larger combustion chamber, and had spent the summer devising an electrical launch system to replace the ropes. He implied that he had made several flight tests, although he admitted to no altitude greater than two hundred feet. Actually, he had had three explosive failures in September and October of 1931. However, they proved the potential of his new 5¾-inch chamber and pointed him back to using nitrogen for pressure.[34]

Goddard next decided that his rockets required gyroscopic stabilization. He redesigned his missile early in 1932, adding a four-inch gyroscope that, through a complicated arrangement of wiring and magnets, operated four small vanes at the mouth of the nozzle. If the missile deviated from the vertical, the blast vanes would dip into the exhaust and correct the error. When he tried it on 19 April, the rocket tipped over as soon as it left the tower and crashed. The blast vanes were too small.[35]

Abbot had lectured Goddard about deviating into new areas of development when it was more important to get a rocket high into the air. Lindbergh succeeded Abbot, urging him to aim high in order to attract financial support.

Even Goddard's crew pestered him about that. "We fellows who worked there often mentioned 'Let's get one way up in the air,'" Charles Mansur recalled. He gave the same answer to all: "Let's get it to go straight first." Mansur realized in retrospect that Goddard never understood that he could get stability as well as altitude so long as he had sufficient acceleration.[36]

Goddard's emphasis on gyroscopes was the most persistent technical mistake of his career. He was not trying to build a guided missile, which would have required steering to reach a destination. Rather, his were sounding rockets, intended to rise straight up and then return to Earth by parachute. Gyroscopes could be useful in steering, but were cumbersome and unnecessary for stabilization; properly placing the rocket's center of gravity forward would suffice, but that also was a ballistic principle that escaped him.

Theodore von Kármán of the Jet Propulsion Laboratory agreed with Mansur. If Goddard had concentrated on increasing exhaust velocity, his rockets would have become self-stabilizing; sounding rockets to this day generally lack gyroscopes. Besides, according to von Kármán, Goddard did not perceive that the real problem was not aerodynamic stability, but the fact that combustion was inherently unstable.[37]

He had more pressing worries in 1932, and he asked Merriam for an early committee meeting, because his two-year funding was coming to an end. Moreover, Lindbergh's baby had been kidnapped at the end of February, distracting his most influential patron. Goddard wanted to know whether he would receive money for the next two years, or should use what was left to shut down the Roswell operation.

Merriam hosted the meeting on 25 May, and the whole group urged that the funding continue. Afterward Goddard and Merriam met separately in New York with Breckinridge and Harry Guggenheim. Goddard, whose self-absorption could even exceed the bounds of common decency, went so far as to send the distraught Lindbergh a set of rocket-launch movies and an appeal for support, as if Slim's family tragedy was secondary to Goddard's rocketry.

Nevertheless, the Roswell enterprise was terminated. "It is a cause of very deep regret to me," Breckinridge wrote Goddard, "that the Guggenheims could not continue their support of the rocket experiments this year. I hope next year things will be better. I think you are a good sport to take such a disappointment with so much courage and philosophy."[38]

Goddard did not know what straits the Guggenheims were in. Daniel's death in 1930 came just as the Depression hit bottom, and it left the family's finances in disarray. The investments that provided income for charitable works were not performing, due to the depressed economy. "They had no eggs to lay," Breckinridge said later. Another issue was a battle with the Internal Revenue Service over how to appraise Daniel's estate for the inheritance tax. The dispute ultimately was settled by compromise, and Harry saved the family fortune.[39]

That happened too late to save Goddard's New Mexico adventure, however. "Owing to the depression there are no further funds available for the rocket project," he told the papers back in Roswell. "There is a possibility that the work will be continued in Roswell next year, but this is by no means certain. I feel that any further statement in the matter should come from the advisory committee."

Dan Wilmot provided storage for the shop equipment, while the crew crated what was going back to Massachusetts and loaded it onto a freight car on 1 July. After everything was cleaned up, Goddard dropped his old work hat on a bench and locked up the shop, while the crew went home. Robert and Esther headed out in their car, making goodwill visits to Cal-Tech and Mount Wilson before turning east. After two years, he was a Clark professor again.[40]

HUMANITY AMONG THE FACULTY

He was hardly back at Clark before he asked Abbot for money. His old patron sent him $250 from the Hodgkins Fund, all the Smithsonian could afford. It was better than nothing, and with it Goddard resumed doing minor experiments on pumps and other components. He also tried to perfect an air-breathing rocket that he had patented in 1930. This got him some publicity, but not much practical success.

He was distracted. His diary reflects increasing time devoted to reading and copying homilies and quotations of a stoical nature. He could be humorously self-pitying: "The rocket is very human. It can raise itself to the very loftiest positions solely by the ejection of enormous quantities of hot air. Emerson says, 'If a man paint a better picture, preach a better sermon, or build a better mousetrap than anyone else, the world will make a beaten path to his door.' I, like many others, have had the misfortune not to be an artist, a preacher, or a manufacturer of mousetraps."

For Christmas 1933 Esther gave him painting lessons at the Worcester Art Museum, and within a few months he thought he could see improvements in his technique. Nobody else could, but his enthusiasm for painting increased.[41]

It was his fund-raising technique that most concerned him. Having waited a decent year after his money was cut off, in May 1933 he wrote Lindbergh an elaborate appeal for renewal of his project. He asked for a meeting with Lindbergh, Breckinridge, and Guggenheim. The Lone Eagle, with problems aplenty of his own, did not reply.[42]

Goddard next went to Washington, where Abbot provided a letter of introduction to the Department of the Navy, which first gave him a runaround and then expressed interest in seeing his reports. Fleming at the Carnegie Institution assured him that his work would resume, and advised contacting Breckinridge, who urged him to write to Daniel's widow, Florence, and ask for $25,000 to resume work at Roswell, or if that was not available, $2,500 to continue working at Clark. Harry answered for his mother, saying that the larger figure was unaffordable, but he would give Clark University the $2,500. Goddard then asked Harry for permission to deal with the navy, and the naval reserve officer complied happily.[43]

He went after the navy in July 1933, sending Acting Secretary H. L. Roosevelt an illustrated report. It offered a history of his work since meeting Lindbergh, and ended with a statement of rocket applications to national defense, especially an "aerial torpedo." The Department of the Navy was taken with the report, and with Abbot's expression of support. Secretary Claude S. Swanson asked Guggenheim for permission to reproduce and distribute the document to various naval bureaus, and Harry agreed on behalf of his mother. He was enthusiastic, but in the end the navy decided it could not afford to develop rockets.

Disappointed, Goddard passed the bad news on to Abbot, doubting that anyone would fund him until he did "something of an especially spectacular nature." That was the first hint that he might understand why Abbot, Lindbergh, and his own crew had pressed him to fly high.[44]

Goddard's doctor chided him for not taking better advantage of his time in the Southwest, because he had missed a chance to improve his health. That amused the rocketeer: "Dr. Fisher said I might have been all well now, if I had lived just as I ought to have, when in the West the last two years, and taken advantage of the opportunity. He apparently thought that the $50,000 grant was made to go West and take a rest cure."

He was healthy enough to sit for an interview and profile in the student newspaper, which called him the "Einstein of Clark University" and a case of "Humanity Among the Faculty." He intimidated most students, but on closer examination he turned out to be "a man intensely human and very much like other human beings." As for Esther: "The lady in this particular picture, Mrs. Goddard, is a charming and capable woman, and she offers an interesting antithesis to her husband, especially in social inclinations."[45]

The students liked him, but Goddard was wearing out his welcome among his colleagues, who regarded him as an unproductive prima donna. He still taught classes and guided theses, but he was more interested in rockets than in professorship. He chafed at continuing to head two departments, and complained to Atwood about his workload.

In fact, he was not doing his job, and the president was not happy about it. Taking prospective students and their alumni fathers on tours in the spring of 1934, Atwood observed what he called a "decline in the Physics Department." The labs were dirty and disorderly, no one was in attendance, and other faculty had complained. Moreover, except for a required freshman course, few students took physics courses at all. "Can we correct all this?" the president asked. He received no reply.[46]

Goddard did not care about his legacy at Clark, because it appeared that he might return to Roswell. Besides, he had paid himself $5,000 per year under the Guggenheim grant, while Clark gave him a fifth less. He had resumed discussions with Lindbergh earlier in 1934, and the two continued talking on the telephone through the spring.

Harry had straightened out his family's affairs, and in the fall of 1933 had reorganized the Daniel and Florence Guggenheim Foundation, first established in 1924. At Lindbergh's urging, in June 1934 Goddard asked Harry whether he could resume the previous level of funding. He wanted at least another $2,500 for further work at Clark. That, he explained, was useful in developing rocket parts, but he wanted to resume flights in New Mexico.

Guggenheim invited him and Lindbergh to Falaise, and the three of them discussed the matter over dinner and breakfast. Goddard's advances and Lindbergh's support paid off. On 11 July the foundation offered a one-year commitment of $18,000 for flight tests at Roswell; further support depended on progress. Atwood would have to accept for the Clark University Research Corporation, but he was on his annual field trip. Goddard tracked him down in Wyoming, and the president agreed to everything. He granted the rocketeer a leave of absence for one year, with Roope to fill in for him as department chairman again. And again, a substitute professor would be required in Roope's place.[47]

The Goddards were already packing and rounding up the old Roswell crew. There was little of the prideful hoopla that had surrounded their previous departure from Clark, however. Now they left behind a faculty and administration considerably disgruntled about Goddard's performance. When he reached Roswell, he found a letter from Atwood announcing that the trustees had declined to contribute the university's share to his annuity plan while he was not on the payroll. The rocketeer could continue to contribute his own part, and in response Goddard sent Atwood a check. "It is signed by Mrs. Goddard," he pointed out for no apparent reason, "as there is only one bank in Roswell, and my own signature is used for Foundation matters only."[48]

The "adventure" was about to resume.

NINE

HE WAS A HOBBYIST

Yes, gentlemen, in spite of the opinions of certain narrow-minded people, who would shut up the human race upon this globe, as within some magic circle which it must never outstep, we shall one day travel to the moon, the planets, and the stars, and with the same facility, rapidity, and certainty as we now make the voyage from Liverpool to New York! Distance is but a relative expression, and must end by being reduced to zero.

—Jules Verne, *From the Earth to the Moon*

Goddard's second tour in Roswell would last eight years, but he did not know that at the outset, because his grant from the Guggenheim Foundation was for one year, and it continued on an annual basis. That it continued at all was owing more to Lindbergh's influence with Harry Guggenheim than to Goddard's own accomplishments. The pattern of annual renewals, often at the last minute, kept Goddard in a state of uncertainty, and it had a similar effect on President Atwood.

The Goddards did not let the door hit them in the back on their way out of Worcester in 1934, and over the years they became more at home in Roswell and less interested in returning to Clark. Goddard avoided burning his bridges, however, in case he had to resume professorship to earn a living, and he retained the chairmanship of the math and physics departments. But he showed little interest in the problems his absence caused others. Percy Roope carried the chairman's responsibilities without the pay or title that went with them, while Atwood could offer only temporary appointments to substitute professors, and those men bolted at the first solid job offer from elsewhere.[1]

I WONDER WHAT IS HAPPENING

In January 1935 Atwood wrote to ask whether Goddard would report anything to the advisory committee—which effectively ceased to exist at about that time—and whether he would return to Clark in the fall. He replied that he intended to continue working, but that he must assume he would return to the university.

Goddard passed on to Guggenheim the president's need for an answer by 1 April, and when Harry invited Goddard and Lindbergh to meet with him at Falaise, Goddard took a train east. Lunch with the Lindberghs and the Guggenheims brought a commitment from the foundation to pay the substitute professor's salary for the next two years. Atwood was mollified. "We are all excited over the conspicuous publicity which you are receiving in the papers," he wrote in August.

The next spring, Goddard invited Atwood to visit during his annual field trip, and the president and his wife stopped by in the fall, when Atwood told him that he was taking heat from the faculty and trustees over his indefinite absence. Goddard climbed onto a high horse, reminding the president that his was really a university project. Atwood again pressed the issue in January, still feeling pressure from other faculty. "Concerning such criticism," Goddard sniffed, "it is perhaps worth mentioning that both Colonel Lindbergh and Mr. Guggenheim feel that the prestige which the University is acquiring as a result of sponsoring a work of this kind should more than offset any disadvantages which might result because of my absence."

When Guggenheim advanced the substitute professor's funding for another year, Atwood invited the Goddards to commencement in June. Lindbergh then stepped in to cover Goddard's flank, sending Atwood a letter praising Goddard's work and the university's support of it, and ensuring that it was read at the commencement and covered by the newspapers. The next winter, however, Atwood complained: "I have not heard anything of your work for so long that I wonder what is happening." Guggenheim provided another year of stop gap funding, and so the friction continued.[2]

FLYING COUPLE VISIT WITH
DR. GODDARD

The Goddards were back in Roswell on 13 September 1934. "Scenery after leaving Raton very pretty," he noted in his diary, "pale blue mountains in west, palest at bottom line.... It was like *two* skies: one near and one distant and very entrancing." They had not realized how much they missed the area. From then on, Goddard habitually rhapsodized about their corner of New Mexico.

"Morning in the desert," he said once, "when the impossible not only seems possible, but easy." Returning from a trip, he observed: "The dry air and the sunlight seemed very good, after the dingy skies, fog, rain, and dampness of the East. It seems as if there were some truth in the saying that there is no air east of the Mississippi."[3]

During his first tour in Roswell he had remained aloof from the town, but now he was entranced by it. "It is real western cattle country here," he wrote to the son of a former classmate, "and on Saturday afternoons the Main Street of the town is filled with cowboys, with wide hats, high-heeled boots, and leather chaps, and with others who look like old prospectors. We have a mountain about 40 miles away at 10,000 feet high, and some higher mountains about 80 miles away, but the country here is very level."[4]

Ljungquist, Kisk, and Charles Mansur reached Roswell soon after the Goddards; Larry Mansur, who had found a teaching job, remained in Massachusetts. They were ecstatic, prancing around as they reopened the shop and cleaned things up. Esther photographed comic rituals—the crew taking down a 1932 calendar and replacing it with one for 1934, Robert ceremoniously donning the hat he had left on the shop bench two years earlier. Their return occasioned the customary blaze of publicity, until something more dramatic occurred.[5]

FAMOUS LINDBERGHS PAY VISIT TO CITY, shouted the *Roswell Morning Dispatch*, then continued: SPEND NIGHT HERE ON TOUR OF AIRPORTS. FLYING COUPLE VISIT WITH DR. GODDARD. Without warning, Charles and Anne Lindbergh landed in Roswell on 15 September. The Goddards had not unpacked, but the Most Famous Couple in America were graciously unimposing. Goddard took them to the shop, out to Eden Valley, then back to town, where he hosted them to dinner at the Nickson Hotel.

The next day, Slim took both the Goddards up in his airplane, circled the launch tower, and landed. When he returned to the Roswell Airport, he discovered that a cactus thorn had punctured a tire. As he and a mechanic fixed the flat, word spread, and the airfield became a mob scene—the usual effect when the Lindberghs appeared anywhere. The crowd most wanted to see the famously beautiful but painfully shy Anne, who retreated to the airport manager's office until they took off that afternoon.

Lindbergh was impressed with Esther's contribution to Robert's work. "I feel sure that Mrs. Goddard deserves more credit than she will ever take for the success of the project," he later wrote to Robert. Anne carried away a lasting impression of "that moon man," as she had previously thought of him. "I never saw a launching," she wrote in her memoirs, "but I remember an evening sitting on a screened porch, while my husband and this quiet intense professor talked of space exploration. Flying was then a new adventure to me; I had just won my pilot's license, and these two men were talking of a step far beyond flying—ascent into space."[6]

The Lindbergh visit impressed Roswell, adding glamour to the mystery that surrounded Goddard. But there was also the man himself, according to Paul Horgan: "His own dignity and his own nature were so appealing and winning that he was taken just as a fellow and liked very much, and Esther was popular and in a lot of things in town. . . . Aside from his work, I would say he was probably the most conventional human being I've ever known in my life. He was absolutely indistinguishable from a man who owned a lumber yard or a lawyer or a vice-president of a bank, highly conservative in everything except his wild interstellar obsession."[7]

Only Cort and May Marley, gatekeepers to Eden Valley, got close to Goddard's work. May remembered: "He'd laugh at things around, and he enjoyed the cattle, he enjoyed watching the men. He and Mr. Marley spent lots of happy minutes together. He got to telling us when he was going to shoot those things off and Mr. Marley would go over and sit on the hill or maybe on horseback someplace and watch them for maybe half the day."[8]

Goddard was a shy, quiet man who had difficulty making new friends, but once he did he blended comfortably, with his deadpan humor and many interests. Because of his ill health, Dr. Charles Beeson and his wife became the Goddards' first friends in Roswell. Esther summoned the doctor a few days after their return, because Robert was in pain and had trouble breathing.

When they visited Worcester in 1937, Esther sent him to his old physician, Edgar A. Fisher, who gave him a complete physical and X rays. Robert told her nothing, so she wrote to the doctor, who complained that Goddard "persistently overdoes it and it is surprising that he does not suffer as a result of it." He was spitting up more than he had in previous years. She was appropriately alarmed, and became ever more protective, but she could do nothing about his self-destructive ways.

His smoking continued, and his drinking increased as the decade wore on. He became famous in his circle as a bartender; his martinis were especially prized. It was not all social, however. He spent his evenings in his undershirt on the screened porch, a cigar in one hand and a drink in the other. George Bode first encountered him that way in 1940, "blowing out big clouds of smoke." Homer Boushey recalled evenings when he and Goddard got drunk on martinis, talking space travel into the night.[9]

After Esther's father died in August 1935, her mother, Augusta, came to live with them, so again he inhabited a household where two mother hens fluttered over him. Esther drove him away from work and into the countryside, often with the Beesons. They motored into the desert or the mountains, watched the Indian Rodeo on the Mescalero Reservation, hunted arrowheads, collected rocks, and shot rattlesnakes. In the summer of 1935 the Goddards and the Beesons went to the mountain resort of Ruidoso, then on to see a bullfight in Juárez, Mexico. For Christmas 1937 Robert and Esther went to Santa Fe. In his diary Robert noted that Esther's "Ma" usually accompanied them, and he ended each expedition's account by saying that he spent the evening smoking, alone or with Dr. Beeson.[10]

Esther threw herself into Roswell society, but Robert only gradually grew more social. She joined St. Andrew's Episcopal Church and became active in its Junior Guild, but there were limits to how far he would go—they remained Easter Christians, attending services only that one time of the year. Robert became a popular speaker in local organizations, talking about rocketry and "stories of the atmosphere" before the high school, the Women's Club, the Civic Club, a joint gathering of the Advertising, Lions, Rotary, and Kiwanis clubs, the St. Andrew's Sunday school, and the New Mexico Education Association. He comfortably addressed such groups, but he joined none.[11]

Esther began building a circle of acquaintances with the NMMI faculty, and their closest familiarity was with Paul Horgan, a congenial bachelor. "You'd think that sundown was a good time," Horgan remembered, "and so you'd go and they'd give you a highball and Bob would smoke a cigar and rock in his chair and flap his foot. That was a characteristic gesture when he would get amused, excited, or interested."

Robert enjoyed music, especially old operas, so he and Esther joined the "Beethoven Society" Horgan hosted at his home every Tuesday evening during the winters. There ten or fifteen people at a time would drink cocktails and listen to records. Horgan remembered that Esther, "whose job it was to see that he was effective and happy," hovered over Robert.[12]

Although Esther ensured that Robert's biography emphasized his piano playing, nobody in Roswell except the attorney Herman Crile remembered his touching a keyboard. His painting was another matter, as he gave his creations away to everyone he knew. Horgan said that Goddard's painting showed more enthusiasm and gobs of paint than talent. Peter Hurd, a more qualified judge, remembered: "Completely without any affectation or pretense to a viable talent as an artist he delighted in the wonderful experience of painting."[13]

Peter's father, Harold, was an attorney from New England who had moved to Roswell for his health in 1902. He acquired twenty acres south of town, and planted dozens of spruce trees in formal rows that earned the property the name Hurd's Folly. The house was full of books, and the elder Hurds and the Goddards became natural associates.

Peter drove down from San Patricio to join them for dinner in the mid-1930s, and afterward Peter and Robert stood in Harold's yard, where the rocketeer pointed out stars and constellations, and they talked about art. This affected the young painter, widening his sense of the wonders of nature.

It also affected the Goddard legend, as the authorized biography quoted Hurd as saying: "Years later, after his death, I realized what he was doing there. He was pointing out landing fields in the solar system." Hurd told biographer Lehman nothing of the sort; Lehman's unanswered question about that possibility became the quotation.

Goddard went to San Patricio to paint with Hurd in 1939. "He loved to spend an afternoon just musing with a piece of canvas in front of him," Hurd remembered, "so he came up here to paint landscapes. We went out together and after about an hour we came in, sat down, had a drink. I can't remember the conversation, but I remember a sense of great well-being, friendship."[14]

The Goddards entertained Roswell by receiving important visitors. Lindbergh was without equal, and he visited almost every year. On 22 September 1935, he flew in, trailed by a press retinue and accompanied by Harry and Carol Guggenheim. Slim wanted to show Harry that Goddard's work deserved a second year of funding. Harry wanted to see a launch, but rain showers limited attempts to fly. Guggenheim's security, meanwhile, kept rubberneckers well back from both the shop and Eden Valley.

The visit turned into an exercise in frustration. Nell failed to ignite on the twenty-third, and on the twenty-fourth it rained again, although Guggenheim was impressed enough to issue a statement affirming his foundation's commitment to continue funding. On the twenty-fifth, the combustion chamber burned out with a loud pop before liftoff, and Guggenheim and Lindbergh departed without seeing the machine leave the ground.

Two days later, the Goddards listened as the radio program *Time Marches On* dramatized his work. The broadcast ended with Guggenheim, at Lindbergh's urging, vowing to support the rocketeer's work "indefinitely."[15]

This visit had a regrettable sequel. Following the trial of Bruno Hauptmann, who was convicted in 1935 of kidnapping the Lindbergh baby, harassment and threats against Slim's family increased, and they left the country. He left behind only three people he trusted: Goddard, Guggenheim, and Dr. Alexis Carrel of the Rockefeller Institute of New York. "But most of all," Lindbergh said in his autobiography, "I missed the opportunity of working closely with Robert Goddard in developing the rockets that I felt might someday carry man out into space." The men continued to correspond, Lindbergh repeatedly assuring Guggenheim that Goddard could carry it off. Eventually, the Lone Eagle made return visits to Roswell, leaving his family safe in England.[16]

Goddard was isolated by the nature of his work, because few people understood enough about it to carry on a conversation. Like most technical enthusiasts, he wanted to schmooze with somebody who could appreciate his interests. Back at Clark, he had Roope to talk to, but there was no one of the sort around Roswell, except when Lindbergh or Guggenheim visited. Far from being the secretive loner of legend, when a fellow technician showed up he could scarcely restrain himself, as when he burst forth on 13 October 1936, the day Lieutenant John W. Sessums, Jr., Army Air Corps, flew in from Wright Field, Ohio.

The lieutenant's mission had started with Major General Archibald H. Sunderland, chief of the Coast Artillery Corps, who had asked Goddard about the possibility of developing rocket-powered drone planes as anti-aircraft targets; he wanted the rocketeer to help him talk the Air Corps into the idea. Sessums arrived to investigate. Esther fed them lunch, then Robert took him on a tour of the shop and out to Eden Valley.

"Goddard was so absorbed in telling me about the status of his work," the officer remembered, "and the future possibilities of it, that he drove in first gear the entire fifteen miles from the launch tower back to the shop." This gusher was set off by Sessums's intelligent question about launch procedures, suggesting the use of a catapult to accelerate takeoff. That was the only positive result of the visit, because his superiors in the Air Corps did not share Sessums's enthusiasm for the rocket-drone idea.[17]

Goddard's most unexpected visitor during the mid-1930s turned out to be his most welcome, and the two of them changed the history of the radio industry. Arthur A. Collins was a radio pioneer who had spent time in prison for stock fraud; now RCA had sued him for patent infringement. He knew the way to Goddard's heart. "It is my belief," he said in 1935, "that your early work with this vacuum-tube oscillator entitles you to recognition as one of the important contributors to the radio art."

The oscillator he referred to, a radio-frequency generator, was the subject of Goddard's first application for a patent, granted in 1915, which he had sold to Westinghouse in 1920. The story began earlier, when Thomas Edison played a role in early radio development. In 1910, Ambrose Fleming went over Edison's work and improved it into a rectifier converting alternating to direct current. Two years later Lee DeForest added a grid to the vacuum tube, producing amplification of radio signals. His patent became grounds for RCA's monopoly in radio transmission equipment.

Goddard's tube did not disappear, however. In 1921 Westinghouse used it to support a lawsuit against RCA, but Westinghouse lost the case in 1928 when the Supreme Court declared that DeForest was the originator of the signal-generating tube as of 12 August 1912. In 1929 Ralph Heintz of Dollaradio used the Goddard patent in another challenge to RCA. He had made enough changes to his "Gammatron" to avoid infringing on the Goddard patent, but in 1935 RCA went after him in court. His interests then became bound up with Collins's.

Goddard met with Collins and his wife and attorney in St. Louis at the end of 1935, and the radio man engaged him as a consultant. The rocketeer had Roope ship his files on the oscillator to him, and assembled a statement on the origins of what Collins was marketing as the "Goddard Tube" (the patent, and Westinghouse's interest, expired in 1932). This distraction from rocketry continued for some time, marked by a very long notarized affidavit sent to Collins's lawyers late in 1936. That gave RCA pause, and Goddard found himself besieged by attorneys from all sides.

The judge found in favor of RCA, but Collins determined to appeal, because the materials Goddard had provided clearly showed that his tube differed from DeForest's, and moreover that he had worked on it before 12 August 1912. RCA abruptly settled out of court for one dollar, granting Collins and Heintz broad licenses.

Most important, the company ceased to monopolize everything involving radio except home receivers, which abetted enormous growth in the electronics industry in the following decades. As Collins told Esther, this could be attributed to the fact that appeals courts would surely recognize Goddard's patent as prior to DeForest's. The whole business had also involved the rocketeer's ego. "His interest as I understood it at the time," Collins said, "was mainly in establishing the integrity of his early scientific work."[18]

HIS MANY YEARS OF EXPERIENCE WOULD HAVE HAD A STRONG INFLUENCE

The Collins case was welcome comfort, because Goddard's ego was under assault. He was no longer alone in the rocket field, and he resented the competition, staying ahead of it in the public eye by generating heavy press coverage. The adjective "lonely" was applied to his workplace at Roswell so frequently that it became a numbing cliché. That isolation sparked the imaginations of eastern reporters, including Harry M. Davis of *The New York Times* and Howard Blakeslee of AP, both of whom visited Roswell several times and produced separate series boosting Goddard and his work. Papers everywhere fretted about whether his foundation grant would be renewed, and cheered when it was.

The amount of technical detail he released to favored reporters was amazing. His patents were greeted with lavish praise, and accounts of his experiments traced his advances in chamber design and cooling, and in gyroscopic stabilization. When Guggenheim praised his work at a conference at New York's Waldorf-Astoria Hotel in 1936, papers all over the country echoed him. G. Edward Pendray was an incessant booster, and over his byline in 1938 *Scientific American* proclaimed Goddard to be the NUMBER ONE ROCKET MAN.[19]

Goddard still received hundreds of letters from foreign countries. REP was active in France, and in 1938 *The New York Times* reported that an aviator in Britain had invented a "torpedo rocket" for use against enemy planes. But always the Germans came first in his mind. Willy Ley arrived in America and pumped him for information, while Goddard suspected that Nazi spies were on his own trail. Lindbergh returned from Germany and told Goddard that he was convinced the Germans were working on military rockets but would not talk about the project. Goddard intervened with publisher Simon & Schuster to demand changes in a book the firm was about to publish, *Rockets Through Space*, because British author P. E. Cleator gave too much credit to the Germans. The publisher inserted a footnote stating that Goddard flew a liquid-fuel rocket before others.[20]

Goddard also faced competition closer to home. At the end of 1935 he sent Abbot affidavits from everybody who had witnessed his rocket work since 1920, for deposit at the Smithsonian. He said he wanted to counter claims from Europe that people there had been first to do things that he had pioneered. What actually bothered him was that other Americans were entering rocketry, and some of them were getting attention.[21]

Rocket organizations were popping up at every hand. They usually promoted cooperation, but Goddard was not about to share his results until he had completed his work. Others, most notably Pendray's American Rocket Society in New York, forged ahead without him, launching their own rockets. That earned them headlines, and Willy Ley drew other notice with his attempts to develop a rocket to carry mail. Robert C. Truax, a precocious midshipman at the Naval Academy who would go on to head up the navy's rocketry programs, wrote Goddard with a long list of sophisticated questions and suggestions on rocket technology.[22]

Looming over the horizon to the west was the Guggenheim Aeronautical Laboratory at the California Institute of Technology—GALCIT. It had been endowed by the Guggenheim Fund in the 1920s and had since become an important research laboratory. The arrival of Theodore von Kármán in 1926 turned the place into a beehive that attracted bright young minds with wide-ranging imaginations. It fell under the general direction of Robert Mil-

likan, a member of Goddard's advisory committee, who wanted to cooperate with Goddard.

One of his advances produced an incident that became part of the Goddard legend. According to Milton Lehman, the Goddards returned to
Roswell from a vacation in August 1936 to find a letter from Millikan introducing a CalTech graduate student, Frank Malina, and Malina himself on
their doorstep. Malina was a GALCIT spy, asking too many questions and
trying to probe into Goddard's secrets. Robert and Esther tensely showed
him around, but sent him away without his having seen anything. Robert's
subsequent refusal to cooperate with GALCIT was a defense against bandits
bent on stealing his ideas.

That Lehman's account was manufactured by Esther is reflected in a
statement he made to Truax and others he interviewed: "The story that the
Goddards [meaning Esther] tell is that Malina came in and demanded to see
everything, and one didn't do this with Dr. G. And Malina kept demanding
and insisting, and Goddard kept refusing flatly, and they said that Malina
declared that he was going to build a better rocket than Goddard did." The
interviewees agreed that Malina was brash, but they disputed the story. Von
Kármán said that Malina returned from Roswell highly enthusiastic.

Millikan did write to Goddard introducing Malina and announcing his
visit. Robert and Esther found the letter on their return home on 30 August,
and together they picked Malina up at the train depot the following day,
took him to dinner at the Nickson, and spent the evening talking rockets on
the back porch. The next day they showed him around the shop and the
launch tower, then put him on a train.

When he got home, Malina wrote to thank them for their hospitality,
"and for your kindness in allowing me to inspect that part of your work
which you considered permissible under the circumstances." According to
von Kármán, the young man returned from Roswell full of ideas, "and we
introduced many changes in the liquid rocket which were due to Goddard's
influence, and using Goddard patents."

Malina took issue with Lehman's account of his visit, which was not
tense, but friendly. He remembered that he saw everything in the shop

except for a few components. However, he also said that the first thing Goddard did when they met was show him a copy of the 1920 editorial in *The New York Times*. "He appeared to suffer keenly from such nonsense directed at him," Malina recalled. This is the only evidence of Goddard's legendary hostility to the press. Not since Sub-Freshman Day in the early 1920s had he even mentioned that editorial.

"The second impression I obtained," Malina said, "was that he felt that rockets were his private preserve, so that any others working on them took on the aspect of intruders. He did not appear to realize that in other countries were men who, independently of him, as so frequently happens in the history of technology, had arrived at the same basic ideas for rocket propulsion."

Malina's further recollections were colored by a letter Goddard sent to Millikan after he left. "I wish I could have been of greater assistance to him," Goddard said, "but it happens that the subject of his work, namely the development of an oxygen-gasoline rocket motor, has been one of the chief problems of my own research work, and I naturally cannot turn over the results of my years of investigation, still incomplete, for use as a student's thesis."

This insulting crack caused Millikan to point out the benefits to be gained from "constant intercourse with men who are working in the same field and from getting a lot of good from a mutual interchange. So far as my experience goes, it always works both ways.... There is a group of pretty able men here at the Institute working on the problems in which you are interested, and I merely thought that this contact might be mutually helpful."

For Malina, who later developed America's first successful sounding rocket, the WAC Corporal, Goddard did not have the influence that von Kármán described in 1956 (and denied ten years later). The GALCIT group, he said, examined Goddard's patents and publications, along with those by Oberth and others, and found them too general to be useful. "As is well known," he pointed out, "patents are not equivalent to know-how and rarely provide the only trial basis for engineering design." In addition: "There is no doubt that had Goddard been willing to co-operate with our CalTech group, his many years of experience would have had a strong influence on our work. As it happened, our group independently initiated the development of liquid and solid propellants different from those that Goddard studied."[23]

Esther may have encouraged Goddard's paranoia about other rocketeers, but he wanted to remain the "number one rocket man" by generating coverage in the daily press. Guggenheim and Lindbergh pressed him to seek out the technical press instead. "As your work progresses," Guggenheim warned, "unless you have a definite plan, publicity may develop along sensational lines that I am sure would be distasteful to you."

Lindbergh had another idea: Goddard should donate a rocket to the Smithsonian, which housed his own plane *The Spirit of St. Louis*. Abbot agreed. Goddard assembled a rocket from parts that had actually flown, coated it with grease, crated it up, and sent it off on 2 November 1935. Robert and Esther together watched Nell's train until it was out of sight. He included a descriptive statement that would have made a fine exhibit label, but required that the artifact remain sealed until he gave permission or, in the event of his death, Lindbergh or Guggenheim did so. The Smithsonian was unable to display it until 1947.[24]

Goddard's patrons urged him to publish an account of his work, preferably in a Smithsonian outlet. He completed a manuscript and sent it to Guggenheim in late December, asking Harry to make any changes and forward it to Abbot. The latter secured him a place on the AAAS program in St. Louis, where Goddard read an abstract of the text. He showed rocket movies along with his talk, with Abbot and Atwood in the audience. Separately, each rose to commend his accomplishments. The result was a nationwide celebration of Goddard's work through the AP and hundreds of newspapers.[25]

Publication by the Smithsonian did not go smoothly, as the proofs for Goddard's review disappeared in the mail. This made the rocketeer think he was being spied upon. When "Liquid-Propellant Rocket Development" appeared in March in the *Smithsonian Miscellaneous Collections*, it offered little that would have satisfied a spy, however.

Ostensibly a reprint of a report to the Guggenheim Foundation, it was not so much a scientific paper as a history of Goddard's rocketry. It included the first formal announcement of his 1926 flight, but it was short on technical details, and even the many photographs were unrevealing. Goddard alluded to a "later paper" and a "later report," for those who wanted more specifics.

The publication earned him another tremendous round of commendatory newspaper coverage, but the first flight of 1926 was scarcely noticed. What caught reporters' eyes was the announcement that a Goddard rocket had risen 7,500 feet at a speed of 700 miles an hour, "or 200 miles greater than the highest airplane speed which engineers expect to achieve," in the words of *The New York Times*. This acclamation stimulated increased enrollments at Goddard's alma mater WPI.[26]

Guggenheim and Lindbergh resumed pressing for a high flight, this time one that was officially registered. Meanwhile, in the aftermath of the Malina visit, Guggenheim had a greater worry. He and Lindbergh had both heard negative remarks from influential scientists about Goddard's uncooperative ways. Guggenheim also wondered why his two rocket projects, at Roswell and at Pasadena, could not get along. Goddard's progress was impressive, but Harry believed it should have been better.[27]

THE MOST THRILLING SIGHT I HAVE EVER WITNESSED

Harry Guggenheim kept Goddard on a shorter leash than had his father. Besides making the grants contingent on progress, he required more frequent and detailed reports, for Lindbergh's review. Harry also made smaller grants—$18,000 the first two years, $20,000 thereafter; Goddard kept his own pay at $5,000 per year, but cut that of his help in half. He wanted to show progress in 1934, but he faced unforeseen problems—scavengers had carried off everything at the launch site but the tower framework. In December 1934, he told Guggenheim that he expected to fly before the end of the month.[28]

Most work took place in the shop, where the Boss was surrounded by his New England Swedes. George Bode, who joined Goddard in 1941, described

ROCKET MAN

the scene: "As you came in the door his desk would be sitting near one end, but he was right in and around all the machines and the welding.... He said that he had to keep the feel of it, the closer he was to the work the better he liked it. In other words, he could have had an office over in his house and been more comfortable but felt he would lose the feel of it."[29]

Robert Truax said Goddard's "primary contribution was in visualizing what had to be done more than in actually doing it. He did enough so he was able to see very clearly what kind of components would be required." Instead of drawings, the way he designed parts was to sit on a high stool, cut up coffee cans, and make things up with a soldering iron.

According to Charles Mansur: "He'd cobble up some of the craziest looking monstrosities—nothing against him, he was a wonderful man—but he couldn't solder, he couldn't weld, he couldn't run a machine, but he did all of it, though. He'd get a big chunk of a thing set together and then it would all fall to pieces.... He liked to do things himself. In other words, he was a hobbyist."[30]

Once things had been manufactured in the shop, and after static tests in the nearby stand, next came flight tests. The rocket went to Eden Valley, and was installed in the tower. Fueling was dangerous, so the whole crew was alert. Everyone kept an eye on the weather, because the approach of a storm or dust or a rise in the wind meant that the launch would be canceled. In that case, Nell's tanks would be drained, and the rocket went back to the shop for cleaning.

If conditions were propitious, the crew scattered to their stations. A launch began when Goddard said: "Press Key One." "For actual launching," said Ljungquist, "we had to watch the pressure gauge through a telescope, and when the pressure came up to the desired point the key was pressed which started the igniter. Then there was a sequence of electrical currents set up which when the rocket pressure came up to a given point, then an indicator would tell us when it was getting full thrust, and then another key was pressed releasing the hold-downs, and then it was on its way."[31]

Prodded by the methodical Lindbergh, Goddard tried to be systematic, organizing a series of experiments designated by letters. The "A" series occupied him from September 1934 to October 1935, while he tried to perfect a gyroscopic stabilizing system. Generally the A rockets were about thirteen to fifteen feet tall and about nine inches in diameter. A third of the length went to a gaseous nitrogen pressure tank, which forced propellants into the motor and powered the gyroscope and blast vanes. The parachute was in a box below the gasoline tank just under the nose cone.

A-1 and A-2 failed to fly in January, and Goddard became concerned about risking his expensive gyroscope in an unproven vehicle. He removed the stabilizing system, and altered the profile of his air vanes for A-3. On 16 February: "The rocket went up straight, and with a rather strong acceleration . . . but there was a pop, and then the chamber burned through. As the rocket began to point downward, the parachute was released; this burned, however, on reaching the ground. The rocket landed on its side."

For A-4, Goddard reintroduced a blast-vane steering mechanism, controlled by a pendulum rather than a gyroscope. On 8 March, this Nell went straight up about a thousand feet, then tilted over and roared across the prairie, trailing fire and exceeding the speed of sound. "It looked more like a meteor passing across the sky than anything else," he told Atwood. He reinstalled the gyroscope.

On 28 March, A-5 "rose out of the tower slowly, less black smoke than before and no trail across the sky. It righted itself moving back and forth a small amount for a few hundred feet and then leaned toward the right, ascending and apparently righting itself somewhat. There were occasional explosions and flashes, perhaps the excess gasoline burning in the air." It got up over a mile high and its speed exceeded seven hundred miles per hour before it smashed into the scrub and cactus two miles away.

A series of violent dust storms hampered testing through the spring of 1935. When conditions were calm, Goddard suffered more failures with A-6 and A-7, because he could not get his blast vanes and parachutes to work right. He sent A-8 and A-10 thundering up over a mile in May and July; A-10 was going so fast when its fuel ran out that it tore off the parachute.

Goddard told Guggenheim that it showed that he had solved the problems of motor design, gyroscopic control, and steering vane design.

Guggenheim wanted to see for himself, but Nells A-11 and A-12 fizzled during his and Lindbergh's visit in September. Finally, A-14 rose over two thousand feet before nosing into the ground. That concluded the A series, except for reconstructing one to send to the Smithsonian. Goddard next set out to develop a more powerful motor to drive a larger rocket.[32]

He aimed at a ten-inch combustion chamber, and eliminated the large nitrogen tank by acquiring a nitrogen liquefier, liquefied gas giving more pressure from less volume. After rigging the tower at Eden Valley to hold the engine down, he began testing the "K" series motors in November, conducting nine static tests between November 1935 and February 1936.

His results were uneven and his chambers suffered from combustion heat, but he started designing a "large" rocket, the "L" series, before K was done with. The new missiles were shorter than before, less than thirteen feet, thanks to the smaller liquid-nitrogen tank, but huskier, at about eighteen inches in diameter. They habitually blew holes in the sides of their combustion chambers; only two of them, L-5 and L-6, got off the ground, each going about two hundred feet before the chamber erupted.

Goddard continually reported success, but he admitted that he was getting nowhere with the burnout problem. It was inherent in liquid-propellant rocketry because of the high combustion temperatures supported by the concentrated oxygen. Other rocketeers had devised "regenerative" cooling, circulating propellant in tubes running around a jacket surrounding the chamber, cooling the chamber while preheating the fuel. Goddard, aware of that alternative but always confident that his way was the only way, stuck stubbornly to what he called "curtain cooling," injecting the gasoline so that it swirled around the chamber walls inside.

Lindbergh had become impatient with Goddard's switching layouts instead of focusing on something simple to make a high flight sooner. Sounding like Abbot, in April 1936 he lectured the rocketeer: "It is extremely difficult for anyone without technical training and who is not a scientist himself, to properly value scientific work and not be influenced, even though unnecessarily, by the general recognition of success which goes with actual demonstration. . . . I feel the morale of everyone concerned would be greatly increased if you would find it possible to obtain a record-breaking flight." Guggenheim agreed.

Still shifting from one idea to another, however, Goddard aimed for altitude by simplifying, but in the process he complicated things further. His last small chamber had worked decently, so he replaced the large one with four small units in a cluster, protruding from tubes at the rocket's tail. They might be more reliable than the big motor, but they offered four chances for failure. L-7 took off on 7 November, and one chamber burned out before it cleared the tower. It made it up about two hundred feet before crashing, and its successors did much the same.

In 1937 the L rockets began to look like the A rockets, some with gyroscopes, others without, and the big gaseous-nitrogen tank was back. Goddard's motors gave more power and made more noise, but he could not get them to fly straight until 26 March, when L-13 worked better. It shot up over eight thousand feet, its guidance system fishtailing it to correct against the winds. The parachute release hung up, however, and the missile was doing three hundred miles per hour downward when the chute opened and tore away. The next flight fizzled, but in April L-15, with a lighter weight due to thin, wire-wound pressure tanks, rose on its fiery tail straight up for about a half mile, then floated beautifully downward.[33]

Nell became longer and leaner during the L series—over eighteen feet by nine inches—while Goddard could not refrain from experiments. In July and August L-16 and L-17 took off sporting swiveling tail sections, with the motor mounted on gyro-controlled gimbals to tilt slightly to maintain straight flight, avoiding the drag induced by blast vanes. Each time Nell climbed to over two thousand feet, correcting herself so well that Goddard concluded that the "movable casing rocket" was the wave of the future. For the moment, however, it had to be set aside as too complicated. With a barometric switch operating the parachute, L-16 had come down fit for another flight as L-17, which broke in two when the chute opened. L-17 had taken off with a boost from a counterweighted catapult, an idea Sessums had inspired and which Goddard did not repeat.

He altered Nell's design again, trying to increase propellant pressure. First he tried a bellows pump to drive the liquid nitrogen, but attempted flights failed. Next he returned to gaseous nitrogen, with disappointing results until L-27 reached 2,200 feet in March 1938.

He had meanwhile been corresponding frequently with Lindbergh and Guggenheim. The former had many interests, including putting Goddard in touch with the National Advisory Committee for Aeronautics, which led to a fruitless exchange about testing rocket models in NACA's wind tunnel. Another issue was the need to promote the military applications of rockets.

Mostly, Lindbergh and Guggenheim hammered Goddard to get something up high. Lindbergh suggested that the best way to get the achievement on record was to have a flight verified by the National Aeronautic Association (NAA), official record keeper in that field. Nobody from NAA wanted to travel to New Mexico, so NAA sent an official barograph to the National Bureau of Standards for calibration, then on to Colonel D. C. Pearson, superintendent of NMMI, who chaired a committee of observers.

Pearson, NMMI physics professor Major John E. Smith, mathematics professor Captain Howard H. Alden, the barograph, and L-28 were all ready at Eden Valley on 20 April. Nell roared aloft to 4,215 feet, according to the theodolite and recording telescope, but there would be no official record, because the barograph was smashed and unreadable after the rocket dived into the prairie.

Goddard rented another barograph from NAA, and on 26 May, L-29 thundered up about four thousand feet before turning over and crashing near the tower; again the smashed instrument gave no readable altitude. The next test was set for 14 June, but high winds prevented a launch, so Charles Mansur stayed overnight at Eden Valley to guard the rocket. The next morning a tornado demolished tower and rocket alike, tossing Mansur a considerable distance, though without serious injury.

The crew reconstructed the tower, making it twenty feet taller, and tried again on 27 July. Howard Alden's wife, Marjorie, also an NMMI math professor, replaced Smith, and with the rest of the committee and Esther watched Nell explode into a storm of fire and shrapnel when the igniter was fired. Finally, on 9 August, L-30 was perfection. Of all the observers, Marjorie Alden was the most descriptive:

> *At about 6:30, Dr. Goddard threw the switch; there was a metallic click followed in a moment by an explosion, and then a steady roar. The flame, first yellow, became white, and in about three seconds after the contact, Dr. Goddard released the anchoring weights and the rocket began to rise. Its rise from the tower was slow, and after clearing the tower, it turned into the wind. Its great acceleration was easily apparent, and after a moment its direction corrected to vertical, altered into the wind, and corrected at least once again, and then remained almost vertical. The rocket continued to move with increasing speed, and left a slight trail of bluish white smoke. Its size diminished so rapidly that by the time the fuel was exhausted it was almost invisible. It continued to rise almost vertically, then turned horizontally, and the cap came off releasing the parachute. The flight, with the parachute release, was the most thrilling sight I have ever witnessed.*

There was confusion about the altitude. The NBS report to NAA set the altitude at 6,565 feet, but Goddard did not understand the difference between altitude above ground level and above sea level. NBS measured the latter and, correcting for Eden Valley's elevation, set the record at 3,294 feet above ground level.

It was the only official flight record he ever obtained, and he objected that it was too low. To exceed it, he believed that next he must concentrate on propellant pumps as the way to "extreme altitudes." He so told Guggenheim, in an uncharacteristic ten-year plan of research that he prepared at Lindbergh's suggestion, as a bid to shore up the millionaire's support for the long haul.[34]

That would wait, however. Goddard was worn out and Esther insisted on a vacation. They wanted to go to Hawaii, but no accommodations were available. Instead, on 11 August they left Roswell by train, planning to see Europe before a war broke out. They sailed from New York on SS *Normandie*, then toured France, Switzerland, and England. They saw but did not enter Germany, which to Goddard "seemed peaceful but, in a strange way, disturbing." By the middle of September they were back in America, guests of Harry Guggenheim, who arranged a meeting that would change Goddard's life.[35]

TEN

HAVE TO CARRY ON ALONE

I am Vishnu striding among sun gods,
the radiant sun among lights;
I am lightning among wind gods,
the moon among the stars.
—*The Bhagavad-Gita* X:21

The Goddards docked in New York on 12 September 1938, and three days later Guggenheim hosted them at his estate. "Mr. Guggenheim," Robert noted in his diary, "has a plan of speeding up a large, light rocket by 'farming out' some of the problems so that the results can be used together and a flight made in a year or so."

He assembled a meeting at Falaise on the eighteenth with Goddard, von Kármán of GALCIT, President Clark Millikan (brother of Robert Millikan) of CalTech, and representatives of NACA. "Guggenheim explained that we were all engaged in working for the country's defense," von Kármán recalled, "so we should pull together and pool our information. We all said we would." The meeting broke up with the understanding that Goddard would farm chamber design to CalTech.

Goddard was not the type to confront anyone directly, and he returned to Roswell fuming. Believing that he had "extreme altitudes" at last within reach, he was furious at the suggestion that he should make room for GALCIT's upstarts to share credit at the end for what had been his work for a lifetime. He wanted to continue in rocketry as he always had—alone, except for his assistants.

He fretted for a month, then wrote Guggenheim to say that he could see no help coming from CalTech because its group was working on small, heavy chambers. In early December he wrote to von Kármán, canceling a scheduled visit to California and with it any assumptions that there had been an understanding between them.

In mid-December the rocketeer told Guggenheim: "The most practicable way of speeding up the work is, I believe, by increasing the number of men and machines here in the shop, as it is the construction phase of the project that requires the most time." He said much the same to Lindbergh.[1]

Goddard viewed newcomers to rocketry as trespassers in his domain. That might have been tolerable, but the duplicitous way he handled the meeting—pretending to agree when he did not, then changing his tune when he was back in New Mexico, where he need not look anybody in the eye—alienated nearly everyone involved. He was fortunate that Guggenheim, Lindbergh, and Abbot remained supporters, although Harry's patience had been tried, but the rocket man had given himself an enduring reputation as a secretive, uncooperative loner.

As if to underscore the isolation to which Goddard had consigned himself, a hurricane tore through Worcester in November. "Cherry tree down," he wrote when he heard the news, "have to carry on alone."[2]

NOT SAYING IT WITH FLOWERS, BUT WITH DOLLARS

The loss of the cherry tree broke Goddard's connection to Worcester, and Esther attached the two of them to Roswell. In the fall of 1940 she threw herself into local activities, joining the Women's Club, the Shakespeare Club, the Junior Book Club, the Bridge Club, and other organizations. She organized crusades to knit for the Red Cross and to raise "Vitamins for Britain," because of the war in Europe that broke out in 1939, and in 1941 she became state president of the Federated Women's Clubs of New Mexico.

At last she drew Robert out. In October 1940 he became a member of the Roswell Rotary Club, the first social organization he had joined since a fraternity in college.

Weekends off, trips to the movies—Robert was a Laurel and Hardy fan—and vacations became more frequent. He no longer worked as constantly as he had earlier, and he enjoyed Roswell more. They had found a home there, so in 1940 they negotiated with Miss Effie to buy Mescalero Ranch, closing the sale the following January.

Goddard falsified his age and his income when applying for credit, claiming he was fifty-three when he was nearly five years older. According to the report of the Roswell Credit Bureau on 24 May 1940, he was on "leave of absence from professorship and an employee of Guggenheim Institute [sic], New York, on a project for testing weather and upper strata conditions by rocket-fired balloons [sic]. Receives appropriation of $35,000.00 on annual basis, renewed each August."[3]

Goddard's Guggenheim grants did not exceed $20,000, and out of that he paid himself $7,500 after 1938. He gave himself a raise to $9,000 when he began receiving military contracts in 1941. Exaggerated statements of income have always been common in credit applications, but this tested the limits.

Goddard's attachment to Roswell paralleled growing estrangement from Clark University. He had never been a good department chairman, and at long distance he was an utter failure. In September 1939 he relayed to Atwood a complaint from Roope that mathematics instructor Cary E. Melville had become rebellious. He refused to take instructions transmitted through Roope, assigned himself the best students, and was too severe a grader. Goddard wanted the president to straighten the situation out.

Instead, Atwood telegraphed a sharp rebuke: "WIRE MELVILLE IMMEDIATELY DEFINITE INSTRUCTIONS REGARDING COURSES. YOU ARE CHAIRMAN." Goddard did so, and Atwood wrote: "I think I have never taken the responsibility of assigning special courses to individual members of the faculty when there was a chairman or acting chairman of the department." He resented being asked to intervene.[4]

The tension between Goddard and Atwood worsened. "What are we going to do about Physics for next year?" the president typically asked. The dean of the undergraduate programs told Atwood "that our standing in Physics was not creditable to the college and that this long period of temporizing was a real injustice to the college." Atwood defended Goddard against that, but with declining vigor. "The situation at Clark is serious enough to warrant a trip East," the rocketeer told Guggenheim early in 1939. He went to Worcester, stayed in Atwood's house, addressed the faculty senate and the professors of all science departments, and the two reconciled for the time being.[5]

"Another year has rolled around, and the same old question comes up," Atwood said in 1940. "What are your plans for next year?" There was growing resistance to extending Goddard's leave. Guggenheim was in Roswell when that letter arrived, and guaranteed $2,500 for the substitute professor's salary for the next year. According to Goddard, Guggenheim "feels that this fact should influence the Faculty and Trustees ... as the Foundation is 'not saying it with flowers, but with dollars.'"[6]

Relations between Goddard and Atwood soured further. Atwood complained early in 1941: "I hear a good many rumblings about this plan of having you away so long and holding the position open for you. We need a stronger set-up in Physics at Clark and one with more permanency." In response, Goddard asserted his own importance to the nation's defense, and how that redounded to the credit of the university. He did not tell Atwood that he had bought the ranch and had no intention of returning to Worcester.

The correspondence between the two of them became acidic, until in early 1942 Atwood invited Goddard to resign. The rocketeer was not to be edged out of his tenure, however. "The problem is not a simple one," he began a disingenuous response, "for I, too, will soon reach retiring age. Clark is my alma mater, and Worcester is my home, and it is a severe wrench even now to contemplate severing relations with the University."

Goddard was interested in protecting his annuities, and not in the least concerned with the difficulties his absence caused. He also had possession of a considerable amount of university equipment, and brushed aside attempts by Atwood to make him account for it. A showdown between the two was only a matter of time.[7]

In contrast to Worcester, at Roswell Goddard enjoyed a rising level of comfort. As Charles Mansur and the other men observed, he became less isolated as he received visitors with whom he could share technical details. Abbot was one. The others were mostly military men, and from them he had no secrets. They included the famous World War I flying ace James H. "Jimmy" Doolittle, who first showed up in 1938, on his friend Harry Guggenheim's recommendation. He was then an executive in Shell Petroleum Corporation's aviation fuels division.

Fuels gave Goddard special problems. Automotive gasolines contained additives that clogged his systems, while aviation gasolines were too light. Doolittle arranged a technical exchange that resulted in Shell's chemists devising a propane-butane combination, to be provided free of charge. Goddard revised his systems to accommodate the new material, but the arrangement ended when Doolittle returned to active duty in the U.S. Army Air Corps in 1940.

The aviator wrote the most thorough report on the rocketeer's work produced by anyone. "Many problems are still being worked on," it said. "Dr. Goddard, rightly, feels that they must be attacked one at a time and each one worked on until a solution, not necessarily the best solution, is obtained. As each problem is solved, he moves on to another, abandoning all work on the one just left. His idea is to get a workable rocket and then refine it at some future date."[8]

During three days in May 1939 Goddard hosted two officials of Linde Air Products, Howard Blakeslee of AP, Herbert Nichols of the *Christian Science Monitor*, and Charles Lindbergh. Each enjoyed Goddard's chauffeur services, Esther's gracious hospitality, Augusta Kisk's Swedish cooking, a tour of the shop and launch tower area, and talks and rocket movies on the screened-in porch.

Lindbergh was, as usual, the most memorable. He was on an official inspection of the Army Air Corps to evaluate that service's readiness to take on the German air force, and had intended Roswell to be only a stopover when he telephoned Goddard from the airport on the afternoon of 10 May.

According to his journal, "it soon became apparent that there were too many things to talk over to cover in half an hour." He continued: "We spent the rest of the afternoon discussing rocket plans and developments—stabilization, pumps, tanks, liquid oxygen, and nitrogen, etc.... Mrs. Goddard's mother is visiting her—a fine old Scandinavian lady. We all had supper together.... Goddard has done good work and had a very successful year."

The two of them also discussed the dangers posed by the Germans, and afterward corresponded about the military applications of his rockets. Military necessities, it turned out, must bend to Goddard's interests, and not the other way around. He saw three applications: a long-range missile to be achieved by completing his current work on high-pressure pumps and means of driving them; a rocket turbine for aircraft to be devised from one of his patents; and the short-range projectiles he had developed in 1918.

Neither Lindbergh nor Goddard knew that they had met for the last time. The Lone Eagle became increasingly devoted to warning his fellow Americans about the German threat, which drew him into the isolationist America First Committee, advising the country to stay away from war in Europe. When the war did arrive on the American doorstep, he joined in, serving unofficially as a pilot and flight-training expert in the Southwest Pacific, then in a postwar review of German aviation. However, America First had distracted him, turning him from national hero into pariah, and leaving less time for him to devote to Goddard's work or to his friendship with Guggenheim.

But in 1939 all that lay in the unknowable future. Slim never lost faith in Goddard, and in the short run he once again saved the Roswell project by persuading Guggenheim that it merited continued support, whether or not Goddard cooperated with GALCIT. He urged Harry to support the rocketeer further by publicizing him, and mentioned the National Geographic Society, where Slim had well-developed connections.

The crucial meeting occurred over lunch at Harry's place in New York on 1 June. As the Lone Eagle recorded: "We also spoke in detail of Goddard's project and his plans and results. Harry was enthusiastic about the last year's progress." Under Lindbergh's influence as always, Guggenheim decided to see for himself what Goddard was up to.[9]

Momentarily distracted by marriage—the third Mrs. Guggenheim was New York socialite and sportswoman Alicia Patterson—Guggenheim arranged to visit Roswell during his honeymoon trip to California; Harry flew his new airplane. "We were delighted that they chose to stop with us," Esther recalled. "They seemed very happy, and deeply interested in both our work and our scenery."

The Goddards picked the couple up at the airport on 20 July, took them to lunch and on a tour of the shop and launch area, and invited Paul Horgan to join them for the evening, which was enlivened by Robert's martinis and Esther's tipsiness. The next day they went fishing on the Mescalero Indian Reservation, then watched rocket movies. The Guggenheims flew out on the twenty-second. The visit proved rewarding to Goddard, because Harry renewed his grant for another year, and mollified Atwood by thanking Clark University for making it all possible. Still wanting to see a rocket fly, he vowed to return.[10]

Of all the visitors, none became as close to the Goddards as Lieutenant Homer A. Boushey, Jr., AAC. He was a courteous young man with a dimpled chin and the brown eyes of a puppy, who held an engineering degree from Stanford and had been an army pilot since 1934. He had followed Goddard's career for years, had invented a rocket-propeller combination for aircraft, and in 1940 asked if he could visit. Goddard invited him for dinner on 28 June, saying "It would be pleasant to talk with someone who is interested in rocket developments."

The Goddards were taken with him so much that, after Boushey married, they became godparents to his children; it was the only such relationship they ever had. Robert and Esther took the young pilot into their home whenever he visited. Inebriated on martinis, Robert and Boushey spent the evenings on the back porch talking about spaceflight. Goddard expanded on all his ideas, including the use of sails to catch the solar "wind" to drive spacecraft. "Of course," Boushey remembered, "New Mexico on a warm summer night is the ideal place to discuss interplanetary flight. You can just look out and almost touch the stars, and having seen the engine, the motor, and Nell with the cloth removed . . . all of this was new to me."

Goddard inspired Boushey to crusade for rocketry in the Air Corps. He was soon promoted to captain and assigned as project engineer in charge of the Corps's jet-propulsion project, for which GALCIT was the principal contractor. Boushey exerted every effort to obtain an army contract for Goddard, and to acquire a surplus LOX generator for his use.

Goddard became unusually open to the younger man. When an attempt to gain an army contract failed, in December 1940 he wrote a disappointed Boushey: "We both seem to have exerted ourselves to the utmost this past summer and fall, and to arrive at a dead end. Even though we are not free to discuss matters as fully as we would like, I hope you will feel free to drop in on us at any time unofficially, for you have earned the right to find the latch string out always."[11]

What drew visitors to Roswell was Goddard's work, but that did not go as well as he would have wished. He had, to his satisfaction, perfected everything he needed for a high-flying rocket, except for a turbine to drive high-speed pumps to force propellants into the chamber. He knew it would be a challenge, and in 1938, saying that he needed a "real physicist" to help him, he tried to lure his old aide C. N. Hickman away from Bell Laboratories. Hickman was tempted, but declined.[12]

Goddard spent the fall of 1938 and most of 1939 working on pumps, testing at least five different models. Using his old small combustion chamber, even the most feeble pumps doubled his exhaust pressures. He experimented with a nine-inch chamber, then settled on the outlines of the P-series rockets—eighteen inches in diameter and twenty-four feet tall.

By spring the rocket man had devised a "gas generator" to drive his pump turbines, "what might be called an internal-combustion boiler for liquid oxygen," he told Guggenheim. "The object was to convert liquid oxygen into warm gaseous oxygen, in a device of small size and light weight, by the combustion of gasoline within this liquid oxygen." The warm gas drove the turbines, and was proven in a series of static tests through the summer.

Goddard developed components through static tests until he thought they worked well enough to try a flight. Flight tests usually failed, sending him back to the shop. In August 1939, he thought his P rocket was almost ready and—irritated by the low altitude figure from a year earlier—made arrangements for another NAA barograph.

He was premature, as failures delayed him into December, when he told Guggenheim that he should be able to fly a rocket after the holidays. "This test will be the first flight attempted with fuel pumps," he warned, "and the chance of mishap is of course greater than in later tests." Nevertheless, he wanted his patron, Lindbergh, photographers from *National Geographic*, and NAA observers to watch.[13]

Goddard optimistically told the NBS to set the barograph to read up to thirty thousand feet, while Howard Alden laid out a series of datum lines at Eden Valley to ensure accurate range-finding. The increased size of the rockets required that a hose-and-pump fueling system replace the former buckets.

The rocketeer told his expected audience to be on hand for a setup on 29 January and a launch the following day. The *National Geographic* people arrived on the former date, received a tour, and followed everyone around, taking pictures. Harry and Alicia Guggenheim arrived on 2 February. The Goddards now had a houseful of people watching snow, ice, and rain fall from the sky, wondering if they would ever see a rocket go up. Weather improved on the sixth, so Goddard tried a launch, but could not get ignition.

By the ninth, the rocket had stood in the tower for ten days, through two big snowstorms and lesser ice and rainfalls. The men knocked ice off the tower, fueled Nell, and at ignition the oxygen pump blew apart in a thunderous blast of shrapnel that shredded the rest of the rocket. Ice had clogged the works.

Goddard drove the Guggenheims to the airport, then went home to bed. He was mortified, but when he told Lindbergh about it Slim reassured him: "I am sure that anyone who has taken part in the early years of aviation, as Mr. Guggenheim did, will thoroughly understand the limitations and unpredictability of weather."

Harry also consoled him, but in March told him to make a flight as soon as possible. *National Geographic*'s photographer returned to watch a promised flight on 21 March, when Nell P-16 hissed, and then her manifolds exploded. This, Goddard told Guggenheim, proved that taking shortcuts was not a good idea. He resumed static tests.[14]

Guggenheim was impatient about the lack of progress. War was approaching, the country was mobilizing, and he proposed to interest the military in rocketry. He therefore wanted something to fly to demonstrate its potential. Goddard, however, soldiered on with his static tests, until 31 July when the NAA observers watched Nell sputter out on ignition.

Mishap followed mishap, until 8 August. After everyone got stuck in the mud on Pine Lodge Road, they discovered that the tower had been struck by lightning the night before, and the launch system had to be rewired. On 9 August Nell P-23 rose slowly, tilted over, performed a loop, and landed on her side four hundred yards away, where the fuel tanks exploded with "a large concussion."

So it continued, with static tests, including a digression into a new chamber design, and explosions and other mishaps. On 8 April 1941, with the Aldens observing for NAA, P-31 rose two hundred feet, tipped over, and landed on her side. On 11 July, with army observers in attendance, P-32 failed because of dust driven by a thunderstorm. On the seventeenth, with Boushey the only observer, P-33 fizzled after thirty-five seconds. P-36 jammed in the tower on 10 October. Military projects had overtaken the work, so that was the last Nell even to attempt a flight in New Mexico.[15]

Goddard had received more money for his research than any other civilian scientist for a single project before World War II—$209,940 in civilian funds. Daniel Guggenheim had provided $50,000, and the Guggenheim Foundation $138,500. The rest came from the Smithsonian, the Carnegie, and Clark University, with a small amount from AAAS. He had forty-eight patents in hand or in the pipeline; they had cost him $530 and the foundation $7,175.

But he was no longer the "number one rocket man." As the P-series pump rockets had demonstrated, he had underestimated the difficulty of the work, and overestimated his ability to do the job alone. Like Icarus, he had challenged the sun, and lost.[16]

A BILLOWING RIVER OF FIRE

Goddard was the most highly publicized American scientist, at least since the death of Thomas Edison in 1931. Increasingly rockets as weapons captured the public's interest, especially after an army officer, echoing Goddard's letters to him, produced a widely reported essay, "What Can We Expect of Rockets?," in 1939. When an Italian newspaper said that Germany was developing "rocket artillery shells" in 1940, AP contacted Goddard. He was "disturbed" at the news, and said: "I have never investigated the possibility of the rocket being used as a weapon, and this may affect my work."

That was a falsehood. He got away with it because he maintained good relations with important reporters. Although Guggenheim cautioned him to treat the entire press equally, he had his favorites, including Howard Blakeslee of AP, Herbert Nichols of the *Christian Science Monitor*, science writer William H. Wenstrom, and a few others. He treated them well, and they returned the favor, dramatizing him and his work.

Witness a December 1940 piece by Blakeslee: "River of Fire. It produces one of the awesome things which strike the eyes of the few persons who have seen one of his rockets taking off. Before it rises, for a second or two, a jet of pure flame strikes down on the valley sand and rolls 50 feet along the surface as a billowing river of fire 10 feet deep. This is the jet of fire which drives the rocket, spreading out as it expands in the air. . . . There is nothing on earth its heat cannot melt."[17]

Lindbergh and Guggenheim appreciated the value of publicity, but also knew the need for prestigious coverage to garner support for Goddard's work. No publication combined popularity and prestige more than *National Geographic*, and when Lindbergh suggested contacting that periodical in June 1939, Guggenheim told him to go ahead. Slim got on the phone, then went to Washington to the National Geographic Society offices, where he always met with a warm reception from President Gilbert Grosvenor.

The magazine's editor, John LaGroce, planned to assign a staff writer to prepare the article, with Goddard providing background and photographs. After Guggenheim visited Goddard that summer, he talked LaGroce into having the article appear under the rocketeer's byline, with professional photography. Goddard drafted the article, which staff writer McFall Kerbey revised into *National Geographic*'s trademark first-person bland. When Kerbey and photographer B. Anthony Stewart arrived in Roswell the following winter, Kerbey and the rocketeer touched up the text.

There were no photos of the rocket in flight, because Nell would not leave the ground when Stewart was around, and because views of a flight were essential, the project was moribund. Interest in publishing the article lingered for two years, until the last P test ended Goddard's efforts to develop a high-altitude sounding rocket. The article never appeared in *National Geographic*, nor did a parallel project Guggenheim had arranged with *Life* magazine see print.[18]

PROBLEMS REGARDING NATIONAL DEFENSE ARE BEGINNING TO ARISE

After the meeting at Guggenheim's estate in September 1938, Goddard concluded that support from the foundation was no longer assured. Looking for other sources of money, he reintroduced himself to General Sunderland in an attempt to revive the old idea of rocket-powered antiaircraft targets. But Sunderland still could not elicit interest from the Air Corps.

Early the next year Goddard dickered with the Glenn L. Martin Company, aircraft manufacturers who expected a naval contract to develop glider bombs. He visited his old associates Louis T. E. Thompson and Nils Riffolt, both at the navy's Dahlgren Proving Ground, and asked them to lobby the navy for him.

Nothing came of that either, and Goddard continued working on his pump rocket. That, he told Guggenheim, was made more pressing by "the fact that problems regarding national defense are beginning to arise, both along the lines of aviation and ordnance, which require for their solution effective liquid-fuel rocket propulsion." He passed on news of his dealings with Martin and the navy, but did not mention that he had asked Abbot to sniff out military opportunities for him.[19]

In 1940 President Franklin Roosevelt appointed Dr. Vannevar Bush—an MIT research engineer and the chairman of NACA who had succeeded Merriam at the Carnegie Institution—to head the National Defense Research Committee (NDRC), which coordinated funding of scientific research for defense purposes. Knowing that Bush regarded him as uncooperative, Goddard felt left out. In April 1940 he asked Abbot to find out what the military was doing about rockets, and what was going on at CalTech. "It does not seem," he fairly whined, "as if the most effective attack could be made by shelving the results I have obtained in the past twenty years." Abbot replied that such matters were very secret, and he had learned nothing about CalTech.[20]

Goddard continued lobbying the navy, via Thompson, on the applicability of his rocket system to such things as armor-piercing bombs. The main order of business, however, was a meeting Guggenheim arranged for 28 May in the office of General H. H. "Hap" Arnold, chief of the Air Corps, with representatives from pertinent branches of the army and navy. Arnold was called away, so the meeting moved to the office of General George H. Brett, the Air Corps's chief of matériel. There Guggenheim offered Goddard's shop, equipment, and people to the defense agencies as a gift from the foundation. Goddard gave a brief history of his work and showed movies of flight tests.

It was not an effective presentation. Except for small rockets to boost aircraft takeoffs, no one could see a use for rocketry. An army ordnance officer said that any development work ought to go to improving trench mortars—this while the fast-moving German army demonstrated that the trench warfare of World War I would not be repeated. The shortsightedness was not all on one side, however. Goddard emphasized his current development without explaining what it might be good for, or evidencing any willingness to alter his work to meet military needs.[21]

Worse, in his view, he had learned in Washington that funding for jet propulsion—the later distinction between air-breathing jets and rockets carrying their own oxidizers had not yet clearly emerged in the language— had gone to CalTech. From this point on, von Kármán and the CalTech group replaced Oberth and the Germans as Goddard's demons.

The Air Corps had started research at GALCIT in 1939, with the rocket-powered target drone that Goddard wanted a piece of. The institute went on to rocket-assisted takeoffs, antiaircraft rockets, antisubmarine rockets, and other developments before it became the Jet Propulsion Laboratory later in the war. With Boushey at the controls, in August 1941 GALCIT would launch America's first rocket-propelled piloted aircraft.

Goddard could not contain his frustration at being so outflanked. As he complained to Abbot:

Apparently my experience and years of work, including early sponsorship by the Smithsonian Institution, were given no consideration by the National Research Council [sic], or at least by those who had charge of this arrangement. This, to my mind, is a serious thing, as it indicates a force which must be overcome before I can hope to work on a really adequate scale, and I would appreciate it very deeply if you would make inquiries among the National Research Council members as to just why I was not even considered in the grant for rocket propulsion.

Still seething, in September Goddard wrote an intemperate letter to Thompson, asking him to pass it on to his superiors in the navy. He accused GALCIT of trying to steal his ideas. CalTech, he said, had started rocket work only four years earlier, when Millikan sent Malina to Roswell. GALCIT was nothing more than a student project focused on powder rockets, at a stage he had passed fifteen years back. Moreover, the participants were inept, and were trying to hoodwink the government.[22]

Goddard emerged from the Washington meeting only with requests from the army to provide an estimate for using a rocket chamber to boost the takeoff and climb of a P-36 pursuit plane; the navy was interested in something similar. "While this is very encouraging," he wrote to General

Arnold, "I cannot help feeling that twenty-odd years of experience on rocket problems, the flight results which we have obtained to date, and the general acceptance of jet propulsion as of real engineering importance warrant having this project undertaken on a much more extensive scale." He asked Lindbergh and Massachusetts senator David I. Walsh to lobby the army and navy for him, and even approached the Canadian government in a search for grant money.[23]

Guggenheim promised that the foundation would fund the project for another year if no government money appeared. Goddard told him that he had heard from Army Ordnance about antitank rockets, although at the meeting in Brett's office the Ordnance people had been interested only in mortars. This, he said, was the result of efforts by his old assistant C. N. Hickman. "I have never had, myself, any great talent for selling ideas," he confessed, "but Dr. Hickman has this in a marked degree, and it may be that only through such efforts as his will the powder phase of the rocket work be carried forward."[24]

While the German army tore across France, Hickman wrote to Goddard to complain that the country was ignoring rocketry in the face of what was going on in Europe. He volunteered to bring pressure to bear in high places. "Go ahead, and God bless you!" Goddard replied, urging him to contact Guggenheim. Hickman did so, and also enlisted his own brother, an army reserve officer, to interest the army in rocketry. He made his own attack on NDRC, but warned Goddard that he doubted any agency would want him to work at Roswell, far from sources of supplies.

In July Bush appointed him chairman of Section H (for Hickman) of Division A of NDRC, under the committee vice-chairman Dr. Richard C. Tolman; Bell Labs granted him leave. Hickman prodded Tolman, who persuaded Bush to appoint Goddard a consultant to Section H. Meanwhile, Hickman connected with Boushey and fellow Clark alumni Thompson and Riffolt.

Goddard's consultancy never amounted to much. His suggestions to Hickman were of limited value, frequently on the level of warning him to watch out for competition from CalTech. More practically, he urged the younger man to use nitroglycerine powder instead of nitrocellulose for armor-piercing weapons. He thereby showed that he was out of touch with the challenges Hickman faced in the new age of armored fighting vehicles, whose technical demands were vastly different from those of World War I.[25]

Hickman began where he and Goddard had left off in 1918, developing rocketry for infantrymen who now faced tanks. Although to the end of his life he credited Goddard with inventing the bazooka, in fact he had to start all over again. The essential problem was in physics—an explosion against a flat surface such as armor plate will go in the direction of least resistance, meaning that much of its force goes outward from the target. That could be overcome through brute force; hardened projectiles moving at high velocity could penetrate armor and explode within. That was fine for artillery and aerial bombs, but of no use to foot soldiers.

Hickman knew that C. E. Munroe had discovered something else in 1888: If an explosive charge contains a concavity, that vacancy becomes the direction of least resistance. This inspired the "shaped" or "hollow" charge— a projectile so designed that it could punch a hole through armor plate and squirt hot gases and molten metal inside. That would unleash hell in a crew or engine compartment.

Hickman sold the idea at NDRC, and he and army experts spent about a year proving its feasibility. The first challenge was to develop a small rocket projectile that would work—selecting the explosive charge and shaping it, followed by the equal difficulty of finding the best solid propellants. Hickman himself devised the shoulder-mounted launching tube.

The bazooka—so called because of its resemblance to a gag musical instrument used by comedian Bob Burns—became Ordnance's proudest achievement in World War II. Hickman, who followed it with a rocket-powered mortar shell in 1944, was the device's chief continuity from Goddard's 1918 work, although the elder man had first crafted a prototypical tube-launched infantry rocket.[26]

Hickman's accomplishments did not result in any grants for Goddard. Despite intervention from Hickman, Boushey, Thompson, Lindbergh,

Guggenheim, and others, the rocketeer made no headway in 1940. Hickman and Tolman repeatedly told him that his proposals lacked details, but he countered that he did not want to give any leads to CalTech. When Boushey and others said that liquid propellants involving LOX had fallen into disfavor, he stubbornly insisted that his liquid pump rockets were the way to go, leaving solid rockets to GALCIT.

He developed proposals for rocket-assisted takeoffs (RATOs) for the Air Corps, which fell into a bureaucratic tangle between the airmen and the navy over which agency would take the lead in development. The navy offered him a well-paid consultantship if he moved to Maryland, or less money if he stayed in New Mexico. He opted for the latter and lost that as well.

Mostly, he brazenly used the war scare to seek money to continue doing what he had been doing, where he had been doing it. His failure to cooperate with his potential clients, by adapting himself and his rocketry to their needs, caused them to turn down all his proposals.[27]

Goddard had tried to get money out of a military still mired in peacetime ways forged in decades of minuscule defense budgets. What he proposed was novel, and he did not sufficiently explain how his rockets might be useful in warfare; only RATOs were understandable to airmen. Despite expanding defense appropriations, military bureaucrats remained cautious about experimental ideas. The fall of France and the Battle of Britain changed that, and money began to flow, especially for ideas related to aviation.

When Boushey and Thompson reignited their respective agencies' interest in RATO early in 1941, Goddard decided to become more adaptable. The Army Air Corps reexamined his RATO proposal from the previous July, and in April asked him to follow up by demonstrating his pump rocket at Roswell. If that went well, further tests would follow at Wright, which could lead to a contract.

Boushey talked to Goddard on the sly all through the negotiations. The rocketeer told him that he liked the proposal except, CalTech on his mind, "I should have some assurance as to my own status and protection while the tests at Wright Field are in progress, should the chamber here interest the Air Corps after the demonstration."

Two demonstrations on 11 and 17 July failed, however. Goddard complained to Guggenheim that he thought the military services were trying to pump him for information, and intended to do their own rocketry without him. Moreover, he had lost the services of Al Kisk, and was almost out of money.[28]

Guggenheim promised to keep the rocket man in funds until the government came through. Goddard thereupon sent the Army Air Forces, as the Air Corps was now called, a revised proposal late in July. Meanwhile, the navy suddenly looked more promising. Driven by Ensign Robert C. Truax, who had started experimenting with rockets while a midshipman in 1936 and had become a Goddard fan, the Bureau of Aeronautics decided to get into the field.

Responding to a request, in June Goddard sent in a "description of a simple form of assisted-takeoff device," which "may serve at least as a basis for discussion." Rear Admiral J. H. Tower, chief of aeronautics, liked the submission, and in July told him: "The Bureau feels that your investigations in this field during the past years have eminently qualified you as an expert on the subject. Accordingly, it is desired to enlist your aid as an engineering consultant to render part-time services, consisting of consulting services, design studies, and other special studies and cooperation, in connection with the development of jet reaction motors." Most work, the admiral said, could be performed in Roswell, 120 days or fewer per year, with a maximum fee of $25 per day.

Such flattery, not to mention the money, was welcome after so long in the professional wilderness. Goddard sent Tower a detailed proposal, identical to the one he had sent the Air Forces, including options for pressure-tank and pump-rocket motors.

The navy sent Truax's superior, Lieutenant (jg) Charles F. Fischer, to Roswell in August, and the Goddards welcomed him as warmly as they had Boushey. "According to [Fischer]," Goddard told Guggenheim, "there is a real need for the jet motor that has been developed under the Foundation auspices for Navy planes." He proposed to deliver a "jet-reaction motor" thrusting eight hundred pounds for thirty seconds within six months, for $18,000, payable in six installments.

Fischer told him on 15 August that his proposal was approved, and to expect a "letter of intent" soon. He continued: "I will continue to enjoy for many years the stunning portrait of your Nell, which I deeply appreciate receiving. The following busy months promise to be very fruitful ones for the Navy and ourselves. The seed now germinating, if properly tended, can easily mature to a commercial giant during the next decade. I know of no man better fitted than yourself to give this new child to the world. It is with extreme eagerness that I look toward this next year with its work and its possibilities."[29]

Goddard missed that letter, because he took Esther to the East Coast to push the navy contract along. He also obtained a first installment of what would become $10,000 lent by Guggenheim to tide him over until government money arrived.

Back in Roswell by the middle of September, he hired Lowell Randall and George Bode as assistants, and received the draft contract from the navy. It included a wrinkle—an option to have work after the first six months performed at Annapolis.

If he thought that he had kept secret his separate negotiations with the army, he was quickly disabused of the notion. The Army Air Forces had been slower than the navy, and was miffed that he had negotiated with the other agency, but in October it awarded him a contract for $13,000 to deliver a duplicate of the navy RATOs. At last the "number one rocket man" would have a chance to show his stuff for the national defense.[30]

IT IS A CASE OF A SQUARE PEG IN A ROUND HOLE

Boushey was disappointed in the way things worked out, because if the army had been the lead contracting agency he would have been the project manager. "Perhaps, sometime, not too far away, I'll have the privilege of working with you," he told Goddard.

The older man consoled him with the thought that the side contract with the Army Air Forces was something "we can put down as a miracle of sorts that you helped to bring about." He also was "sorry at the turn of events," but wrote: "It is a pleasure to know that you would still like to work with me sometime, and who knows but what it may come to pass? There is certainly no one with whom I should enjoy working more."[31]

Dealing with the hard-core bureaucracy, Goddard was a fish out of water. He had no contract, just the letter of intent; the former took months to obtain the proper signatures in Washington. The delivery schedule stood, so he spent his own money along with the foundation's to double the size of his shop, hire people, set up test facilities, and acquire materials.

When he finally saw the contract in November, he was appalled that the government procured his services as if he were any other commodity. "It is a case of a square peg in a round hole," he complained to Fischer. On the other hand, the navy had raised his contract price to $21,000 because of a technical change. By Christmas Eve, he and the naval bureaucrats had everything sorted out, and he sent in the contract with his signature, his first progress report, and an invoice for $4,000.[32]

Goddard's crew built a test bed at the ranch to avoid trips to Eden Valley. To dampen the noise, they installed an "underground muffler," a hole through a concrete block into a tunnel lined with oil drums covered with concrete, L-shaped and twenty-six feet long in all. This was all behind a heavy steel frame on which rocket motors were anchored. It did not work well, according to Lowell Randall, because the racket of test firings nearly ruined the nearby dairy herd. Consequently, tests at full thrust could not be conducted at Roswell, and the first low-thrust test, itself noisy enough, took place on 6 December. By that time, Howard and Marjorie Alden worked as volunteers, making graphs and tables of test data, relieving Goddard of a heavy burden.

The first RATO units were to be installed on PBY seaplanes, so Goddard asked for the hull of one to use in fitting design. When one arrived from San Diego in February, it was riddled with holes. "It evidently came from Pearl Harbor, judging from the bullet holes," Goddard told Guggenheim. "I can think of nothing that would give me greater satisfaction than to have it contribute something to the inevitable retaliation."

He was adapting his pump rocket to work in a horizontal position aboard aircraft. Just as he had it almost perfected, the navy switched from pumps to pressure-tank rocketry. That would be simpler to manufacture and operate, and sufficient for the need, but it also meant that he must abandon pump rockets for the duration. Nevertheless, after ten bench tests, he was near completion of a variable-thrust RATO.

On 25 March Fischer tossed him another curve—the navy would exercise the option to move him to the Naval Engineering Experiment Station at Annapolis. He should be in place by May to prepare for flight tests in June. Goddard countered that the change to pressure tanks would cause delays and required amendment of the contract.[33]

He alerted Guggenheim to the pending move, and obtained clearance to move the foundation's equipment. He nevertheless continued to drag his feet, nearly done with the Air Forces unit but giving the navy a laundry list of complaints. He had still not moved by June, and now badgered Fischer for information about employee housing and tire and gasoline rations for the trip. To eliminate his excuses, the navy granted him an extra $5,000 to cover the move.[34]

He surrendered, told Fischer they were going to Annapolis, then rented out Mescalero Ranch and began packing. On 2 July, he gave the local draft board the names of the eight men who would accompany him, saying they would be gone on a military project for at least six months. With Marjorie Alden looking after the house, Robert and Esther drove out of Roswell on 4 July, and within a week they were settled in Annapolis. Soon the freight cars and the crew were on hand, the shop was set up, and the Air Forces had been alerted to the change of locale.[35]

The great adventure in New Mexico was over.

ELEVEN

ONE OF THE PITIES OF THE PRESENT WAR TIME

I am Death the destroyer of all,
the source of what will be . . .
I am the scepter of rulers,
the morality of ambitious men;
I am the silence of mysteries,
what men of knowledge know.
—*The Bhagavad-Gita*, X:34, 38

Settling into the Annapolis area, Goddard told his patent attorney, Charles T. Hawley, that the place was not as comfortable as the Southwest. "The humidity here," he said, "which you can cut with the proverbial knife, is 'guaranteed not for years, not for life, but forever.' " Esther, writing to Marjorie Alden, eschewed advertising slogans. "He is coughing quite a bit," she reported, "but I see that he gets 2 shots of halibut liver oil each day. . . . Bob has a quart of milk and individual box of cereal for lunch, so he is eating gargantuan meals at night, much to my delight."[1]

In Roswell, Goddard had maintained his office in the shop, spending the days among the men and machines. Now he presided over a grander establishment, with a private office where he spent a growing share of his time, the door shut. The red tape overwhelmed him. Auditors examined every expenditure, and when they were told to punch a time clock he and the whole Roswell crew were offended.

Truax, who had looked forward to the great rocketeer's arrival, was disappointed by the man in the flesh. "He could have had us sitting around his knee," he recalled. "But he would never talk rockets. I think he was afraid someone would steal his stuff. . . . He had the vision, certainly, and there was almost nothing that was later developed successfully that he didn't try, at least once. But he was very much a loner."[2]

BOB NEEDS HOWARD MORE DESPERATELY THAN EVER

The first order of business was to complete the RATO—by then generally renamed jet-assisted takeoff, or JATO—begun for the army and navy at Roswell. The Air Forces job was the simpler one. Four army representatives arrived on 14 October to watch a successful, bone-shaking demonstration of a thruster giving 750 pounds for 20 seconds, and in early December Goddard shipped it to Wright Field.

That did not complete the contract, however, because without drawings, wiring diagrams, and an operating manual, the thing was useless. Delays in the drafting department at Annapolis held up the "Instruction Manual for the Goddard A-1 Jet-propulsion Motor Unit," as he grandly called it, until late March. The army then let it gather dust, favoring solid-fuel JATOs instead.[3]

The navy wanted a variable-thrust model, and by the end of July Goddard was testing it on a stand next to the Severn River. Chamber burnout persisted, but daily tests showed improvement, while Lieutenant Fischer operated the controls by September. He was an impetuous young man, eager to get a rocket-driven PBY into the air, so Goddard and his crew had one mounted in a plane and ready to try, with the lieutenant in the pilot's seat, on 23 September 1942.

A succession of progressively longer burns went from dead stop to taxi to the sixth test, when: "The plane took off satisfactorily." Watching from a crash boat, Goddard thought he detected excess oxygen supply. He wanted to halt for adjustments, but Fischer took off again. This time the JATO burst into flames, which engulfed the rear of the aircraft. Fischer landed without anyone being hurt, but the plane was badly mauled. A liquid-fuel rocket-assisted takeoff was not attempted again, and Goddard resumed static tests.

In November Fischer transferred to other duties, to be replaced by Lieutenant J. S. Warfel. The departure of his friend increased Goddard's sense of isolation among uniforms and bureaucrats; he had already lost Howard Alden, who had returned to NMMI in August. "Bob needs Howard more desperately than ever," Esther wrote Marjorie in November. Goddard pressured Superintendent Pearson for Alden's return, and NMMI released him at the end of the semester. The Aldens were in Maryland by Christmas, and remained till the end of the project.[4]

The work went forward under Ljungquist's direction, with Alden providing engineering expertise. In 1944 Henry Sachs emerged from retirement to rejoin the team, which stayed together after they transferred to the Curtiss-Wright Corporation in New Jersey in 1945. The principal outgrowth of their work under Goddard was the XLR-25-CW-1 variable-thrust motor, which later drove the Bell X-2 rocket plane at 2,100 miles per hour to an altitude of 126,000 feet.

Goddard spent the war trying to continue the pump-rocket work he had begun at Roswell. He insisted on staying with gasoline and LOX, when other propellants proved superior, and solid fuels gained increasing favor as being safer and more convenient in storage and handling. Moreover, with liquid fuels there was a trend toward acids and other exotic fluids, culminating in the use of hypergolic (self-igniting) fuels. Nitric acid proved more portable and storable an oxidizer than LOX, while aniline, an oily derivative of nitrobenzene, was sometimes used as an additive, until Frank Malina had an inspiration and used it straight as a fuel. It worked.

Goddard resisted such innovations, beginning with orders from the navy to experiment with nitric acid. When a tank blew up in January 1943, he objected: "I do not wish to assume responsibility for this or any other accidents with nitric as stated in my reports.... I believe that spontaneous ignition and toxic qualities of both nitric acid and aniline constitute a hazard which is not present with the liquid oxygen and gasoline."

Acid-aniline motors were what the navy wanted, so his men had to work on them. On his own, he continued working on pump rockets with the old fuels. In 1944 he developed a cooling method by mixing water in the chamber with the gasoline and LOX, and tried to sell the idea to the navy.

Instead, the navy diverted him into developing pumps for nitromethane. In 1945, he finally received a go-ahead for his pump rocket, and was working on it at his death.[5]

Goddard's technical accomplishments were modest during the war, compared to his achievements of the previous two decades. He had reached the limit of his professional development in the P-series, and came up with few new ideas. He was also not as healthy or positive in outlook in the humid East as he had been in the sunny Southwest.

Mostly, he was confined to his office, contending with the bureaucracy, feathering his nest, and in the last months strolling down memory lane. The first two consumed the most time. From 1942 to 1945 he negotiated four contracts with the navy, along with two negotiations with the Curtiss-Wright Corporation and two with Linde Air. Not all these dealings were successful, but each generated a mountain of enormously complex correspondence.

Goddard had met M. B. Bleecker of Curtiss-Wright's propeller division on an earlier trip east. When he told Bleecker he was at Annapolis, the man visited him on 6 August, offering to retain the rocketeer as a consultant for assisted takeoffs and long-range missiles, with Goddard to license his patents to the company for postwar application.

Goddard told Guggenheim, who was on active duty in the navy: "It seems to me that an arrangement with the Curtiss-Wright people might solve a number of problems. It would make possible my turning over parts for them to manufacture during the war, so that they would be in a good position to carry out work on my developments afterwards. They would also be influential enough and interested enough to look after my interests even before the end of hostilities, which might be better than to wait until the war is over and then look around for someone interested in the patents."[6]

The rocket man ground out about thirty-five patent applications between 1942 and 1945. A two-year contract he was negotiating with the navy, he told Hawley in 1942, would allow the government to use his patents but give him exclusive rights to sell them to others. He knew that Uncle Sam could take his inventions for defense use, compensation to be determined later, and allow competing firms to manufacture them, so he needed "a backer powerful enough to do the fighting after the war." That, he said, was what he expected from Curtiss-Wright.

In September, with the navy concurring, he recommended that the Patent Office put all his applications under secrecy orders. He thereby protected himself from prying eyes foreign and domestic.[7]

Goddard further complicated his situation by playing Linde and Curtiss-Wright against each other to see which would offer the sweetest deal. Guggenheim referred him to Washington attorney Clarence M. Fisher, to represent his interests in military and industrial contracts. Most of his worries were about the navy's claims to his patents, and the navy soon appointed its own team for negotiations that went on after Goddard's death.

Fisher enlisted the aid of Bleecker. He suggested that to limit a government stranglehold Goddard should grant a nonexclusive and nontransferable license to use only those devices covered by patents and applications dating before the contract, paying him a royalty. The government would have the right to royalty-free license to patents developed under its sponsorship, but this also should be nonexclusive and nontransferable. "In no event," Bleecker concluded, "should the option or the license in any way give the government the right to sublicense."[8]

With Fisher doing the talking, by December Goddard was in heavy negotiations with Curtiss-Wright and Linde. He had meanwhile completed his earlier navy contract, and negotiations for a new one also were underway. Fisher advised taking the best offer from the commercial sector, then telling the navy that he could no longer devote all his time to its work. At about the same time, Curtiss-Wright proposed to fold the Goddard team into its own organization, which would offer his services to the navy under its own contract.

Fisher and Goddard worked out a new contract with the navy in January, backdated to the previous September. It made him the technical director of the laboratory at Annapolis to the following June, with options to renew for one or two years. With that in hand, Goddard continued to dicker with the corporations.[9]

Curtiss-Wright was the more aggressive of the two, although Goddard's price was too high for them. Linde also said the rocketeer wanted too much money. Goddard was undismayed. "I personally feel," he told Guggenheim, "that an arrangement whereby I can have the assistance of a large manufacturing and engineering concern is essential if the device or devices are to be of practical use in this war."

Curtiss-Wright took the lead, and Bleecker pressed him for a meeting to iron out a contract, the negotiations for which took place in April 1943, in Breckinridge's New York office. They involved Guggenheim and his lawyer, Goddard and Fisher, and Curtiss-Wright officials.

Fisher, conferring with Breckinridge, uncovered a new wrinkle. He opined that Goddard was not bound by the 1930 agreement to give partial royalties to Clark University because the "consideration for that agreement" was Guggenheim's grant of $100,000, only half of which had come forth. Because the estate was not liable for the balance of the grant, according to Breckinridge, all bets were off. In conclusion: "Therefore, I recommend that you proceed with the Curtiss-Wright contract without advising Clark University either before or after the contract has been executed by all the parties thereto."

By following that advice, Goddard and Guggenheim together neatly sidestepped any obligations to Clark. Goddard now reassigned half his royalties to the Daniel and Florence Guggenheim Foundation, which could not reassign its interest without his permission.[10]

While this Byzantine maneuver went forward, the navy exercised its option to extend Goddard's contract until June 1944. Within a month, Fisher also had the details worked out with Curtiss-Wright, and on 20 July Goddard, accompanied by Esther and her mother, went to the lawyer's Washington office to sign consultant and licensing agreements. The navy was pleased that Curtiss-Wright assisted its project by making parts and helping with design, while Goddard's resources and products served both employers.

With his patents licensed to Curtiss-Wright, and thus protected, he thereupon secured permission from the Patent Office to show clients the applications held under secrecy orders. He had become potentially a wealthy man, and in the process had enlarged the assets of the Guggenheim Foundation.[11]

There were limits to Goddard's acquisitiveness, however. He traveled often to the Curtiss-Wright plant at Caldwell, New Jersey, submitting an expense voucher after each trip. Robert L. Earle, a vice president of Curtiss-Wright, chided him in 1944: "Upon seeing how much you spent for meals I greatly wonder how you exist on so little food."

That brought out the rocket man's impish sense of humor. "Thank you for your solicitude regarding my meals ... ," he replied. "I will admit, however, that being a professor for a number of years has developed the habit of living on short rations, besides keeping me a novice in the technique of making out expense accounts. I will see to it that this fault is corrected in the future."[12]

THE UNIVERSITY WOULD LIKE TO HAVE MY RESIGNATION

Goddard was not a reckless man. For years he maneuvered to retain his tenure at Clark University, securing a refuge if his rocket funding ran out. The running joust with Atwood continued after the move to Maryland, with three sores festering between them. One was Goddard's double-dipping in retirement plans; he was now in his sixties and almost eligible for annuities that had been building since 1914. Another was the substantial amount of university property he had with him. The third was the repeated leaves of absence.

Since he was earning more money from the government, in January 1943 Goddard wrote to TIAA, which managed Clark's retirement plan, asking whether he could increase his contribution. TIAA replied that that raised the issue of Clark's matching contribution. Since he was on unpaid leave, there was none, and for several years Goddard had been contributing both shares.

He approached Atwood, saying that he had also checked on his

Carnegie Foundation annuity. It projected a nice figure, he said, but it would not be enough for his retirement needs. Atwood consulted the payroll office, and told him he could send his enlarged contribution to Worcester, and the university would pass it on to TIAA. As long as Goddard held on to his tenure, he stood to receive two annuities.[13]

The question of the laboratory and shop equipment he had taken to Roswell and Annapolis was more thorny. The president demanded that the navy purchase the Clark equipment in Goddard's hands, adding a claim that objects shown on Goddard's inventories as belonging to the foundation and the Smithsonian actually belonged to the university, which nominally received the donations.

Goddard consulted with Breckinridge, and replied that Smithsonian grants and loans of machinery were to him personally. Carnegie and Guggenheim Foundation purchases probably belonged to Clark, but the intention was that he use them as long as he needed, while equipment purchased under navy contracts belonged to the navy. As for Clark equipment taken in 1930, Goddard said that on receipt of an appraisal he would ask Guggenheim if the foundation would buy it. Finally, he sniffed, the machine tools were not useful to students anyway.

The dispute ended in July 1943, when the foundation paid Clark University $4,600 for its property, without an appraisal. Guggenheim decided that that was the "simplest" course of action, and Atwood assented.[14]

Goddard would not give up his tenure as a professor or department chairman until he felt secure in private industry or the government. In January 1943 he slyly told Atwood that if it appeared unlikely he could return to Clark for a "reasonable number of years of teaching before retirement," he would let the president know. "On the other hand, should events occur that make it very desirable to appoint a permanent head of the Physics Department, I shall of course step aside, if circumstances do not then permit my return." That was so qualified as to be an empty promise, and Atwood knew it.

"As affairs are developing here on the campus," the president wrote in April, "I am certain that it is wise for me to make it clear to you that I must not count on your holding any longer a responsibility in the administration of either Mathematics or Physics during your period of absence." The "men on the ground" must have authority to plan their courses.

Goddard blew up at that: "If I understand the first paragraph of your letter . . . correctly," he rejoined, "the University would like to have my resignation at the present time. If this is so I am hereby tendering it."

Atwood boldly penciled at the bottom of this letter "YES!", underlined four times. But he wanted Goddard only to give up the chairmanships, not sever his ties with the university, and advised: "I am shocked at your reaction. I have not asked for your resignation from your professorship at Clark."

Goddard in turn said, "I am happy to discover that I have misinterpreted your meaning, and shall be glad to continue in my former status as outlined in my letter of January 9, 1943." He retained both professorship and chairmanship.[15]

The showdown came in August, after Goddard had secured his contract with Curtiss-Wright, and his annuities were safe. When yet another temporary professor left, Atwood wrote: "We have another emergency to meet. . . . We simply cannot keep a good man here without any prospect open to him for the future." With Clark taking on "Army emergency work," the trustees demanded that Goddard return to take over the physics program.

Goddard telegraphed that he was needed by the navy, and "HENCE I MUST REGRETFULLY RESIGN IN ORDER TO HELP YOU MEET THE PRESENT EMERGENCY." He followed with a letter. "As I stated," he wrote, "I feel that I can be of enough service in the war effort to make my decision to remain here the only course to follow. The fact that I am near the retiring age is also a factor of definite weight. A third point is that I doubt if I could lecture as soon as September. I had a severe cold last spring which settled in the larynx, and a specialist tells me I ought not to talk above a whisper for about two months." Atwood accepted his resignation without hesitation.[16]

The physics department was a hollow shell, thanks to Goddard's negligence and Atwood's plundering. As soon as Goddard was out, Atwood announced that Clark would accept no more graduate students in physics.

The two were not quite done with each other, however, and in 1945 they competed in a cynical display of mutual affection. Atwood persuaded the trustees to offer Goddard an honorary degree of Doctor of Science and invited him to the commencement to receive it. The rocketeer accepted, "and will naturally feel highly honored by this distinction. It will be no small additional pleasure to receive the degree from the hands of one under whom I worked so long at Clark, and whom I esteem so highly as a friend."

The ceremony took place on 3 June, with Dr. Benjamin S. Merigold of the chemistry department presenting Goddard to Atwood to receive the degree, and reading a citation that honored his scientific experiments. Percy Roope recalled that Goddard was "tremendously pleased" by Merigold's words, but could not remember his reaction to the degree itself.[17]

Atwood might have gloated had he known that the rocketeer had not found a gravy train at Curtiss-Wright after all. His problems began with both the army's and the navy's loss of interest in JATO. In 1944, Boushey, by then a colonel, and Commander Guggenheim tried to drum up interest in a rocket-powered glider project, but got nowhere. Goddard fretted that Cal-Tech would steal such an undertaking anyway. Curtiss-Wright required every technical detail relating to seventy-nine existing or pending patents, but lost interest in Goddard's first love, high-altitude rockets.[18]

Goddard was overwhelmed by changes in business and technology. Members of the American Rocket Society formed Reaction Motors, Inc. (RMI), so they could be eligible for government contracts, while on the West Coast the CalTech crew formed Aerojet Engineering. Jet engines were coming into their own, and in the fall of 1944 the German V-2 campaign reawakened interest in long-range rocketry, the very field that Goddard regarded as his own. Some of the new competitors were more interested in his brand of rocketry than was Curtiss-Wright.

Boushey urged Army Ordnance to develop an American V-2 program, hoping to involve Goddard, and Project HERMES was underway by early 1945, with General Electric (GE) the lead contractor. GE offered Goddard a consulting contract, but Curtiss-Wright refused to let him share the company's proprietary information, including his patents. The same happened with RMI. Goddard felt himself shut out of high-altitude rocketry by an employer that now tied up inventions it had no intention of using.[19]

Back in Annapolis, throughout 1944 and 1945 he was plagued by rumors sweeping through his shop that he and his men soon would be terminated. The stories were usually caused by delays in navy contract extensions, and ultimately he and the crew were secured through June 1945. But in April he discovered that Curtiss-Wright and the navy had been dealing behind his back, making arrangements to move him, his men, and the equipment to Caldwell. To relieve him of paperwork, his navy work would be under contract to Curtiss-Wright.

He assembled his crew and explained what was afoot, pointing out the advantages of staying together at Curtiss-Wright, at least to the end of the war. All eight agreed to go along, although Alden soon accepted an offer from the University of Wyoming. As they prepared to move, Goddard entered the hospital, leaving Ljungquist in charge.[20]

VERGELTUNGSWAFFE

After critic Edmund Wilson offered a favorable review of Willy Ley's book *Rockets*, on 10 June 1944 Goddard wrote him to counter any implications that the Germans were first in anything in rocketry. Three days later the first *Vergeltungswaffe*—"vengeance weapon," as Nazi propaganda minister Joseph Goebbels called it—struck London.[21]

Known as V-1 after Goebbels promised a second vengeance weapon in August, the flying "buzz bomb" or "doodle bug" or Fi-103 "Cherry Stone" was an air-breathing pulse jet powering what was later called a guided "cruise missile." Its guidance and timing system cut the engine off at the proper moment so that it would fall on the target city. Although it was a shock at first, it proved vulnerable because of its low speed and altitude.

Goddard was beset by reporters asking about the buzz bomb, but he did not know enough to offer details until August, when he told the North American Newspaper Alliance: "The V-1 flying bomb as a flight-controlled plane has already been shown to be an American idea, through patents to [Charles F.] Kettering and [Elmer A.] Sperry. The jet-propulsion engine is another," referring to his 1934 patent on "Propulsion Apparatus." He went on: "Features of the patent that appear in the bomb engine are shutter-type valves in a fixed grill; fuel injection orifices incorporated in this grill; combustion chamber; spark plug; and nozzle. . . . Anyone could have read it during the last ten years."[22]

The V-1 engine was actually an outgrowth of a patent issued to Paul Schmidt of Munich in 1930, and developed by the German military since 1934. As Walter Dornberger, head of the V-2 project but not involved in the V-1, described its action, it was a pulse drive involving about five hundred fuel explosions per minute. Air was sucked into the forward end "by means of a grid valve fitted to the duct head and provided with many rows of single flap plates opening inward. Fuel oil was injected into the compressed air and ignited. The resultant combustion pressure closed the valve flaps of the grid forward and forced the combustion gases and the air contained in the duct astern. This was accompanied by powerful expansion of the gases, and reaction propulsion took place." There were coincidental similarities to Goddard's design, but it was entirely a German production.[23]

Unlike the V-1, the *Aggregat-4*—what Goebbels called the V-2—was a true rocket similar to what Goddard had spent a lifetime trying to perfect. As reports of its construction came in, he was ever more agitated by the conviction that the Germans had stolen it from him. It "is reported to be almost identical with the rocket we were working on in New Mexico at the time we changed over to war work, except that it is longer," he told Guggenheim in December. He asked for a meeting to discuss it.[24]

Guggenheim summoned him to Mercer Field, New Jersey, where he was commanding officer. There the rocketeer presented him a memorandum comparing the V-2 with his own rocket. In his view, the two missiles were substantially the same in all important elements, while differences were minor. To illustrate his points, he gave Guggenheim a photograph of his pump rocket resting in the shop at Roswell, inscribed on the back: "Rocket produced in New Mexico in the Spring of 1941, under the Daniel and Florence Guggenheim Foundation. It is practically identical with the German V-2 rocket. Robert H. Goddard. Mercer Field. December 18, 1944." The foundation later distributed the photo and caption by the thousands.

With Lindbergh away at war and no longer available as his intermediary, Harry abandoned his previously cool distance from the rocket man. Goddard was plainly at a loss as to how to confront this blatant challenge to his and the Guggenheims' investments in rocketry. Harry took the shaken rocketeer in hand and told him exactly what to do to get the word out, while he made telephone calls to grease the way. He sent him to the National Geographic Society.

On 4 January Goddard met with LaGorce and Kerbey, and the latter drafted an article claiming that the V-2 was a theft of Goddard's designs. When Goddard gave him a list of his patents allegedly copied in the German craft, Kerbey revised his text, adhering to Goddard's statements, and it appeared in the society's *News Bulletin* on 19 January. The "some physicists" it cited were actually Goddard. Thus was born the legend of Goddard and the V-2.[25]

Goddard and his crew examined a V-2 at Annapolis in spring 1945, and as with Lindbergh when he first saw one, they agreed that it resembled their own work. However, as the years passed all reached the conclusion that the Goddard rocket and the V-2 were coincidental developments. Goddard himself went to the grave believing otherwise, despite major differences in fuel and cooling methods. He would not concede that his sounding rocket and a long-range ballistic missile were separate species connected by elements common to all liquid-fuel rockets.[26]

The legend fed on itself. If the V-2 was a theft of American patents, use of its technology by the United States Army would invite legal complications, and the military bureaucrats did not want that. When they began testing the German missile in this country, they avoided all mention of Goddard's name.

Thanks to Guggenheim, however, the word got out. Articles in newspapers and magazines trumpeted the twin themes of German theft and the American government's failure to appreciate the value of Goddard's work. J. F. McAllister, head of the GE rocket program, made the latter point explicit in a speech in 1947; he was echoed by the editor of *Flying* magazine. The next year, when the Guggenheim Foundation established Goddard professorships at Princeton and CalTech, it proclaimed baldly: "The German development of the V-2 rockets was based on Dr. Goddard's work in this country, as has been much of the progress in jet-propelled guided missiles the world over in recent years."

By 1949 the legend had entered the Cold War. L. T. E. Thompson and Jimmy Doolittle proclaimed that if the United States had supported Goddard the way Germany had supported its rocketeers, there would be nothing to fear from the Soviets. After the Commies launched the first satellite, *Sputnik*, in 1957, that theme became a chorus ringing from the lungs of such authorities as Gilbert Grosvenor, Charles Abbot, Homer Boushey, and even the head of the real V-2 program, Walter Dornberger, now an American citizen working in the aerospace industry.

Meanwhile, as early as 1949 the myth acquired a submyth—that when the German rocketeers were asked about their work in 1945, they rejoined: "Why don't you ask your own Dr. Goddard?"[27]

Neither the myth nor its interrogatory offspring was true, as an abundant literature has demonstrated that the German rocket program produced the technology on its own. It was an effort whose roots sprouted in civilian rocket clubs of the 1920s, inspired by Oberth, although without much technical input from the theory-minded German "father" of rocketry. But the vengeance weapons acquired considerable psychological and propaganda value for Hitler and his minions, for the German people, and for the Allies.

It was easy for the casual observer to miss the fact that the V-2s were relatively inefficient when compared to their promise, and appeared too late to make a difference in the war. An American program based on Goddard's work would not have accomplished much except, as in the German case, to divert resources from more conventional ways of delivering explosives.[28]

As for the "Why don't you ask" question, that was probably a case of spontaneous generation. The first American and British interrogation of the German rocketeers was inept, earning the contempt of the prisoners. Dornberger and his technical director, Wernher von Braun, manipulated the process in order to pursue a transfer to America, but they never mentioned

Goddard. Later interrogators from Aerojet and CalTech were qualified to ask intelligent questions, but Goddard's name never came up for them, either. Major General Leslie E. Simm and Lieutenant Colonel James P. Hamill of the American missile program, familiar with the interrogations, so told Goddard's widow in 1957.[29]

The "question" had by that time become a matter of faith for Goddard boosters, and it was hard for Esther and others to let it go. Milton Lehman sweated von Braun on the subject in a long interview in 1956, and elicited the following:

> You're asked why did you do this and why did you do that? Why you do this thing? Why don't you ask Goddard? you would say. I wouldn't be surprised if I said that, because you know, because it was a very obvious thing to ask. This situation they ask how did it ever occur to you to build rockets? Why did you build rockets? So wouldn't it be a very logical counter-question?... Everybody knew at Peenemunde that there was a man by the name of Goddard working in the United States who was trying to do the same thing we were....I don't remember saying this, but I wouldn't be a bit surprised if I had.[30]

That was remarkable, because in that interview and others von Braun maintained that he had scarcely heard of Goddard before 1950.

Esther learned in 1958 where the "question" may have originated, but she told no one about it. She received a letter from G. Edward Pendray's partner in the public-relations business, who said that in 1953 Pendray had told Harry Guggenheim about a meeting he had had a few years previously with Merian C. Cooper. The Hollywood director and old aviator friend of Guggenheim's claimed to have been in Germany when the Americans seized the German rocketeers. He said: "When we took them, the head of the delegation said, 'You have the man in your country who knows all about rockets and from whom we got our ideas—Robert H. Goddard.'"[31]

Esther inadvertently helped to discredit the V-2 legend by swallowing it whole and then driving Robert's authorized biographer to verify it. He ran into a wall, because his interviews with knowledgeable Americans, including Goddard's crew and Lindbergh, revealed that none believed the story any longer. Truax simply said: "The laws of physics and chemistry are impartial. They apply equally to Germans and Americans." The Germans he contacted, including Oberth, von Braun, and Dornberger, all denied any Goddard connection to the V-2. "The basic question, of course," biographer Lehman complained to Esther, "is whether the Germans have 'sent out word' to downplay any Goddard influence and whether the von Braun interview is deliberate evasion or not."

Esther compiled pre-war American and European literature on rocketry, doing her own translations, but could not find any direct connection between her husband and the German programs. Lehman was reduced in the biography to using misleading quotations and a third-hand statement that a German general not connected to the rocket program had posed the "Why don't you ask" question.[32]

The German rocketeers always denied gaining any information from Goddard. Dornberger said repeatedly that such complicated undertakings required teamwork, not solitary inventors like Goddard. They were proud of their achievements, but keeping some kind of Goddard story going after they moved to America was useful to them. Having an American doing the same things they did somehow made their own complicity in Nazi horrors less serious.

Hanging over their heads was the butchery of the slave-labor camps where the V-2 rockets were manufactured—twenty-five thousand people worked to death or murdered outright in a year at the Dora camp. One of the Germans, Arthur Rudolph, was forced to leave the United States in 1984 when his role in the atrocities became public knowledge.[33]

The Germans in America countered the Goddard V-2 legend with a legend of their own—that they had never heard of the man. Taking a cue from Willy Ley, they maintained that he had been unknown in Germany before the Nazi period, that his publications had not circulated there, and—the grossest lie of all—that his patents had been classified. Dornberger testified before a congressional committee in 1958: "At least ninety-eight percent of Dr. Goddard's patents were classified." In fact, Goddard's patent applications first went under a secrecy hold in the Patent Office after 1942. Only in the 1980s would some of the Germans admit that Goddard had been well-known when they entered rocketry.[34]

The most interesting German was Wernher von Braun, who hid his Nazi past as part of an effort to become an American hero. Interviewed in 1956, he expressed bitterness toward Harry Guggenheim for spreading the tale that the Germans stole the V-2 from Goddard. But he contradicted himself, saying in the same breath that he knew nothing of Goddard before 1950, and also that Goddard had been his "boyhood hero."[35]

The claim that the German rocketeers never heard of Goddard was made more absurd by the fact that Dornberger had used him before and during the war as a straw man in arguing for budgets and priorities for the V-2 program. He placed the German effort within an international rocket arms race, implying to his superiors that because of Goddard the Americans might win.

Saying the Germans could have known nothing of Goddard's patents was also ridiculous. Commercial and military attachés routinely acquired patents in their host countries, supplementing those exchanged under international agreements. Thanks to the military attaché in Washington, during the 1930s the Germans spied on Goddard, although they were not very good at it.

On 7 January 1936 General Friedrich von Boetticher fowarded to the German army's General Staff in Berlin the 4 January issue of *Science News Letter*. It reported Goddard's presentation at AAAS the previous week, including an account of the 7,500-foot flight, with a picture of Nell in her launch tower on the cover. Goddard, the attaché said, "has worked for decades with the problem of rocket flight," and enjoyed "considerable support" from the Guggenheims and Carnegie.

The General Staff forwarded this material to Army Ordnance, supervisor of the V-2 program. That elicited questions from the German rocketeers, which the foreign intelligence service tried to answer by directing a spy in

New York, Gustav Guellich, to learn more. He sent his masters several reports during the following year, claiming to have gone to Roswell and watched a launch, and to have obtained the schedule for future launches (there was none).

The creative spy seldom failed to answer a question, but he made up details. On 27 January, for instance, he reported the source of Goddard's funds as the army, the navy, and "DuPont Old Hikory [*sic*], Tennessee." His reports were of limited value, to say the least, but they were reviewed by the German rocket scientists. Dornberger and von Braun also saw a copy of Goddard's 1936 Smithsonian paper.[36]

It was reasonable that the German rocketeers would deny—and they could do so honestly—that they had stolen their technology from a dead American, but not so easy to figure out why they told so many lies about it. The Goddard V-2 legend was untrue, albeit at the beginning more an honest mistake than a fabrication. As to why it was propounded, the story has to be followed back to its real author—not Goddard, but Guggenheim.

ALMOST SINGLE-HANDED, DR. GODDARD DEVELOPED ROCKETRY

During their first two months in Maryland, the Goddards rented a place in Annapolis. They moved to Tydings-on-the-Bay, a small resort community, in October, renting a neglected house with few neighbors. It was cooler than Annapolis, nearer to Goddard's work, and it offered a beautiful view of Chesapeake Bay. "We are simply loving the place," Esther wrote Marjorie Alden, "tho I am still scouring in spots."

They had left most of their furniture in Roswell, so she replaced their mismatched possessions with finery shipped from New York. "But the view is magnificent, the quiet is perfection, and the house is *gemütlich* [cozy], in spite of decorative gaps," she said. That was small compass for someone with Esther's energy, and the house became confining after her mother moved to her brother Al's house in Massachusetts.

In October 1944 they moved to a larger place, with a yard big enough for the Goddards to take up archery. Esther loved the house, but Robert thought it was too big and gloomy, and he told a friend he wanted to go back to Roswell. Esther enrolled at Johns Hopkins University in 1944, Augusta returned to keep house, and Robert bought his wife a second car to commute to the train station.[37]

Their social life was more restricted than it had been, because they had few neighbors and wartime rationing limited travel. Christmas, shared with guests including the Aldens, was the high point of the year; Robert's gifts typically included cigars and Scotch. When possible, they vacationed in New England, or traveled to Baltimore or Washington to see the sights and visit old friends.

One such friend was Abbot, who in 1944 retired as secretary of the Smithsonian. His departure from that post left the rocketeer feeling increasingly isolated. Lindbergh was away at war, and only Guggenheim looked out for him.[38]

Goddard's work schedule at the navy laboratory was demanding, typically forty-eight hours per week. By 1944 he spent most of his time in his office, organizing the records of his rocket work during the 1930s. In addition, because of wartime hush-hush, he no longer enjoyed widespread publicity, except for a few pieces that said he was helping with the war effort. When the papers announced the development of Frank Whittle's jet engine, a project of the American and British air forces, in 1944, Goddard was both pleased to see himself credited as a contributing pioneer and annoyed to share the credit with Tsiolkovsky.[39]

While at Annapolis Goddard enlarged his relationship with G. Edward Pendray of the American Rocket Society, a long-standing Goddard fan. After the German rockets began to fly, Pendray became an energetic booster, writing letters to editors extolling Goddard as the real pioneer of astronautics. When Pendray proposed that the society republish Goddard's two reports for the Smithsonian, the rocketeer was delighted, and composed an introduction.

When he told Roope about that, the man who had been "time keeper and distance measurer" for the 1926 flight wrote him: "I am glad that, at last, proper recognition is being given to your work on rockets, and also that the United States is finally waking up to the vast importance of rockets and jets.... Let us hope that after the war the eyes of America remain open and that peacetime applications continue apace."[40]

Goddard would not see that happen, because he was dying. Beginning early in 1943, a diary entry stating "Stayed in bed all day" became increasingly frequent. In June he visited Baltimore tuberculosis specialist Dr. Charles R. Austrian, who referred him to throat specialist Dr. Harry L. Slack. Slack advised him not to talk at all until whatever was weakening his voice improved. It seemed to help for a while, but by winter Goddard was often bedridden.

When he was on the job his voicelessness interfered with everything. Unable to talk by late 1944, he tried writing, but his scrawl was so crabbed that few could read it. Finally, he gave his crew a typewritten copy of the Morse code, and tapped out his words with a pencil on a table. That did not work well, either.

In March 1945 Robert and Esther went to dinner with Commander Fischer and his wife at the Army-Navy Club in Washington. There he suffered an upset stomach so severe that an alarmed Fischer drove them home. By May Dr. Slack was flatly ordering him to take time off and sit in the sun. The navy obliged by giving him and his crew a two-week vacation before the move to New Jersey.

The vacation did not help, and in June Slack referred him to a surgeon, Dr. Edwin Looper, who on the sixteenth discovered a growth in his throat and advised immediate surgery. That took place in the University of Maryland Hospital on the nineteenth, when Looper removed his larynx and upper trachea.[41]

Those who knew Goddard were used to his weak voice, his thinness, his respiratory distress, and the fact that he never complained about his condition. Those who worked with him every day did not suspect that he was in such bad shape, while other acquaintances who were at a distance were shocked at the news of his hospitalization. Even Atwood, who had long since ceased to be a friend, offered warm encouragement.

Guggenheim—by then executive officer of a carrier group in the western Pacific—wrote twice after the surgery, once promising to talk about future work, the second time saying: "The thing for you to do now is get a complete rest somewhere, and make a full recovery. In the meantime, don't worry about jet propulsion or anything else. Your experiments have been many a year in the making, and there is a long road ahead, so a few months more or less will really make no difference. Your job is to get your mind off anything but your health, and when that has been recovered, we'll make a fresh start on the great future ahead."[42]

Goddard's iron will had seen him through tuberculosis, his smoking and drinking, and the frustrations and long hours of his work. But it could not conquer throat cancer, the disease that had claimed his father. He lingered in an oxygen tent, sustained by transfusions, tended by Ljungquist's wife, a registered nurse. He communicated by writing on a notepad, and refused to admit that he was in pain.

Esther was there day and night, but on the morning of 10 August 1945 she needed to go home for a few hours. While she was gone, he quietly died. "A lot of people you have a fair idea," Mandy Ljungquist recalled, "but with him there was no warning at all. He just passed away."[43]

He was buried on the thirteenth in the family plot in Hope Cemetery, among his grandparents and parents and not far from his brother, Richard. His passing was noticed in the hundreds of papers that had formerly charted his progress toward the Moon; many of them said he had invented the V-2. Pendray penned an obituary for the AAAS journal, *Science*, presaging the Goddard the world would come to know:

Even more impressive than Dr. Goddard's technical skill, insight, and ingenuity were his extraordinary perseverance, patience, and courage. He carried on many of his investigations in the teeth of public skepticism and indifference, with limited financial resources and in spite of heartbreaking technical difficulties—a combination of obstacles which might have baffled and disheartened a less stout-hearted pioneer. Almost single-handed, Dr. Goddard developed rocketry from a vague dream to one of the most significant branches of modern engineering.[44]

Out in Roswell, the editor of the *Morning Dispatch* remembered a Robert Goddard who had especially endeared himself to newsmen. "It is hard for a writing person to deal briefly with the passing of Dr. Goddard," he said, "who got close to all of us during the years the Goddards lived on the Berrendo farms. One could make a story about his work fully as interesting as that of the atomic bombs and much more easily understood. . . . Only 62, it is one of the pities of the present war time that Goddard should have been taken. His passing was certainly regretted by every newspaper man in the southwest."[45]

Goddard the man was gone. Goddard the legend now took his place.

TWELVE

WE HAVE COLLABORATED IN SERVING BOB'S MEMORY

Women have served all these centuries as looking-glasses possessing the magic and delicious power of reflecting the figure of man at twice its natural size.

—Virginia Woolf, *A Room of One's Own*

On 25 June 1876, Lieutenant Colonel George Armstrong Custer led 700 men in an attack on a large Indian encampment on the Little Bighorn River in southern Montana. When it was over, he and 265 other soldiers lay dead.

He left behind a widow, Elizabeth Bacon Custer, who had few resources other than her wits and energy. Those she had aplenty, along with a winsome charm. Over the next fifty-seven years, she went from impoverished widowhood to financial security by becoming a successful author and lecturer, and a friend of the rich and powerful. She did all that by remanufacturing her late husband's image into a myth of Custer the "boy hero," a brilliant soldier, solid Christian, patriot, and faithful husband. Not much of this was true, but so formidable was she that only after her death did historians and military scholars reexamine Custer's career and character.[1]

Esther Goddard faced a problem similar to Libbie Custer's. Her husband's maneuvers to secure two annuities bore little fruit because the widow's allotment was small. She began substitute teaching in the Worcester schools, but mostly she concentrated on how she wanted Robert to be remembered. She already had powerful friends, and wits, energy, and charm in abundance.

Within a year of Robert's death, Esther wrote a speech that she gave without important alteration for the next thirty years. "The Life and Achievements of Dr. Robert H. Goddard" extolled the man she knew, not the scientist. It presented him the way she wanted the world to know him— as a boy of exceptional brilliance, of humble origins and poor health, who dreamed great dreams and pursued them throughout a dedicated life. He was a distinguished but absentminded professor, a saintly man of rich humor, an enthusiastic piano player and painter, loved by everybody who knew him. Although his own country failed to appreciate the importance of what he did, he continued in his work despite widespread ridicule and the attempts of others to steal it. He never complained, never evinced discouragement or frustration. Above all, he never gave up.

It was a successful performance. When she first trotted it out in 1946 before the American Rocket Society, the American Society of Mechanical Engineers, and the delegates to the United Nations, she received standing ovations. It was the opening shot in a campaign to secure Goddard's place in history, and to ensure that his patents paid off.[2]

THE PATENTS WERE *MY* PRIMARY GOAL

Esther's most powerful friend was Harry Guggenheim. Two weeks after Goddard died, Harry contacted her to make plans to deal with the rocketeer's patents. He proposed that she assign to the foundation half interest in any patents not already assigned, and help secure patents for inventions not yet applied for. He also wanted to liquidate the foundation's interest in mutual property, beginning with the sale of the shop equipment to Curtiss-Wright. Ljungquist handled that, but clearing out of Roswell was more complicated.

With Harry paying expenses, in the fall of 1946 Esther returned to New Mexico and sold her ranch to friends. The shop building, which belonged to the foundation, went for $3,000 to someone who planned to move it and remodel it into a home. "I am glad," she told Guggenheim, "it is not to be somebody's barn. As you so generously suggested, I am keeping this sum as a gift from you, and I thank you again for this useful backlog. It will eventually go toward a memorial to Bob." The contents of the shop brought $1,200, which paid to ship the rockets and parts east; she suggested that they be donated to the Institute of Aeronautical Sciences (IAS), in New York.[3]

Another influential friend was Lindbergh, who felt some guilt about not providing more help to Goddard during times when he was devoted to his own interests, including the antiwar America First movement and then the war itself. Although Esther remained unaware of his activities for some time, he began working in her interest immediately after the war.

Still believing that the V-2 had copied Goddard's work, Lindbergh thought that, during his prewar visits, the Germans had manipulated him for propaganda and to learn about the state of rocketry in America. He was angry at the Germans, but furious at the American military, which had failed, in his view, to recognize the importance of Goddard's work even after the attack on Pearl Harbor. He encouraged the pursuit of an American rocket program, and he wanted it carried out with due regard to Goddard's contributions and his estate's interests.[4]

By early 1946, Esther was feeding material for patent applications to Hawley, with the foundation covering all expenses. The result was 131 new Goddard patents, raising his total to 214. This enormous and highly expensive undertaking went forward in a blaze of publicity. The aim was to establish Goddard once and for all as the inventor of everything in rocketry, and to dramatically assert his successors' right to payment for use of his ideas.

The project soon required full-time management by a knowledgeable rocket expert. Esther and Guggenheim tried to lure Boushey away from the Air Force, but he would not give up his uniform. Instead, he became her private informant in the Pentagon. In the fall of 1946, the foundation hired Pendray as its rocket consultant. "Of course," Harry told Esther, "Mr. Pendray will be the spearhead for the various plans that we have discussed and which we may carry out to stimulate further rocket development and in publicizing Bob's and the Foundation's contributions to rockets and jet propulsion."[5]

Guggenheim also may have felt some guilt, as unwarranted as that harbored by Lindbergh, at not having done more to support Goddard, leaving the American rocketeer to be upstaged by the Germans. He certainly had a personal stake in vindicating his and his family's investments in Goddard's work. Moreover, starting with a scheme to establish passenger-rocket service between New York and Buenos Aires, he believed that there was money to be made in rocketry, and that Goddard's patents could put a lock on it. Curtiss-Wright still controlled many of them under its 1943 contract with Goddard, and Harry wanted to get those license rights back.

First, there was Esther to look after. The final element in Harry's motivations for helping her establish Robert's legacy was growing, albeit platonic, affection for the widow, and his desire to secure her financial future. He certainly became far more indulgent toward her sentiments than he ever was to any of his wives. Nor could he resist her charming manner. To Harry, Esther always presented the image of the needful, and grateful, widow of a neglected national hero.

He put Esther on the foundation accounts without her thinking it was an act of charity. As he told her in July 1947: "It is not the intention of the Foundation that any charges should be made against you in the development and protection of our patent rights unless, and until, these expenses should be offset by returns from licenses or other profits derived from the patent rights, in which event such expenses would be a first charge against such profits."[6]

Guggenheim had concluded that, via government contracts, other companies were using Goddard patents without paying royalties. He engaged two law firms to explore the range of questions involved. "When this service has been completed," he told the foundation's directors, "we will be in a position to negotiate with the Army, the Navy, and other potential licensees for our patents."

He hoped that Esther would agree to the following proposal: "It is obvious that the Foundation from this point on will actively have to market and protect its patents. This will necessitate the creation of a small organization for this purpose and will involve the expenditure of appreciable sums of money. The alternative is to abandon any hope of returns for our patent rights. Obviously, at this stage, such a course would be folly." Esther and the directors agreed.[7]

A visit to Esther from Curtiss-Wright representatives in December 1947 provoked a crisis. The company was frustrated because the government was contracting to other firms and giving them the patents to which it held rights. Curtiss-Wright did not want to bear the expense of suing the government, and would surrender its exclusive claims if the foundation developed a licensing agreement with the federal agencies.

Esther and Pendray agreed that it would be best to find a way to pressure the government short of a lawsuit, but from then on the government was the target. Harry assembled a large meeting in his office on 17 December, to launch a campaign to protect the Goddard patents from the United States.

He shifted from general exploitation and defense of the patents to going after the government because of a report from patent investigator Donal F. McCarthy, who had gone through all the Goddard material. He declared that there was no income potential in any of it, for three reasons: Some of the patents had expired; there was no use for them in any field but rocketry, and that would take time for commercial development; and developers in the field would work on parallel lines to avoid infringement. Any money to be made out of Goddard's patents depended on a claim that Uncle Sam was violating Esther's and the foundation's interests in them.[8]

The chink in the federal armor, Guggenheim knew, was the V-2. If he claimed that the agencies were using Goddard's patents without license, they could reply that they had seized German technology as spoils of war. He would have none of that, asserting that the V-2 was stolen American property.

In 1948 he claimed that there was "evidence" that German rocketeers followed Goddard's work from 1919 until his death: "They were familiar with his patents, all his published material, and any additional data which they were able to obtain by correspondence directly or through engineers of other nationalities. It will also be apparent that the much-discussed German V-2 rockets, though larger, were almost identical versions of some of Dr. Goddard's missiles tested in flight.... All of the Goddard experiments and patents were made available to our Government before Pearl Harbor."[9]

The latter point was especially strong. The government might claim that its rocketry was based on the German program, and the Germans, as Harry anticipated, could be expected to deny that they had stolen their ideas from an American. But to most Americans they were nothing more than lying Nazis, not to be believed in any case. The United States government, however, could not deny that Guggenheim had offered Goddard and his rocket program to the federal agencies before the war, nor that Uncle Sam had had full access to Goddard's material, including patents, applications, and design specifications, since the fall of 1941. As Guggenheim pictured it, one way or another the American postwar rocket program led back to Goddard's "Guggenheim Rocket Research Project," so-called on the rocketeer's Roswell letterhead.

Whether Guggenheim had thought in those terms when he launched the Goddard V-2 legend in 1945 is not apparent. By 1950 it seemed the best way to fill a tall order. He had been around the government long enough to know that it would resist any challenge to its activities. Because the patent workload was about to increase even further, he dropped the charade of compensating Esther as a consultant and essentially put her on the foundation's payroll full-time.[10]

When the first patent infringement claim against the United States was filed in 1950, Lieutenant Colonel James P. Hamill of the American rocket program gave Wernher von Braun a pile of Goddard patents and applications, and asked him to review them from an engineering standpoint. Von Braun professed himself amazed that Goddard had addressed every problem the Germans had faced, often earlier.

His conclusion—remarkable because the United States Army still held him on a short leash as a suspect German—was that the United States government, in developing rocket and jet technology, inevitably infringed on Goddard patents. If his technical opinion translated into a legal one, the federal agencies would be liable to Goddard's successors for damages. To hide that exposure, the bureaucrats classified everything secret and dug in for a fight.[11]

Esther naïvely expected things to go quickly. "The end is now in sight," she told Boushey, "and I shall be free to leave Worcester, where I have stayed in order to be near the patent attorney, and look for an exciting job somewhere, perhaps in New Mexico again." She enrolled at Clark and received her M.A. in history in June 1951, but she was soon bogged down in the patent case.

In April 1951 she visited Curtiss-Wright to gather information. In September she provided power of attorney to Charles H. Murchison, the foundation's lead in negotiations with the government, so that he could represent both parties. By December her involvement in the work with Murchison and his successor, Henry G. Beauregard, was so heavy that Guggenheim gave her a retainer of $150 per month plus expenses. She turned down a $5,000 per year job with the National Research Council in Washington to stay with the patent case.[12]

Esther had many admirable qualities, but patience was not one of them. As negotiations with the government dragged out over years, frequently she blew her temper. Harry or Pendray, sometimes Boushey, calmed her down. In 1956, Guggenheim raised her pay to $5,000 per year and expenses, and gave her the wherewithal to run a wide-ranging campaign that included a biography of Robert and other propaganda.

Nevertheless, she felt betrayed. The lawyers had told her in 1950 that an administrative claim would bring faster results than a lawsuit or a private bill in Congress, yet there was no settlement years later. Thanks to Boushey, she knew that the Air Force especially was dragging its feet in negotiations. She even threatened to break with the foundation and pursue her own suit or private bill, until Boushey pacified her.[13]

After the last Goddard patent had been awarded, Charles T. Hawley retired to Vermont in 1957. Esther told him: "I count myself most fortunate to have been able to work with one so experienced and sympathetic in pursuing Bob's ideas to so satisfactory a conclusion. The teamwork of yourself, Mr. Guggenheim, and myself has truly made to Bob the gift he would consider the greatest and most important of all." He replied that it had been "the longest and most interesting experience in my legal career. I feel as if I had been a first-hand observer (and oft-times a participant) in a scientific research project which has rarely been equaled in efforts or results. And where may it all end?"[14]

The patent case picked up steam after 1958, when the National Aeronautics and Space Administration (NASA) was established. NASA shortly announced that its unmanned-spacecraft facility would be called the Goddard Space Flight Center. Clearly losing a public-relations war, the military services responded when Beauregard reduced the infringement claim to $1 million—he had been asking for several times that. In September 1959 the army and Air Force surrendered. Then the navy, whose share of the total would be only $10,000, balked.

In January Esther and Beauregard asked General Boushey, as he now ranked, to intervene. He did, and the dam broke in June. "Mr. Beauregard has been very busy indeed," Esther told Boushey. "He telephoned me Wednesday and in a shaky voice said he had in his hand a *settlement*. Would I meet him in HFG's office the next day? I would. HFG and I signed a lot of papers . . . and now Beau is in Washington, getting Govt. signatures, filing documents, etc. etc."[15]

It was more complicated than that, and Esther never understood just what happened or why. The case focused, in practical and legal terms, on two patents issued in 1946, although the terms of the settlement required that the foundation license to the government all Goddard patents since 1926. Esther resented that, believing there should be acknowledgment of infringement and monetary compensation for all the patents.

Moreover, she was technically not a part of it at all. On the advice of Harry's lawyers, when the government offered a deal, Esther assigned her entire interest in all the patents to the foundation. In return, she would receive from the foundation a total of $400,000, at $20,000 per year for twenty years, in quarterly installments of $5,000, which she should declare as capital gains. If she accepted a lump sum of a half million dollars, she would have only $62,000 left after taxes.

The settlement was therefore between the government and the foundation, which received the money. However, the federal attorneys insisted that Esther also sign the paperwork, voiding any further claim she might make.

The next problem was that NASA insisted on making thunderous news about it, while Harry feared that publicity would expose Esther to cranks and beggars. Fights within the government and between NASA and the foundation consumed over a month, until on 5 August NASA issued a short release covering the bare facts.

United Press International (UPI) went further in its own release, emphasizing the government's infringements on Goddard's patents and saying of him: "Now he is generally credited with being 'the father of modern rocketry,' the German V-2 missile, the American bazooka of World War II, and virtually the entire family of U.S. space and military rockets." Esther could not have said that better.

The press detailed the breakdown of liability within the government: the Air Force $765,000, the army $125,000, NASA $100,000, and the navy $10,000. The NASA release stressed—and most papers reported—that the infringement involved only two patents, not wholesale theft of Goddard's legacy.

Kenneth S. Davis, writing for *The New York Times Magazine*, thought otherwise: "When the Government recently announced payment of $1,000,000 for infringing patents by the late Dr. Robert H. Goddard, it specifically acknowledged that the Air Force's Atlas and Thor, the Army's Jupiter and Redstone, and the Navy's Vanguard had used what were essentially Goddard rocket engines and Goddard liquid propellants. Tacitly it acknowledged numerous infringements of other Goddard patents by acquiring rights to more than 200 of them, many so basic that every modern rocket, every guided missile, must use them."[16]

Once it was over, Esther and everyone else lightened up. She told Hawley that it had not all been grim: "Mr. Beauregard told me a story, during his elation over the outcome. . . . At a meeting of patent counsel from all three services some time ago, the Navy representatives started in, as they had several times previously, by saying the Goddard patents were 'no good.' Finally one of the Army patent counsel sprang to his feet and said heatedly, 'The hell they aren't any good! I was a Patent Examiner when those patents were going through the Patent Office, and *I* passed on them! I *know* they are perfectly valid patents!' "[17]

Above all, Esther proclaimed her gratitude to Harry Guggenheim for making it happen. As she told him, for Robert "the patents were only New England coming out, as something of an afterthought, to protect his inventive victories and support his claim to them. . . . But after Bob's death, Harry, the patents were *my* primary goal. To me they have been the bulwark against any inclination to neglect or downgrade the Goddard contributions to the jet problem. To me they are quite simply the whole foundation upon which you and I have built the shining image that is Goddard today."[18]

EVERYTHING IN OUR POWER TO INFORM THE PUBLIC

In Esther's view, vindicating Goddard's legacy involved far more than winning acknowledgment that his patents had been important to the American rocket program. There was a collateral implication that her man had been unfairly treated by an ungrateful government that first ignored him and then shamelessly exploited his work. The patent claim was part of a larger effort to turn him into the Mount Rushmore of rocketry. That involved getting his story out through his own writings and by means of a biography that would let him say from the grave: "I had it first!"

With Guggenheim money, in the 1940s she hired two typists to help her transcribe his research notes in an office in the Worcester house. They copied 5,500 pages of notes, mounted and labeled 2,600 photographs, and produced 500 photostats of his drawings. These were assembled in twenty-two volumes—five more were added in following years—of illustrated typescript, with copies donated to such repositories as the Smithsonian, the Library of Congress, Clark University, and the Roswell Museum and Art Center.

Next she transcribed Robert's diaries, and with Pendray's help completed a project Goddard had started in 1944, compiling his laboratory note-books from 1929 to 1941. They published this record of the development of rocket components in 1948, with an introduction by Guggenheim, who sub-sidized the publication.[19]

In 1953, with the patent negotiations going at their glacial pace, Esther determined to produce an acceptable biography. "There has been a ten-dency in two or three writings," she told Boushey, "to make him out as secretive, shy, and even bitter about the V-2—all about as far from the truth as one could imagine, as you know." She persuaded Harry to in effect hire Wes-ley Price, assistant editor of *The Saturday Evening Post*, to prepare the text.

When he saw the material Esther had available, he submitted a book proposal to Farrar, Straus, and Young. Roger Straus, Jr., son of Guggenheim's sister and head of that publishing house, passed it on to his uncle Harry, who urged approval. There was no resistance—three members of the Straus family were directors of the Guggenheim Foundation. In 1954 Price began conducting interviews with the rocketeer's associates, while Esther selec-tively lent him documents, retrieving each before giving him another.[20]

Guggenheim told Hickman about the project in 1956: "As you know, we are doing everything in our power to inform the public about Dr. Goddard's achievements," he said. "We have made some headway, but not enough. A book has been in preparation for the last two years, but, unfortunately, the author appointed to do the job has had a physical breakdown and a new author has recently taken over the work."

The new man was freelance writer Milton Lehman, a publicist at the Motion Picture Association of America (MPAA). He had first come to Esther's attention in 1954 with an article about White Sands Missile Range that mentioned Goddard, followed by others about the man himself. Lehman started interviews in 1956, even before financial arrangements were worked out. One of his first subjects was Lindbergh, who told him that "he feels that this book should be timeless. It should be a permanent reference." The Lone Eagle became so intimately involved in the project as almost to earn the status of coauthor.

Lehman wanted a grant from the foundation, but Guggenheim arranged a contract with Straus instead. It gave him a $9,000 advance, at $600 per month for fifteen months, together with expenses and royalties; he continued working for MPAA part-time.

Esther pledged to help out, but did not want to interfere with the patent claim by offending someone in the military. Her preference was to ignore the whole Annapolis period. "Dangerous, while claim pending?" she asked herself. Besides the question of Goddard's influence on the Germans, she reserved to herself the exclusive right to do Lehman's research in her husband's papers. She fed him what she wanted him to see, withheld the rest, and gave her own gloss to the story.

Lehman, meanwhile, carried on an extensive correspondence with Lindbergh. Slim again made common cause with Guggenheim on Goddard's behalf, during his frequent stays at Harry's homes.[21]

Lehman began producing a draft by early 1958, behind schedule. Esther, impatient as ever, asked Guggenheim to lean on him. Lindbergh was favorably impressed with the text, and urged Esther to lay off the man. What he had seen so far was incomplete, because Lehman's frustration at not being able to connect Goddard to the V-2 caused him to dodge the subject. Lindbergh, who now knew there was no connection, suggested "that you make some mention of the German and Russian work that was contemporary with Goddard's. I think this will strengthen the book, and leaving it out leaves too obvious a gap in the history of rocket development."[22]

Lehman was in a difficult position. Esther prodded him on the one hand to shape the story in a manner the facts would not support, and on the other to get the job done yesterday. Personal difficulties also plagued him, including his father's death and his own hospitalization after an automobile accident.

By 1959 Lindbergh was firmly at his side, and the two spent a considerable amount of time together. Slim's chief contribution was to persuade Esther and Harry to cease browbeating the writer, opining that he was being pushed too fast. He pointedly reminded Esther that it had taken him fourteen years to write his own most recent book. He prevailed, Esther backed off, and by 1960 Harry was reading and commenting on the drafts.

Still the project dragged on, and Lehman's difficulties continued to multiply. Lindbergh, over dinner, talked Guggenheim into giving the writer more money, and arranging for more editorial assistance from the publisher. Thanks to his constant encouragement and positive attitude, Lindbergh supported Lehman to completion at the end of 1962.

He also erased Esther's frustration at the delays. When Lehman sent her a copy of the book's front matter it contained a wonderful surprise—Lindbergh had written a preface. She was ecstatic. Slim next prodded Guggenheim and Straus to promote the book. He and Guggenheim both bought large numbers of copies, but he missed the grand publication party in the fall of 1963 because he had to go to Europe.[23]

Despite the delay, Esther was delighted with *This High Man*, its title derived from a poem by Robert Browning. It presented Goddard the way she wanted the world to know him, which was how she had described him in her speech since 1946.

That accorded also with the Goddard that Lindbergh, the V-2 issue aside, and Guggenheim, still upholding the V-2 legend, wanted to remember, a man whose greatness vastly exceeded any questions about German rockets. Said Slim in the preface: "Sitting in his home in Worcester, Massachusetts, in 1929, I listened to Robert Goddard outline his ideas for the future development of rockets—what might be practically expected, what might be eventually achieved. Thirty years later, watching a giant rocket rise above the Air Force test base at Cape Canaveral, I wondered whether he was dreaming then or I was dreaming now."[24]

In 1958 Esther looked for the next project to further Robert's reputation. Paul Garber of the Smithsonian's Air Museum suggested a more comprehensive publication of the rocketeer's reports than had been done previously. She pitched the idea to Guggenheim, expanding it into a general publication of Robert's papers. Within a year Harry had worked out a deal with reference-book publisher McGraw-Hill to edit and publish the Goddard papers, with Esther in charge. The foundation would reimburse McGraw-Hill for all costs.

"This is a project that gives me great pleasure," Guggenheim told Esther, "as I hope it also does you. It should result in a book of great importance and interest." She was delighted. "For a long time now," she told him, "I haven't been able to find the words to say what should be said to you for Bob and myself. Such sponsorship borders on the legendary."[25]

She hired a typist and got to work. Pendray visited twice a month, and they checked with each other on selections. She wrote "Keep?" on doubtful documents, and he answered yes or no. He usually deferred to her judgment, and with her temperament she would prevail in any event.

Polishing her husband's image was her chief guiding criterion. She discarded all or parts of items that she regarded as too personal, or which put Robert in a bad light, such as his correspondence with Atwood, his poor performance as department chairman, or his relations with GALCIT.

The project grew to two volumes by 1962, and to three by 1963. Guggenheim and McGraw-Hill quietly shuddered at the increasing cost, but Harry stood by her. Other than her public appearances, the papers became Esther's chief activity, and she slaved at it. She completed the job and sent it in late in 1967, and after the page proofs arrived she prepared an excellent index. Awaiting publication in 1970, she organized her own papers.

Meanwhile, in 1964 she had heard of plans to build a Goddard Library at Clark, and the university asked for Goddard's papers. She had promised them to the Library of Congress, but now wanted to change her mind. When she asked Guggenheim, he said: "I feel that you should determine where you would like these Papers to go, and send them there. I will heartily concur in your decision." They went to Clark.[26]

However much it reflected Esther's legend building, *The Papers of Robert H. Goddard* was a magnificent achievement and a valuable contribution to the history of technology. The foundation purchased hundreds of copies, widely distributed in influential circles, generating the kind of publicity Guggenheim and Lindbergh had wanted thirty years earlier. The publication was received with favor by the press, and enjoyed by researchers ever after. It was designed as another monument to Robert Goddard, but it was equally a monument to the skill, energy, hard work, and uxorial dedication of Esther Goddard.

Slim Lindbergh recognized that. He returned home from a trip at three in the morning to find the *Papers* awaiting him, and immediately sat down and started going through them. Then he wrote to Esther: "What wonderful accomplishments these volumes mark, not only your husband's but your own, and Harry Guggenheim's, and Ed Pendray's, in producing them. I think that, more than most people, I realize the comprehensive and painstaking effort you put into these volumes, year after year, the selfless interest, the love and dedication. What a monumental work results—timeless in the records of history! It rounds out, in a way that only you could have directed in its present stage, one of the greatest achievements of mankind."[27]

THERE ARE STILL SOME TAG ENDS TO DO FOR BOB

Esther told a reporter in 1963 that she was proud of the honors showered on Robert, but she hastened to say that she had not instigated them. "My husband wouldn't have liked me to do that," she said. "He didn't like aggressive women." She could be as disingenuous as he.

They were married for twenty-one years, and she outlived him by thirty-seven. During the earlier period she hovered over him and protected him and generally ran his life except for the rocket experiments. During the later period she spent nearly every moment building her man into a national monument, the unassailable giant of rocketry.[28]

An important early element of that effort came in 1948. The foundation paid the American Museum of Natural History to prepare a traveling exhibit of Goddard rockets and components, for display at the Smithsonian when it finished touring. It was dedicated on 21 April 1948, with Esther sharing the stage with Guggenheim, Lindbergh, and Jimmy Doolittle.

Guggenheim emphasized Goddard's purported role in the V-2, and promised that the foundation would continue to support the research he had pioneered. Doolittle described the prophet without honor in his own land who would become the core of the Goddard legend: "Dr. Goddard's rockets were flown successfully in New Mexico while the Germans were just beginning their rocket work on the shores of the Baltic. In that fact lies a tragic lesson. . . . We had in this country before the war the most know-how and the leading developments in the field of rockets. We neglected them until our enemies demonstrated their value."

Guggenheim was true to his word—never in the quest for patent compensation was there any sign of greed, because all proceeds went to support research. Late in 1948 the foundation endowed Daniel and Florence Guggenheim Jet Propulsion Centers at Princeton University and CalTech, each headed by a Robert H. Goddard Professorship. By the spring of 1949 Esther and Harry were well into a campaign to shame the government into paying for the use of Goddard's ideas. Over the years, the man who invented the V-2 became the man who put America on the road to the Moon.[29]

Goddard artifacts that did not go to the Smithsonian Esther donated to the Institute of Aeronautical Sciences in New York. Under a sign proclaiming "The Father of Rocketry," and paid for by the Guggenheim Foundation, "one of the most complete collections of literature and records dealing with rocket propulsion" opened at the Institute in 1950. A grand banquet was attended by Esther, Harry, Laurence Rockefeller, Abbot, dignitaries from Clark and WPI, C. N. Hickman, and others associated with Goddard's life. Esther formally transferred the property to the institute.[30]

She was optimistic that her crusade would show early results, as she told Boushey: "I am still living with my mother in my old house, but keeping right busy. . . . There are still some tag ends to do for Bob, but my long task is drawing to a close, apparently, and I shall be thinking about what to do with the rest of my life. Everyone is most kind, so I'm not worrying particularly."[31]

However, she was beginning to doubt that New York was the right place for what she had come to conceive as the "Goddard National Memorial." Her heart was in Roswell. The town gave her a tremendous reception in 1949 when she dedicated a new exhibit on the grounds of the city's art museum. She had donated the launch tower to the city along with $1,000 from the sale of the shop to pay for moving it to the museum.

After its dedication, she could not stay away from the place, stopping by during a visit to White Sands to watch the launch of a V-2 the next year. She began sending publications to the museum, and gathering artifacts for a Goddard exhibit there, which opened in 1953. She asked Harry for money to add a wing to the museum building.[32]

She had given the best stuff to the Institute of Aeronautical Sciences, and that did not leave enough for the grand wing that she encouraged the museum director to design. With Harry's backing, in 1957 she approached IAS regarding a loan of the artifacts, and within months she had it. The foundation then pledged $7,500 to build a wing. Esther helped select photographs and labels for the exhibit, and talked Clark University into donating some of Goddard's laboratory equipment and the army into lending a collection of rockets. In April 1959, after she visited Mexico with Peter Hurd and his wife, the Goddard Wing was ready for dedication.[33]

It was a grand show. Guggenheim could not attend because of an injury, so Esther represented the foundation. She was joined by her husband's former helpers, and by high-ranking military officers and civilian dignitaries. The star of the day was Wernher von Braun, who there launched his career as a Goddard booster. It was part of his lifelong effort to make himself an American hero, hoping his adopted country would forget his Nazi past.

Over the next several years, whenever Esther dedicated a new honor to her husband, von Braun was beside her, as close as a barnacle. Beginning at Roswell, he delivered a stock talk about his "boyhood hero." Those present were unaware that when his audience was German that title went to Oberth.[34]

Esther never really left Roswell. "I know what you mean by 'what might have been,' as I drive past these familiar streets and places," she told Boushey in 1962. "The Mescalero Ranch looks better now than it did a few years ago, and Roswell itself seems more attractive than it used to be, after the war. But there is an ache, especially when more old friends are no longer here to greet me." On that trip, she gave the museum a copy of the Congressional Medal recently awarded her husband. In 1964 she attended the first-day ceremonies celebrating a Goddard commemorative postage stamp. A year later she dedicated the new Goddard High School.[35]

For a while Esther wanted some sort of a monument, even a national historic site, at Eden Valley, but removal of the tower had destroyed the integrity of the place. She focused on the museum, which she had long since selected as the "Goddard National Memorial." In 1964 she talked the foundation out of $5,000 to install improved exhibit cases, and for years she continued to round up artifacts as donations or loans from other institutions. When the Roswell Rotary Club planned to reconstruct Goddard's workshop in the museum, she talked Curtiss-Wright into donating the machinery it still had on hand, and the Smithsonian into lending one of its Nells.

It was estimated that the project would cost $15,000, and the Rotary began fund-raising. The club enlisted help from old Goddard hands including Ljungquist, who still had drawings of the shop's layout before 1941. When it appeared that the fund-raising would be short by about $5,000 the club did not ask for Esther's help, but went directly to the Guggenheim Foundation. Observing that the museum had already received $12,500, Harry answered: "In view of these contributions, we do not feel that the Foundation should make further grants to the Roswell Museum at this time." Local sources found the money, and in 1969 Esther and von Braun dedicated both the shop reconstruction and a new Goddard Planetarium attached to the museum.[36]

As Esther aged into her seventies, her correspondence with the museum slacked off somewhat, although she continued a minute supervision of details begun in the 1950s. When astronaut Harrison Schmitt offered his spacesuit for exhibit in 1974, she objected to displaying it in the Goddard Wing, which she wanted to focus strictly on her husband. Told that the exhibit would be temporary, she withdrew her objection. During her last trip to Roswell in 1975 she helped Schmitt dedicate the display and presided over a separate tribute to Goddard.[37]

While Roswell was important to her, Esther did not neglect other opportunities to glorify her man. She had plenty of help. On the tenth anniversary of Goddard's death, Guggenheim, Lindbergh, and Doolittle together paid tribute to the rocketeer, calling him "the almost forgotten man behind today's international developments in long-range rockets, earth satellites, and space flight."

Esther had not known that was coming, and wrote to Harry to thank him: "In these last ten years, as we have collaborated in serving Bob's memory, I have expressed gratitude to you very many times, for very many things. That you should have continued your interest in my welfare after Bob passed away, so far beyond what might reasonably have been expected, has been not only a source of courage to me, but also an indication of your own fine and steadfast spirit."[38]

Esther did whatever she could to gain what she regarded as appropriate recognition for Robert, while Harry cultivated his political connections to the same end. She provided photographs and other materials to any publication that would do a spread on the "father" of rocketry, as the two of them went after suitable posthumous honors. The American Institute of Aeronautics and Astronautics (AIAA) presented the Louis W. Hill Space Transportation Award in 1958.

The following year Esther turned her attention to NASA, which wanted to make Goddard a national hero for a space program dominated by Germans. "Let me assure you," Deputy Administrator Hugh Dryden told her, "that I have not forgotten our project to secure additional recognition of the accomplishments of your late husband." He suggested that she ask Guggenheim to pressure the regents of the Smithsonian. "The project of making Bob better known to the general public has received great impetus," Esther told Harry.[39]

She pressed her congressman and others to obtain medals for Robert, and in 1959 Guggenheim exerted his influence in favor of a Congressional Gold Medal. Esther was miffed, thinking that something from the president would be superior. Saying she spoke for Harry, she urged Boushey to use his connections to gain a White House ceremony. The savvy general gently set her straight: "I would recommend that you let things take their natural course and accept the honor which it appears Congress is quite willing to bestow. The Congressional Medal is a higher honor than a Presidential.... As regards Mr. Guggenheim's suggestion that 'someone' persuade the President to present a medal, this is of course a matter which must be decided by you and Mr. Guggenheim. But again my advice (for what it is worth) would be not to press the Presidential issue at all and feel very fortunate that things have taken the Congressional turn."

Esther calmed down, and in September 1959 Congress ordered that a medal be struck for Robert H. Goddard. The next year the Smithsonian presented him its Langley Medal. In 1964, on Lindbergh's strong recommendation, the Guggenheim Foundation presented the Daniel Guggenheim Medal, the highest honor in aeronautics, first awarded to Orville Wright in 1929. It was governed by an independent board, which said that it was "virtually impossible today to design, construct or launch a rocket without using ideas or devices covered by Goddard patents." So it continued.

In 1965, President Lyndon Johnson proclaimed 16 March, the anniversary of the first flight, to be Goddard Day, a time for nationwide tributes to the "father" of rocketry. "It is twenty years now since my husband died," Esther told Guggenheim. "Looking back, I am deeply proud and grateful for what we have done for his memory. It is far more than he, or I, might have dreamed of."[40]

Medals were not enough for Esther, however. There were also places to name for Goddard, and monuments to him to be dedicated by her and the ever-present von Braun. They included a Goddard monument at WPI in 1959, a monument in 1960 in Auburn to commemorate the 1926 flight, the Goddard Space Flight Center in Greenbelt, Maryland, in 1961, the new Robert H. Goddard Library at Clark University in 1969, and in 1978 a Goddard memorial paid for by Congress at that institution. In 1969 Goddard was inducted into the Space Hall of Fame in Houston.[41]

Esther continued to enjoy Lindbergh's support. He was shy to begin with, and as he aged he became more reluctant to speak in public. Nevertheless, as the Apollo moon program advanced in the late 1960s, he told his "Goddard story" to selected audiences. They included the Apollo VIII astronauts the night before they took off in 1968, a White House dinner honoring the astronauts, and in 1969 the Airline Pilots Association. The "story" was always the same, about how Goddard had said in 1929 that it would be possible to reach the Moon, but it would cost a million dollars and that was out of reach.[42]

Esther, Slim, and Harry were all growing old by that time, but they had accomplished what they had set out to do in 1945. Goddard was a famous man, the no-longer-neglected "father" of rocketry and, in general opinion, the true author of the V-2. Lindbergh called him the "first astronaut."

A measure of their success first appeared in 1962, a juvenile biography by Anne Perkins Dewey called *Robert Goddard: Space Pioneer.* A succession of tracts aimed at youngsters followed over the next four decades. In keeping with their genre, they had a "Little Engine That Could" tone, emphasizing Goddard's hard work and perseverance, even when he was laughed at. Later in the century, with the growing emphasis on "self-esteem" for children, the biographies tended to focus on the cherry tree dream that Goddard remained true to throughout his life. The "father" of rocketry joined Abraham Lincoln, Thomas Edison, and Booker T. Washington as a model for children wanting to overcome adversity.[43]

HE TOOK HIMSELF TOO SERIOUSLY

Esther and her fellow memorialists succeeded too well, and an inevitable reaction set in. Overstating Goddard's prominence in the history of his art elided the contributions of others. Some specialists believed that Tsiolkovsky and Oberth deserved comparable notice, and the Goddard V-2 legend was proved false, while NASA's Germans and other American rocketeers felt themselves slighted by the glare focused on the professor from Worcester. For Theodore von Kármán, Lehman's biography was the last straw. The founder of the Jet Propulsion Laboratory penned his own memoirs, and vented his spleen:

I believe Goddard became bitter in his later years, because he had had no real success with rockets, while Aerojet-General Corporation and other organizations were making an industry out of them. There is no direct line from Goddard to present-day rocketry. He is on a branch that died. He was an inventive man and had a good scientific foundation, but he was not a creator of science, and he took himself too seriously. If he had taken others into his con-

fidence, I think he would have developed workable high-altitude rockets and his achievements would have been greater than they were. But not listening to, or communicating with, other qualified people hindered his accomplishments.[44]

Esther dismissed such talk. She had always dedicated herself to the well-being and reputation of her husband, and now had little life apart from Robert's memory. She continued working for the Guggenheim Foundation until 1976, still on the propaganda campaign. Payment for her transferred interests in the patents continued until 1980. But Harry Guggenheim died in 1971 and Slim Lindbergh in 1974. She was alone in Gram's old house in Worcester, surrounded by her memories.

As her loss of temper over the patents, the biography, and medals had demonstrated, Esther Goddard had an irascible streak. With advancing age, she became isolated, haunting the house where she received few visitors. On 5 June 1982, she passed away alone, at home. Two days later, she joined Robert in the family plot at Hope Cemetery.

She would have been pleased to know that her obituaries focused mostly on her husband and his career. That was what she had lived for.[45]

EPILOGUE

INTO FIELDS OF THE IMMORTALS

If a man does not keep pace with his companions, perhaps it is because he hears a different drummer. Let him step to the music which he hears, however measured or far away. It is not important that he should mature as soon as an apple tree or an oak. Shall he turn his spring into summer? If the condition of things which we were made for is not yet, what were any reality which we can substitute?

—Henry David Thoreau, *Walden*, 18

Before Apollo XI took off for the Moon atop a multi-stage pump rocket in 1969, the history of spaceflight began to trace its origins to three "fathers"—Tsiolkovsky, Goddard, and Oberth. This diplomatically ended nationalistic boasting in the Soviet Union, the United States, and Germany, with each claiming its own "father" who outshone the others. Historians and museum exhibit designers have followed this new tradition in the decades since. It bears reexamination.[1]

Placing the origins of spaceflight on the shoulders of one or three "fathers" might be likened to the old legend that Abner Doubleday "invented" baseball. Not only was that untrue, it begged the question, because the idea of baseball is as old as the idea of hitting something with a stick. The spaceflight idea and rocketry also have ancient origins. It is possible to identify milestones and important contributors, but no one "invented" either.

Defining a complex technology around one or three characters dismisses the contributions of others. Like radio, automobiles, computers, and other wonders of the twentieth century, progress in rocketry depended on enthusiasts, some professional, others amateur. Max Valier, Robert Esnault-Pelterie, G. Edward Pendray and the American Rocket Society, Nikolai Tikhomirov, Mikhail Tukhachevsky, Sergei Korolev, Theodore von Kármán, Frank Malina and GALCIT, C.N. Hickman, Rudolph Nebel and the German rocket clubs, Johannes Winkler, Wernher von Braun, Robert Truax, Charles Mansur, and others made valuable contributions. Some labored under a spell cast by a national "father," but they deserve credit in the general project.

The lives of the three "fathers" reveal some interesting similarities. All were inspired by Jules Verne. All were teachers oriented toward applied mathematics. All produced treatises demonstrating that projecting objects into space by means of liquid-fuel combustion was theoretically possible. Each had a streak of mysticism, and indulged in fantasies about carrying the human species to the stars. Each had a flawed personality and difficulty getting along with others, and yet each was likable enough to turn acquaintances into devoted admirers. Each had a monumental ego that approached the narcissistic, Tsiolkovsky less so than the others. All three lived to see the field each believed he had invented pass them by.

Such coincidences notwithstanding, it is fair to assert that Goddard stands out among the three "fathers." If he had not made high-altitude rocketry both famous and respectable in 1920, the world might never have heard of either Tsiolkovsky or Oberth. Moreover, neither of the others ever launched a rocket, so only Goddard turned theory into practice. Fatherhood, it might be observed from biology, begins not with conception but at birth.

In his own thinking and that of his boosters, Goddard's achievements paraded as "firsts," and they made a formidable list. He first applied a De Laval nozzle to a rocket, and so redefined the rocket motor. He first proved that a rocket would work in a vacuum. He turned multi-stage and liquid-fuel rocketry into mechanical designs, not just ideas. He redefined battlefield rocketry, and provided the conceptual foundation for the bazooka. His 1919 publication on reaching "extreme altitudes" was original, monumental, and elegant. He was the first inventor to launch a liquid-fuel rocket, the first to produce a rocket with an inertial guidance system, the first to use thrust-vector control (blast vanes) in a rocket, the first to use a gimballed engine, the first to build turbopumps for a rocket, the first to assemble liquid-fueled rocket motors in clusters, and the first to send a powered vehicle faster than the speed of sound.

The list of achievements was longer than that, and was reflected by the 214 patents in his name. As the federal government poured millions into rocket development after World War II, it used his inventions in violation of his rights, hiding behind secrecy classification to avoid retribution for that shameless exploitation of the man's creativity.

Thanks to Harry Guggenheim, the government's gambit failed. Admitting prior infringement and then buying license to Goddard's patents, in 1960 the federal agencies tacitly concurred with Guggenheim's claim that no rocket or jet aircraft could take to the sky without Goddard's inventions.

Robert H. Goddard the man was succeded by Goddard the legend—or more precisely, by two competing legends. One was created by Esther with the help of Guggenheim and Lindbergh. The other was a counterlegend created by historians and other rocketeers in reaction to the extreme claims of the first.

Goddard the man was a complex and sometimes inscrutable individual. He had many admirable qualities, chief among them the patience, persistence, and iron will that helped him to overcome tuberculosis, then to pursue rocketry for three decades. Seldom expressing frustration or discouragement, he accepted failure as part of invention, and kept on working. That did not make him unique among inventors—Thomas Edison was famous for his perseverance.

On the other hand, Goddard sometimes turned persistence into stubbornness, and was loath to give up on an old idea. He wasted years on the multiple-charge solid rocket and gyroscopic control where it was unnecessary, and he resisted adapting his experiments to the needs of the military. He could be disorganized, and was often distracted.

Away from work, his most admired quality was his likability. He ranged widely in conversation, had a sly sense of humor, and was a congenial host.

He was also a boy who never grew up, who could not cease expounding his ideas to those who understood his technology and his vision. The Goddard of Roswell was an overgrown boy playing with complicated toys; one of his men described him as a "hobbyist." He had a magnificent intellect, but it was governed by a puerile mind.

He was a case of arrested maturity. Isolated in childhood, he became dependent on women who protected and cared for him, so that without his mother, then Gram, then Esther, he could not have fended for himself. That dependency translated into a shy sort of helplessness that attracted a succession of guardian patrons, until at the end, Guggenheim almost led him by the hand. After his death Harry and Slim heroically promoted his reputation and his posthumous interests, including Esther's well-being.

He would not have gained such support, or held it as long as he did, without being a brilliant and visionary man. His patrons—Duff, Webster, Abbot, Lindbergh, and Guggenheim—all believed that he could accomplish great things in an unexplored field.

But he had an enormous ego. He believed himself the originator of everything, and resented intruders in his scientific preserve, earning himself a posthumous reputation as a secretive loner. In 1938 he placed his self-interest above the national interest, at least as it was defined by others. Believing himself about to perfect high-altitude rocketry at last, he declined to share the stage with GALCIT. Next he used the approach of war to seek money from the military to pursue his own interests. When that did not work, to get any funding at all he ended up at Annapolis where, his health fading, he became withdrawn.

There were some deplorable aspects to the man's personality, and some questionable ones, such as the self-destructive smoking and drinking. Most unfortunate was Goddard's narcissism. Urged on by Esther, he believed that he had been born a Great Man who need not be burdened by the needs of others. Witness the abuse of his tenure at Clark University, disregarding the damage it did to the physics department. He was not above lying, whether in his dealings with Atwood or when applying for credit. There was also a cowardly streak. He shrank from confrontation, and found indirect ways to stay his ground, as after the 1938 meeting at Guggenheim's estate.

Goddard the legend was created by Esther, Harry, and Slim after his death. It began with the V-2 story, and placed him at the beginning of everything in rocketry. The publicity hound who spent a quarter century courting the press and bathing in its attention became in their vision a shy man privately wounded by ridicule. Two sentences from an otherwise laudatory editorial in *The New York Times*—erroneous comments about Goddard's claim that a rocket could work outside the atmosphere—became a torrent of venom, which did not deter him from continuing his work. He was a saintly, quiet inventor unappreciated by his country before the war, and abused by it afterward. Robert H. Goddard deserved recognition as the one true "father" of rocketry, and all the medals and monuments that went with it.

Beginning in the 1960s, at the legend's height, a reaction set in. First Goddard had to make room for rocketry's two other "fathers." Then the V-2 story was put to rest, as it should have been. It was also absurd for his boosters to accuse the United States of blindness for failing to pour money into his work, because he did not sell his product effectively, and always insisted that he have complete control. Moreover, this technically myopic nation was the same one that produced the C-47, the B-17, the Jeep, the Sherman tank, DDT, mass production of penicillin, storable blood plasma, radar, the modern computer, advanced code breaking, and the atomic bomb.

In the last quarter of the twentieth century, a new legend emerged: Goddard's career was mostly a succession of failures, and he was so reclusive that, whatever his achievements, he had little influence. Because he allegedly denied the details of his work to others, they had to reinvent rocketry on their own.

This second legend is as flawed as the earlier one, and the truth is somewhere between them. Goddard, alone except for his crew, did pioneer liquid-fuel rocketry in America, although rocketeers in other countries followed their own traditions, however much Goddard did or did not inspire them. He invented many of rocketry's components, and he began before anyone else, launching the first liquid-fuel rockets to fly anywhere. But he did not solve all problems, and he did decline to cooperate with others.

The American "father" of rocketry is entitled to his due—not the hagiography that Esther, Harry, and Slim wanted, but a fair appreciation of the man and his contributions. Whatever his flaws, he forged ahead and made modern rocketry practical. He did not keep everything secret, because he patented his ideas, and others learned from them. When younger fellows with their own concepts entered the field, he behaved like the dog in the manger, but he had popularized rocketry and attracted those youngsters into it. Given what he had accomplished before them, the new rocketeers started out standing on his shoulders.

Finally, for nearly five decades Goddard held the unwavering devotion of the world's most famous aviator, Charles Lindbergh. The Lone Eagle believed he knew where Robert H. Goddard stood in the pantheon of the world's rocketeers. As he once told Esther:

> *Probably no figure in the history of science had a greater vision than that of Robert Goddard, or more courage and tenacity in translating his vision into fact. The determination of a youth to conquer the universe in spite of doctors' warnings that he had a few weeks more to live, the vision of a man physically projecting himself off his Earth and into space, the design and construction of prototypes by which this project would shortly be accomplished, what more is needed to carry mortal man into fields of the immortals?*[2]

NOTES

COMPLETE DATA ON WORKS CITED IN THE CHAPTER NOTES ARE PRESENTED IN THE BIBLIOGRAPHY.

NOTES TO PROLOGUE

[1] King and Kutta, *Impact*, 242–43.

[2] Kerbey, "Germany's Flying Vengeance Bomb."

[3] Mosley, *Lindbergh*, 343–44.

NOTES TO CHAPTER ONE

[1] The Daedalus story is a staple of Greek mythology. A handy summary of the various versions is Tripp, *Meridian Handbook*, 185–87.

[2] RHG to Edward F. Bigelow 19 JUN 01, Bigelow to RHG 26 JUN 01, and RHG, "Autobiographical Statement" 6 FEB 21, GP.

[3] Sources of Worcester's history include *Massachusetts: A Guide*, 392–401; Brooke, *Heart of Commonwealth*; Brown and Tager, *Massachusetts*; Farnsworth and O'Flynn, *Story of Worcester*; Kolesar, *Politics and Policy*; Roberge, "Three-Decker"; Lincoln, *History of Worcester*; Erskine, *Worcester*; Cohen, *Worcester's Ethnic Groups*; Chasan, *Civilizing Worcester*; and Rosenzweig, *Eight Hours*. Subscription histories include Crane, *Historic Homes*; Nelson, *Worcester County*; and Nutt, *History of Worcester*. See also True, *Worcester Area Writers*. Population figures are available in the decennial census reports; Wells and Perkins, *New England Community Statistical Abstracts*; and the *Manual for the City Council* accessible at the Reference Desk of the Worcester Public Library.

[4] ECG developed a great interest in RHG's genealogy shortly after their marriage in 1924, and kept at it for the rest of her life; this is reflected in THM. See also Charles Goddard, *Samuel Goddard Families*; William Goddard, *Genealogy*; and Magee, "Two Goddards."

[5] For reasons to be explained shortly, this account follows THM, 11–14.

[6] RHG "Material for Autobiography," 1927 with interpolations made 1933, GP; THM, 14–17. RHG "Material" was published posthumously in *Astronautics*, April 1959. RHG was confirmed in the Episcopal Church 21 MAR 1897, at St. James Church in Roxbury; confirmation certificate, RGP Box 1.

[7] Nobel Prize–winner Richard Feynman's account of his youthful tinkering is remarkably similar to RHG's. Feynman, *Surely You're Joking*, early chapters.

[8] RHG "Material"; THM, 15–21; Diary 19 FEB 1898.

[9] THM, 16.

[10] Ibid., 21–22.

[11] Top and others, *Communicable and Infectious Diseases*, 356–95. See also Dajer,

"Sleeping Giant." An estimated 20 percent of Americans were infected with tuberculosis in 1900.

[12] THM, 23–24.

[13] RHG "Material."

[14] Ibid.; Diary 19 OCT 1899.

[15] Holme, *Bullfinch's Mythology*, 187–88; Burrows, *This New Ocean*, 6–7. There are several fine summaries of the history of the space-travel idea in literature, and of rocketry. Most of them follow the three editions of Ley, *Rockets*, who pioneered the subject. Among them are Burrows, *This New Ocean*, ch. 1; Clarke, *Exploration of Space*, 1–8; and WVB and Ordway, *History of Rocketry*, 2–39. Works by Rabelais, Poe, Verne, Wells, and Lowell discussed below are listed in the bibliography.

[16] Barzun, *From Dawn to Decadence*, 112.

[17] RHG "Material"; Diary 1899–1901. During this period RHG showed his first interest in patents, corresponding with attorneys regarding a "Projecting Apparatus," of which nothing more is now known. Munn & Co. to RHG 7 MAR 01, GP.

[18] RHG "Material"; Diary 1901–04.

[19] RHG "The Navigation of Space" 25–29 DEC 01, Editor *Popular Science News* to RHG 31 DEC 01 and 27 JAN 02, GP; Diary 25–29 DEC 01.

[20] S. P. Langley SecSI to RHG 7 MAR 02, Calvin DeWitt USA to RHG 12 MAR 02, F. W. Hodge SI to RHG 11 NOV 03, and Editor *Scientific American* to RHG 28 JAN 04, GP.

[21] RHG "Material"; Diary 2–4 JUL 03.

[22] RHG Notebook titled "Suggestions" 12 JUL 03, GP.

[23] RHG "On Taking Things for Granted," GP; Diary JUN 04; *Worcester Gazette* 3 JUN 04; *Worcester Post* 24 JUN 04. Coverage of Class Day 3 JUN 04 was the first time RHG's name appeared in a newspaper, while his speech was printed 24 JUN 04.

[24] RHG "Material."

[25] Diary 24 JUN 04. The following account of his first romance necessarily follows THM, because ECG attempted to delete all record of Miriam from RHG's life.

[26] Diary 1904–09 passim; THM, 36–38, 40–41, 44–45, 50–52.

NOTES TO CHAPTER TWO

[1] The standard history of WPI is Tymeson, *Two Towers*. On Duff, see Masius, "A. Wilmer Duff," and examples of his many textbooks in the bibliography.

[2] THM, 41–42.

[3] RHG "Who I Am" 22 NOV 04, GP.

[4] Diary 1905–08 passim; Tymeson, *Two Towers*, 91, 109; *Aftermath, 1908*; RHG "On Some Anomalous Electrical Conductors"; RHG "On Some Peculiarities"; *Worcester Telegram* 10 and 11 JUN 08.

[5] Diary end 04, 2 and 16 FEB 05, 26 JUL 06, 8 NOV 08, end 08.

[6] RHG "Material." The summary mentioned is in the Green Notebooks, RGP, discussed below.

[7] Green Notebooks, RGP; Diary 4 MAR 06.

[8] RHG "Use of Gyroscope"; Diary 28 JUN 07; RHG "Material"; GP 1:16–17, 77–79.

[9] RHG "The High-Speed Bet" JAN 06, RHG "Limit of Rapid Transit" 20 NOV 09; Editor *Scientific American* to RHG 26 NOV 09; RHG to Editor *Boston Sunday American* 27 MAY 12, GP; RHG "Material," RHG "Bachelet's Frictionless Railway," RHG "Limit of Rapid Transit." The patents are, respectively, No. 2,488,287, issued 15 NOV 49, and No. 2,511,979, issued 20 JUN 50.

[10] RHG "On the Possibility of Navigating Interplanetary Space" 3 OCT 07; Editor *Scientific American* to RHG 9 OCT 07; William W. Payne to RHG 15 JAN 08, GP; RHG "Material."

[11] Diary 15 and 25–29 MAR 09. RHG rejected European study because "I can only see $800 at most coming for the two years." Diary 15 MAR 09.

[12] RHG "The Last Migration" 14 JAN 18, excerpt GP 1:95n.

[13] Green Notebooks, especially 4, 24, 25, and 31 JAN, 2 FEB, 11 JUN and 28 DEC 09, RGP.

[14] Winter, "Who First Flew?" and Burrows, *This New Ocean*, 21–22. Histories of rocketry generally follow Ley, *Rockets*, 54–78, and besides Burrows include WVB and Ordway, *History of Rocketry*, 2–39, and Clarke, *Exploration*, 1–8. See also Winter, *First Golden Age*.

[15] Baker, *Rocket*, 8–9; Burrows, *This New Ocean*, 8–9; James and Thorpe, *Ancient Inventions*, 238.

[16] Burrows, *This New Ocean*, 8–12; James and Thorpe, *Ancient Inventions*, 78–79; Baker, *History of Manned Space Flight*, 8–13.

[17] Ley, *Rockets*, 74–75; WVB and Ordway, *History of Rocketry*, 30–34; Burrows, *This New Ocean*, 23–24.

[18] Gartmann, *Men Behind*, 11–25; Ley, *Rockets*, 91–98.

[19] McDougall, *Heavens and Earth*, 17–25; Riabchikov, *Russians in Space*, 90–97; Burrows, *This New Ocean*, 36–44; Gartmann, *Men Behind*, 26–34; Ley, *Rockets*, 90–91, 98–103. Tsiolkovsky's autobiography appears in Clarke, ed., *Coming of Space Age*, 100–04. See also Tsiolkovsky, *Works*.

[20] Diary 7 MAY 09. The photo appears in GP 1:101. RHG's cigar smoking is discussed below.

[21] Koelsch, *Clark University*. Also informative are Veysey, *Emergence*; Kirkpatrick, *American College*, 142–51 (which calls Clark "an adventure in efficiency"); Ryan, *Studies in Early Graduate Education*; Barzun, *Dawn*, 641, 659; and Pope, *Colleges That Change Lives*, 25–32.

[22] Koelsch, *Clark University*, passim. On Hall's career, see Pruette, *G. Stanley Hall*;

Ross, *G. Stanley Hall*; and Rosenzweig, *Freud, Jung, and Hall*. Examples of Hall's works are in the bibliography.

[23] Koelsch, *Clark University*, 66–69. Examples of Webster's works are in the bibliography.

[24] Diary 7–10 SEP 09.

[25] Diary 9–16 JUN 11; RHG "Theory of Diffraction" and *On the Conduction*, RGP.

[26] Diary MAY–SEP 11; THM, 57–58.

[27] Diary 16 AUG, 15–22 SEP and 5–31 DEC 11, and 27–29 APR 12.

[28] Diary 30 JUL 12; RHG "Material"; GP 1:14–16 and note; Shiers, "First Electron Tube."

[29] Diary 29 APR–3 JUN 12; Magie to RHG 24 MAY and 8 JUN 12.

[30] Diary 10–11 and 23 SEP 12; RHG "Material."

[31] Diary 22 OCT–3 DEC 12; RHG "Material."

[32] Green Notebooks 4 and 5; RHG "Material."

[33] Diary 3 MAR–19 APR 13.

NOTES TO CHAPTER THREE

[1] RHG "Material."

[2] Ibid.; ML Int Stimson, nd 1956, MLP; Hubbard to ECG 15 SEP 45, in "Our Professor Goddard."

[3] RHG "Material"; Diary 14 MAY–2 JUL 13, 3 APR 34.

[4] RHG "Material"; Diary 21–23 MAY 14; THM, 63–72; Sanford to RHG 23 MAY 14, RGP.

[5] ML Int Horgan 6 JUL 56, MLP. Waterhouse's poem is in most quotation dictionaries.

[6] Practices that may be benign for most people can be risky for a tubercular.

[7] Green Notebook 5, 19 OCT 13, RGP.

[8] Since Goddard's day, the measure of rocket efficiency has become "specific impulse," the velocity of the exhaust gases divided by gravitational acceleration, measured in "seconds." Regarding efficiency of early twentieth-century rockets compared to more modern ones, Alway observes: "The most efficient, the Coston ship rocket, had a specific impulse of about 32 seconds (by contrast, the cheapest model rocket engines today have a specific impulse around 80 sec, and the Space Shuttle main engine has a specific impulse of 445 sec)." Goddard's chamber-and-nozzle combination, he says, converted "over half the heat produced by burning into exhaust velocity. The resulting specific impulses approached 250 sec." Alway, *Retro Rockets*, 10.

[9] Diary SEP 13–JUL 14; RHG "Material"; Goddard Patent Files, GCR; Ley, *Rockets*, 188; Alway, *Retro Rockets*, 10; Alway, *Rockets*, 80–81.

[10] RHG "Material."

[11] RHG "Past Revisited."

[12] See for instance Jardine, *Ingenious Pursuits*.

[13] Howard, *Wilbur and Orville*, 319–51; Crouch, *Bishop's Boys*, 407–16.

[14] RHG "Outline of Article on 'The Navigation of Interplanetary Space,'" 10 SEP–11 OCT 13, GP.

[15] RHG "Material"; Magie to RHG 8 APR and 20 JUN 14, Sanford to RHG 23 MAY 14, RHG to Sanford 28 MAY 14, GP. The *Physical Review* article was RHG "On Ponderomotive Force."

[16] RHG "The Problem of Raising a Body to a Great Altitude Above the Surface of the Earth," 21 AUG 14, GP.

[17] Diary 21 AUG–12 SEP 14; Green Notebook 5, 19 OCT 14, RGP; RHG "Material."

[18] Koelsch, *Clark University*, esp. ch. 3; RHG "Material"; Diary 25 JAN 15.

[19] RHG "Material."

[20] Diary 6 and 8 JAN and 1 and 2 FEB 15; *WPI Tech News* 12 and 26 JAN 15; *Worcester Evening Gazette* 2 FEB 15.

[21] Diary through spring 1915; RHG "Material"; William T. Foster to RGH 19 MAR 15, GP; Diary 25 MAR 15. The correspondence with various manufacturers is in RGP.

[22] Diary 3 Feb–23 OCT 15.

[23] RHG "Material." The earlier ion patent was No. 1,137,964, "Method of and Means for Producing Electrically-Charged Particles," filed 26 APR 13, granted 4 MAY 15. The other five in order were No. 1,154,009, "Apparatus for Producing Gases," filed 4 JAN 15, granted 24 NOV 15; No. 1,191,299, "Rocket Apparatus," filed 8 NOV 15, granted 18 JUL 16; No. 1,206,837, "Rocket Apparatus," filed 28 JUN 16, granted 5 DEC 16; No. 1,341,053, "Magazine-Rocket," filed 12 NOV 17, granted 25 MAY 20; and No. 1,363,037, "Method of and Means for Producing Electrified Jets of Gas," filed 4 OCT 17, granted 21 DEC 20; Goddard Patent Files, GCR.

[24] RHG "Material"; RHG "Method of Reaching"; Adams, *Space Flight*, 15.

[25] Sanford to RHG 28 MAY 15 and 1 MAR 16, GP.

[26] David L. Murphy Int Roope 20 APR 66, Roope File, GCR. The Clark College Class of 1920 dedicated that year's yearbook to Goddard, rather than to either of the retiring presidents Sanford and Hall.

[27] "Our Professor Goddard."

[28] Riffolt to ML 1 OCT 56, MLP.

[29] THM, 77–78.

[30] Diary 19 and 29 JUL and 8 AUG 15.

[31] Diary 25 FEB, 1 MAR, 10 APR and 3 MAR–27 SEP 16 passim; RHG to SecSI 27 Sept 16, and Maxim to RHG 22 FEB 17, GP.

[32] Diary 2 and 21 DEC 15, 2 OCT–16 NOV 16 passim.

[33] Diary 17–19 OCT 16.

NOTES TO CHAPTER FOUR

[1] RHG to SecSI 27 SEP 16, GP. Earlier quests for funding are mentioned in RHG "Material."

[2] CGA, *Adventures*, 100; CGA to CDW 2 Oct 16, GP. Examples of CGA's publications are in the bibliography.

[3] CDW to RHG 11 OCT 16, RHG to SecSI 19 OCT 16, GP.

[4] CDW to RHG 29 NOV 16, GP.

[5] RHG to SecSI 4 DEC 16, GP.

[6] CDW to RHG 9 DEC 16, CGA to CDW 18 DEC 16, and Buckingham to CDW 26 DEC 16, GP.

[7] CGA, *Adventures*, 100; GP 1:190n.

[8] CDW to RHG 5 JAN 17 and RHG to SecSI 9 JAN 17, GP.

[9] Diary 8–22 JAN 17; *WPI Tech News*, 13 MAR 17; *Worcester Evening Post*, 16 MAR 17.

[10] Tymeson, *Two Towers*, 122–23; RHG to SecSI 24 MAR 17, GP.

[11] Diary 17–21 MAR 17; RHG to SecSI 24 MAR 17, GP.

[12] Diary 9–30 JUL 17; RHG to SecSI 6 AUG 17 and Acting SecSI to RHG 16 AUG 17, GP.

[13] RHG to SecSI 5 DEC 17, GP.

[14] ML Int Riffolt JUN 56, MLP; RHG to SecSI 9 JAN 18, GP.

[15] Diary 1 APR 14; RHG to Daniels 25 JUL 14, and Daniels to RHG 12 AUG 14, GP.

[16] RHG to Daniels 16 OCT 14, Roosevelt to RHG 24 OCT 14, and RHG to SecNav 26 OCT 14, GP.

[17] Green Notebook 5, 19 OCT 15, RGP.

[18] RHG to SecSI 11 APR 17, CGA to CDW 14 APR 17, and CDW to RHG 20 APR 17, GP.

[19] Koelsch, *Clark University*, 119–22; RHG to SecSI 24 MAY 17, GP.

[20] RHG to SecSI 20 AUG 17 and RHG to Chief of Ordnance 20 AUG 17, GP.

[21] Diary 27–28 AUG and 27–28 NOV 17; RHG to SecSI 5 DEC 17, GP.

[22] Diary 11–13 NOV 17. Dunwoody resigned as a colonel in 1919 and went into private life until he became a civilian adviser to the Chief of the Transportation Corps in 1942, during World War II. His father, Henry Harrison Chase Dunwoody, USMA class of 1866, was also an artillerist who went into the Signal Corps, became a meteorologist and served in the Weather Bureau in the 1890s, and retired as a brigadier general in 1904. He held the first patent for a crystal radio receiver. Register.

[23] Diary 15, 22, and 31 DEC 17; RHG to SecSI 5 DEC 17, RHG to Rockwood 26 DEC 17 and 24 JAN 18, GP.

[24] Webster to Rear Adm. Ralph Earle 16 JAN 18, GP; Diary 17–23 JAN 18.

[25] CDW and Stratton to Squier 22 JAN 18, and "Report on Dr. Goddard's Device by C. G. Abbot and Edgar Buckingham" 22 JAN 18, GP; Durant, "Goddard and Smithsonian"; Register; DAB.

[26] RHG to SecSI 7 and 20 FEB and 1 MAR 18, and CDW to RHG 15 FEB 18. The story of the Langley lathe was related by Charles Mansur, later RHG's longtime crewman, in Mallan, *Men, Rockets*, 65. It accompanied RHG to New Mexico in 1930, Annapolis in 1942, and New Jersey in 1945.

[27] Diary 16–19 MAR 18; CGA to CDW 19 MAR 18, GP.

[28] RHG to SecSI 1 APR 18, CDW to RHG 19 APR 18, and CDW to Squier 30 APR 18, GP.

[29] RHG to SecSI 12 MAY 18 and CGA to CDW 15 MAY 18, GP.

[30] RHG Memorandum re Haigis 11 APR 18, RHG to SecSI 12 and 18 MAY 18, H. W. Dorsey to RHG 18 MAY 18, and RHG to CDW Tel 18 MAY 18, GP.

[31] RHG to SecSI 22 MAY 18, GP.

[32] Shinkle to Rockwood 27 MAY 18, Saltzman to Acting Chief of Ordnance 27 MAY 18, and RHG to SecSI 29 MAY 18, GP. Shinkle remained in Ordnance. He retired a brigadier general in 1942, was recalled to duty to command an ordnance depot for some months, then became the general manager of a defense manufacturer for the remainder of the war. Register.

[33] CGA, *Adventures*, 100–01; CDW to Squier 31 MAY 18, GP.

[34] RHG, "The Last Migration," 14 JAN 18, RGP; RHG "Material."

[35] RHG "Material"; Sanford to RHG 4 JUN 18, GP.

[36] CDW to Woodward 1 JUN 18, Woodward to CDW 3 JUN 18, and CDW to RHG 3 JUN 18, GP.

[37] RHG to CGA 4 JUN 18 and CGA to RHG Tel 5 JUN 18, GP; GP 1:234n.

[38] Diary 5–23 JUN 18; Nahum D. Goddard to CGA 10 JUN 18 and RHG to CGA 12 JUN 18, GP.

[39] DAB; Diary 11–23 JUN 18; RHG to CGA 12 JUN 18, GP.

[40] RHG to SecSI 1, 8, and 15 JUL 18; Hale to CGA Tel 10 JUL 18, GP. Diary 9–10 JUL 18.

[41] Diary 7 AUG 18; RHG to Sanford 15 JUL and 5 AUG 18, Squier to Chief of Ordnance 19 JUL 18, CGA to RHG 22 JUL and 5 AUG 18, RHG to CGA Tel 1 AUG 18, Hale to CGA Tel 7 AUG 18, CGA to Hale Tel 8 AUG 18, GP.

[42] RHG to CGA 8 AUG 18, GP.

[43] Ibid.; THM, 92–94.

[44] Diary JUL–AUG 18; Robert Goddard Photo Scrapbook, CUA.

[45] RHG to CGA 12 AUG 18, Sanford to RHG 13 AUG 18, Hale to CGA 17 AUG 18, RHG to SecSI 19 and 24 AUG 18, H. W. Dorsey to CGA 22 AUG 18, Squier to Chief of Ordnance 22 AUG 18, RHG to CGA 28 AUG 18, H. W. Dorsey Memorandum 29 AUG 18, GP.

[46] Diary 13–14 SEP 18; RHG to CGA Tel 16 SEP 18 and RHG to SecSI 20 SEP 18, GP.

[47] Correspondence involving RHG, CGA, CDW, Maxim, Squier, and others, 20 SEP–18 OCT 18, GP. See also RHG to SecSI and OCT 18, GP.

[48] Diary 26 OCT–9 NOV 18; Gen. George W. Burr to CDW 4 NOV 18, "Program for Tests at Aberdeen Proving Ground" 6 NOV 18, RHG to Sanford 9 NOV 18, RHG to SecSI 15 NOV 18, GP and RGP; CGA, *Adventures*, 101–02; Weeks, *Men Against Tanks*, 165.

[49] Diary 18–20 NOV 18, 10–13 JAN 19, Lieut. Col. Herbert O'Leary to SecSI 19 NOV 18, RHG to Chief of Ordnance 19 NOV 18, Walter S. Adams to RHG 16 NOV 18, RHG to CGA 28 NOV 18, RHG to Maj. W. A. Borden 29 JAN 19, RHG to CGA 7 FEB 19, and CGA to RHG 16 MAR 19, GP.

NOTES TO CHAPTER FIVE

[1] Diary 16–24 FEB 19, 13–15 APR 20, CNH to RHG 13 MAR 19, CGA to RHG 8 APR 20 and Tel 12 APR 20, RGH to CGA Tel 12 APR 20, W. deC. Ravenel SI to RHG 23 JUL 20, CGA to RHG 26 NOV 10, and Baldwin to RHG 8 DEC 21, GP; Mackenzie, *Alexander Graham Bell*, 325–26, 337–38.

[2] *Worcester Evening Gazette* 28 MAR 19, *Boston Globe* 29 MAR 19, *Greenville (South Carolina) News* 15 APR 19, *Indianapolis News* 15 APR 19, *Salt Lake City Tribune* 30 MAR 19, *New York Sunday Tribune* 30 MAR 19, and many others.

[3] CGA to CDW 8 MAR 20, GP.

[4] Diary 18–23 MAR and 1–7 APR 20, 8 JAN–19 FEB 21; Program of 1920 Annual Meeting of Eastern Association of Physics Teachers, RGP Box 75; RHG "Possibilities of the Rocket"; *Chicago Tribune* 3 APR 20, *Chicago Herald Examiner* 3 APR 20, *Boston Globe* 11 APR 20, *Worcester Sunday Telegram* 11 APR 20, *Washington Star* 27 APR 20, *Washington Times* 28 APR 30, *Boston Advertiser* 27 APR 20, *Worcester Post* 27 APR 20; SI Annual Report 30 JUN 21, excerpt in GP.

[5] Diary 4–10 APR and 3 JUN 21; RHG "Material"; RHG to Trustees 11 APR 21 and CDW to WWA 11 APR 21, GP; RHG statement to trustees and related material, WWAP.

[6] Earle to RHG 27 MAR and 28 APR 20; RHG to Earle 7 APR and 11 MAY 20; RHG to Lieut. Cdr. O. M. Hustvedt 21 JUN, 21 AUG, and 21 SEP 20, and 6 JAN, 24 MAR (2 letters), and 15 SEP 21; RHG Report to BuOrD on Preliminary Experiments Performed at Indian Head, Maryland, 9–20 AUG 20; RHG "Application of the Rocket to Armor-Piercing Projectiles" 24 MAY 21; Chief BuOrd to Solicitor DepNav 11 JUL 21; and RHG to Lieut. Cdr. T. S. Wilkinson 17 MAY 22, all GP. Reports to BuOrd 1920–23, RGP. Tymeson, *Two Towers*, 133ff.

[7] Fries to RHG 7 and 26 MAY and 10 JUN 20; RHG to Fries 18 MAY and 8 JUN 20; Maj. A. Gibson to RHG 6 NOV 22, 6 APR, and 24 MAY 23; Rear Adm. Charles B. McVay, Jr., Chief BuOrd to RHG 12 MAY 23, GP. Brig. Gen. Amos Alfred Fries, USMA class of 1898, was a Corps of Engineers veteran of the Philippine Insurrection. Moving to the Chemical Warfare Service in World War I, he became its chief in France, earning the Distinguished Service Medal and in 1918 promotion to brigadier. He served as chief of the service from 1920 to retirement in 1929. Register.

[8] RHG to CGA 11 SEP 19 and 29 SEP 22, RHG Report on Rocket Development nd SEP 19, and CDW to RHG 8 JUN 22, GP; Diary 23 SEP–23 OCT 20; RHG "Material."

[9] Quoted in THM, 127.

[10] Diary 8–9 JAN 19; RHG "Report to Smithsonian Institution Concerning Further Developments," MAR 20 with notes added 1929, GP. This was actually followed by two supplements, MAR 24 and AUG 29, RGP.

[11] Diary 8 JAN–3 FEB 21; RHG "Material."

[12] *Worcester Evening Gazette* 28 MAR 19; CGA to Lieut. Col. Herbert O'Leary 31 MAR 19 and RHG to CGA 1 APR 19, GP.

[13] CGA to RHG 4 APR 19, GP.

[14] RHG "Material"; RHG to CGA 7 and 15 APR 19, CGA to RHG 10 and 28 APR 29, C.F. Marvin (chief, Weather Bureau) to RHG 24 APR 19, RHG "Outline for Paper Read before American Physical Society" 26 APR 19 and "Points of General Interest in Paper Entitled 'High-Velocity Jets of Gas, Applicable to Rockets' Read before Meeting of American Physical Society" 26 APR 19, Fulcher to RHG 28 APR and 17 MAY 19, and RHG to Fulcher 12 MAY 19, all GP. The various earlier versions of the Smithsonian publication are in RGP.

[15] Diary 4 MAY 19; RHG to SecSI 2 JUN 19, GP; GP 1:337n.

[16] RHG, "Method of Reaching."

[17] Ibid., emphasis in original.

[18] He had learned that meteors were usually too small and spread too far apart to threaten a spacecraft, ending a fear that had obsessed him since his student days.

[19] Quoted in *Boston Herald* 12 JAN 20, and other newspapers. The original Smithsonian press release and the identity of its author disappeared decades ago. The AP bulletin copied it verbatim, and many newspapers followed suit, although the press generally focused on the paragraph quoted. See also Durant, "Goddard and Smithsonian."

[20] RHG "Material"; *Boston Herald* 12 JAN 20.

[21] There were hundreds of others, most based on the AP account. ECG tracked down all she could and placed them in scrapbooks, noting in the margins those press accounts for which she knew the date but had no copy. RGP.

[22] Hundreds of letters from volunteers, uninvited commentators, and just plain crackpots survive in RGP, with a selection in GP. Press reports after the initial JAN 20 rush include *Pittsburgh Press* 28 MAR 20, *New York American* 21 MAR 20, *Chicago American* 22 MAR 20, *Boston Sunday Advertiser* 21 MAR 20, *Boston Post* 4 APR 20, *Boston Globe* 4 APR 20, *Chicago Daily News* 3 APR 20, *Springfield (Massachusetts) Journal* 25 APR 20, *Boston Herald* 28 APR 20, NYT 28 and 29 APR 20, *Chicago News* 29 APR 20, *Boston Sunday Globe* 13 JUN 20, *Worcester Sunday Telegram* 19 SEP 20, *Boston Post* 16 and 17 SEP 20, *Boston Herald* 19 SEP 20, NYT 28 JAN 21, *Boston American* 27 JAN 21, *Washington Times* 28 JAN 21, *New York Sun* 28 JAN 21, AP story dated 20 JUL

21 in NYT 29 JUL 21, *Atlanta American* 14 AUG 21, *New York American* 21 AUG 21, and others in a slightly diminishing stream into 1923.

[23] RHG Statement for Newspapers 18 JAN 20, and RHG to SecSI 19 JAN 20, GP.

[24] *Worcester Gazette* 22 JAN 20.

[25] RHG, "On High-Altitude Research."

[26] RHG "Autobiographical Statement"; NYT 28 JAN 21. The editors of GP could not determine whether the "Autobiographical Statement" was ever published. GP 1:43n.

[27] RHG, "That Moon Rocket Proposition."

[28] *Worcester Daily Telegram* 6 and 19 MAY 21.

[29] THM, passim.

[30] Winter, *Prelude.*

[31] The correspondence with Macmillan and Company (publishers of *Nature*), together with a wide selection of foreign inquiries in the early 1920s, is in GP. On REP, see Lise Blosset, "Robert Esnault-Pelterie: Space Pioneer," in Durant and James, *First Steps*; REP to RHG 31 MAR, 10 MAY and 16 JUN 20, and RHG to REP 20 APR, 1 JUN and 8 JUL 20, GP.

[32] ML Int von Kármán 3 JUL 56, MLP.

[33] Burrows, *This New Ocean*, 48–53; WVB and Ordway, *Space Travel*, 55–62; Ley, *Rockets*, 108–10. See also Rauschenbach, *Hermann Oberth*, a biography that for unmitigated fawning exceeds THM.

[34] Oberth to RHG 3 MAY and 12 JUN 22, RHG to Oberth 17 JUN 24, RHG to Editor *Nature* 9 SEP 24, GP; ML Int Riffolt, ML Int Oberth 30 JUL 56, MLP; Rauschenbach, *Hermann Oberth*.

[35] Merigold quoted in THM, 108; NYT 13 JAN 20. This editorial has been reproduced frequently, including in Clarke, *Coming of Space Age*, 66–67, and as a sidebar to RHG, "Past Revisited." On page 43 of the 17 JUL 69 issue, with the Apollo 11 moon mission underway, NYT printed a "correction," saying that "it is now definitely established that a rocket can function in a vacuum as well as in an atmosphere. The Times regrets the error."

[36] Diary 29 JAN–1 FEB 20, 9 FEB 21; *Worcester Telegram* 2 FEB 21.

[37] Diary 26–28 MAY, 22–30 SEP, and 5–19 OCT 19; Industrial Research Laboratories letterhead, GP; GP 1:328n; ML Int Riffolt, MLP.

[38] Diary 9 and 18 OCT 19, and passim 1919–24; ECG birth certificate, baptismal and confirmation certificates, grade reports, and other material, EGP; THM 106–07, 120; quotation from Donald Shaughnessey Int ECG for Columbia Oral History Project 1960, EGP. ECG allowed ML to point out her resemblance to RHG's mother, and was proud of it. THM, 137–38.

[39] ECG, "Life and Achievements." THM makes much of the piano playing, but examination of ML's interview transcripts reveals only one person who could remember it. MLP.

[40] ML Int William Cole 14 JUN 56, MLP.

[41] Shaughnessey Int ECG, and Mr. and Mrs. Kisk's wedding announcement, EGP; *Worcester Evening Post* 21 JUN 24; *Boston Globe* 22 JUN 24.

[42] The photo file labeled "Robert Goddard Photo Scrapbook," CUA, reveals this.

[43] THM, 137–38; Feynman, *What Do You Care*, which offers this intimate detail regarding his marriage to his first wife, a nonpulmonary tubercular. See also Top and others, *Communicable and Infectious Diseases*, 356–95. The "photo scrapbook," CUA, shows that Esther documented every room of every house they lived in. The bedroom views reveal the sleeping arrangements described.

[44] THM, 145–46, summarizes this. Photos in RGP provide more information. Except for utility upgrades and cyclic maintenance, the house is nearly identical to what it was after Esther finished in 1924.

[45] ECG, "Life and Adventures."

[46] THM, 6; ML Int Horgan, MLP.

NOTES TO CHAPTER SIX

[1] *WPI Journal* 1920; Koelsch, *Clark University*, 112; Clarkson, "Ah, Distinctly"; "Our Professor Goddard"; Murphy Int Roope, GCR; RHG to CGA 12 JAN 29, GP.

[2] Sanford to RHG 7 JUN 20, and RHG to Sanford 12 and 16 JUN 20, GP; Tymeson, *Two Towers*, 131, 133ff; Masius, "A. Wilmer Duff."

[3] Koelsch, *Clark University*, 123–35; Diary 23 SEP 20; DAB; Murphy Int Roope, GCR. Examples of WWA's publications are in the bibliography.

[4] Koelsch, *Clark University*, 131–35, 143–45.

[5] WWA to RHG 12 MAY 21 and 9 APR 28, WWAP; RHG to WWA 14 APR 22 and 16 APR 28, Magie to RHG 23 NOV and 7 DEC 26, and RHG to Magie 30 NOV 26, GP.

[6] WWA Memorandum for Trustees Business nd (probably early 1926), WWAP; Veysey, *Emergence of the American University*, 313, 387. See the TIAA-CREF Web Center, www.tiaa-cref.org.

[7] Diary 21 OCT and 2 NOV 20 and 4 SEP 22; RHG to WWA 7 JUN 21 and 15 MAR 26, RHG Memorandum Regarding Radio License 15 MAR 26, and Acting President to RHG 15 JAN 23, WWAP; RHG to C.S. Thompson 30 OCT 22, GP.

[8] THM, 147; Diary 6 JUN 16; Goddard photo collection, mostly the product of Esther's work, CUA; genealogy and clipping files, RGP.

[9] ML Int Cole 14 JUN 56, MLP.

[10] Murphy Int Roope, GCR.

[11] The seven patents granted were No. 1,661,473, "Accumulator for Radiant Energy," filed 10 JUN 24, granted 6 MAR 28, and shared with Nils Riffolt; No. 1,700,675, "Vaporizer for Use with Solar Energy," filed 27 MAY 27, granted 29 JAN 29; No. 1,809,115, "Apparatus for Producing Ions," filed 16 JUL 26, granted 9 JUN 31; No. 1,809,271, "Propulsion of Aircraft," filed 28 JUN 29, granted 9 JUN 31, one-half

assigned to George Crompton of Worcester; No. 1,834,149, "Means for Decelerating Aircraft," filed 28 MAR 30, granted 1 DEC 31, covering an awkward idea for tail-mounted speed brakes; No. 1,860,891, "Apparatus for Pumping Low Temperature Liquids," filed 23 APR 30, granted 31 MAY 32, an early approach to what would become his pressure-tank fuel feeds for his rockets; and No. 1,879,186, "Apparatus for Igniting Liquid Fuel," filed 30 OCT 30, granted 27 SEP 32.

[12] ML Int Cole, MLP; Hawley quoted THM, 152. On stopping colleagues in the hall and having them sign as witnesses, see as examples Diary 4 and 21 JUN 26.

[13] GP 1:547, 548.

[14] See for example the oath required of assistants by Robert Boyle (1627–91), for whom Boyle's law is named, in Jardine, *Ingenious Pursuits*, 334: "Whereas I ___ being now in ye service of Mr. Boyle he is pleas'd to imploy me about ye making of divers Experiments that he would not have to be divulg'd; I do hereby solemnly and faithfully promise & ingage myself that I wil be true to ye trust repos'd by my sayd master in me, that I will not knowingly discover to any person whatsoever, whether directly or indirectly, any process, medicine, or other Experiment, which he shal injoin me to keep secret & not impart; without his consent first obtain'd to communicate it. And this I promise in ye faith of a Christian, witnes my hand this ___ day of ___."

[15] Jardine, *Ingenious Pursuits*, especially ch. 8. See also Barzun, *Dawn to Decadence*, passim. Academic hostility to scientific secretiveness is fierce. Witness the proof of Fermat's Last Theorem by Princeton mathematician Andrew Wiles in 1994, which earned him celebrity status for something that no one but theoretical mathematicians could understand, but it also brought him a hailstorm of criticism from his colleagues for working on the problem in secret for eight years. Hoffman, *Man Who Loved*, 119–217.

[16] ML Int Riffolt, MLP.

[17] RHG to SecSI 1 AUG 23 (quotation), Supplementary Report to Trustees, Clark University . . . 1 AUG 23, and Report to the Trustees, Clark University, on the Principles and Possibilities of the Rocket Developed by R. H. Goddard 1 AUG 23, GP.

[18] RHG, "On the Present Status of the High-Altitude Rocket," 27–30 DEC 23, GP. An abstract of this paper appeared in *Physical Review* FEB 24.

[19] RHG, "High-Altitude Rocket."

[20] *Boston American* 7 JAN 24, *Boston Traveler* 7 JAN 24, *Buffalo News* 10 JAN 24, *Washington Star* 11 JAN 24, *New York Tribune* 13 JAN 24, and scores of others during JAN 24.

[21] NYT 22 NOV 25.

[22] Editor *Industrie-und Handels-Zeitung* to RHG 31 OCT 25, RHG to Editor 16 NOV 25, Editor *Der Berliner Westen* to RHG 12 DEC 25, RHG to Editor 1 FEB 26, Otto Willi Gail to RHG 12 APR 26 and RHG to Gail 19 JUL 26, GP. The novel was *Der Schuss ins All* (*A Shot into Space*).

[23] RHG to Robert Lademann 14 NOV 26, German Consul General to RHG 22 DEC 26, RHG to Consul General 29 DEC 26, RHG to CGA 29 DEC 26, and CGA to RHG 14 JAN 27, GP.

[24] The voluminous correspondence with foreigners during this period is retained in RGP, with a selection published in GP 2:583–86. The latter includes Rynin to RHG 9 JAN and 23 FEB 26 and RHG to Rynin, 11 DEC 26 and 6 JAN 27. See also McDougall, *Heavens and Earth*, 27, and Stoiko, *Soviet Rocketry*, 39–40.

[25] NYT 31 DEC 23, *Boston Herald* 2 JAN 24, *Review of Reviews* JUN 24, *New York Herald Tribune* 17 JUN 24, *London Daily Express* 27 MAR 25, *Literary Digest* 21 MAR 25, *Radio News* MAR 25, *Philadelphia Bulletin* 28 MAR 25, *Scientific American* JUL 26, and others. See also Pollard, "Professor Goddard's Rocket" and Deisch, "How High Up?" and clippings for the period in RGP Boxes 75 and 76. Watson Davis of the Science Service approached RHG during this time to produce a scientifically correct new version of Verne's *From the Earth to the Moon*. The project came to naught because SI had become impatient with the untoward publicity. RHG to Davis 29 MAY 24, GP. THM, 138, states that this project went forward.

[26] CGA to RHG 3 OCT and 27 DEC 22 and 25 JAN 24, RHG to CGA 18 and 27 DEC 23 and 22 and 28 JAN 24, GP.

[27] RHG to CGA 17 and 23 JAN 24, and CGA to RHG 21 JAN 24, GP, and GP 1:524n.

[28] Hamerschlag to CDW 31 JAN 24, CDW to Hamerschlag 1 FEB and 31 MAR 24, GP. CGA drafted CDW's letters.

[29] RHG to WWA 23 JAN 24, WWAP; Burton E. Livingstone Sec AAAS to RHG 23 FEB 24, RHG to Livingstone 26 FEB 24, RHG to CGA 26 FEB 24, and CGA to RHG 29 FEB 24, GP.

[30] CGA to RHG 10 MAR 24, RHG to CGA 12 and 26 MAR 24, CDW to Hamerschlag 17 MAR 24, and RHG Supplementary Report on Ultimate Developments MAR 24, GP.

[31] CGA to RHG 26 APR 24, RHG to CGA 29 APR 24, CDW to RHG 16 FEB 26, RHG to CDW 24 FEB 26, and CGA to Hamerschlag 20 MAY 26, GP.

[32] ML Int Truax 4 JUL 56, MLP. This book aims at a biography of the man rather than his machines. Those wanting complete technical detail may refer to the notebooks, memoranda, affidavits, documentary photographs, and lab or shop notes in RGP; condensations in RHG, *Rocket Development*; or the twenty-seven volumes of experimental notes from 1921 to 1941, with copies lodged in CUA, D&FGF, SI, RMAC, and other institutions. The largest collections of RHG artifacts are now at RMAC and SI, with smaller collections at CUA and WPI. Patents are housed at the National Archives, with copies at RMAC and CUA. Correspondence and other background on patents and applications are in RGP, with documents on posthumous patents and the patent-infringement claim against the United States in EGP.

[33] Murphy Int Roope, GCR.

[34] Thompson, "Robert Goddard and the Weapons Program."

[35] Diary 11 JUL–13 SEP 21.

[36] Diary 5 OCT 21–3 MAY 22; RHG to President and Trustees 29 MAR 22 and RHG, First Report on Rocket Development . . . 1 APR 22, GP.

[37] Diary 7 MAY 22–1 NOV 23; RHG to President and Trustees 21 MAR 23 and RHG, Second Report on Rocket Development 21 MAR 23, GP.

[38] ML Int Riffolt, MLP; CGA to RHG 3 and 12 NOV 23 and 11 NOV 24, RHG to CGA 6 NOV 23 and 3 NOV and 22 DEC 24, CGA to CDW 27 MAR 24, CGA Summary of the Present State of the Goddard Rocket 25 MAR 24, RHG to Watson Davis 9 SEP 14, David White AAAS to RHG 19 DEC 24, RHG to White 24 DEC 24, all GP. On Riffolt's departure see GP 1:547n.

[39] RHG "Material." Among the theses supervised by RHG during this period were "A Study of the Elastic Properties of Light Metals and Alloys," Donald E. Higgins, JUN 24; "A Study in the Absorption of Radiant Energy," Nils A. Riffolt, JUN 24; "The Emissions of Electricity from Substances on Incandescent Carbon," Russel B. Hastings, JUN 25; "The Emission of Positive Electricity from Potassium, Heated on a Platinum Filament in a Vacuum," Lewis M. Sleeper, JUN 26; and "A Study in Rapid Transferences of Heat from Metal Surfaces," Clyde F. Benner, JUN 27. RHG listed these in "Material" as being helpful in working out rocket and spaceflight problems, and also in patent applications. Clark University theses and dissertations are in the custody of CUA. Years later Hastings said: "I never suspected that he was planning to use my work to prove ion propulsion for rockets." Quoted THM, 151.

[40] GP 1:548n; RHG "Material"; CGA to RHG 8 and 17 JAN, 5 and 14 MAY, 15 JUN, and 9 OCT 25; RHG to CGA 13 and 24 JAN, 16 MAR, 11 and 18 MAY, and 5 OCT 25; Humphreys to RHG 7 FEB 25; RHG to Humphreys 11 FEB 25, GP. Diary 18–28 NOV 25.

[41] Diary 6 DEC 25; RHG "Material"; RHG to CGA 5 MAY 26, GP.

[42] This account of the first flight and the events leading to it follows Diary DEC 25–MAR 26 (quotation 17 MAR 26); RHG "Material" (1927 quotation); THM, 143–44 (Esther's version); and RHG to CGA 5 MAY 26 and related materials in GP, along with the photographs in CUA. The story has been retold many times; one of the more precise accounts is Alway, *Retro Rockets*, 12–16.

[43] RHG to CGA 5 MAY 26, GP.

NOTES TO CHAPTER SEVEN

[1] Nowlan, "Armageddon" and "Airlords"; THM, 204; Shaughnessey Int ECG, EGP. The texts of the Nowlan stories are on the Buck Rogers Website, www.buck-rogers.com.

[2] *Boston Herald* 24 FEB and 24 and 27 APR and 13 SEP 27, *Boston American* 5 JUN

27, *Baltimore News* 15 JUN 27, *Worcester Gazette* 25 and 26 JUN 27, NYT 4 JUL 27, and others.

[3] *Worcester Telegram* 13 FEB 28; Gray, "Speed"; Gray to RHG 27 MAR and 19 JUL 29, RHG to Gray 2 APR and 23 MAY 29, and RHG to Birchall 12 NOV 29, GP; Diary 17 MAY and 8 NOV 29; NYT 15 DEC 29.

[4] T. O'Conor Sloane to RHG 21 APR 30, RHG to Sloane 24 APR 30, and RHG to Editor *Worcester Evening Gazette* (quotation) 16 MAY 30, GP; Wenstrom, "Rockets to the Moon"; *Boston Traveler* 15 FEB 30.

[5] Lasser to RHG 8 APR 30 and RHG to Lasser 12 APR 30, GP.

[6] F. H. of Calamba, Philippine Islands, to RHG 8 JUL 27; Mary Proctor to RHG 12 SEP and 27 OCT 27; RHG to Proctor 4 OCT 27; W.J. Humphreys to RHG 5 JUN 28; RHG to Humphreys 20 JUN 28; REP to RHG 30 JUL and 7 DEC 28; and RHG to REP 27 AUG 28, GP. Proctor, *Romance of the Moon*.

[7] NYT 8 MAY 27; Three Students to RHG 26 MAY 29, GP; Harford, *Korolev*, ch. 3; scrapbook of Moscow exhibition, CUA.

[8] RHG to CGA 6 JUN 27, GP.

[9] RHG to CGA 21 MAR and 28 APR 27; CGA to RHG 23 MAR 27; RHG to Lademann 28 MAR and 30 JUN 27 and 3 JAN, 12 MAR, and 25 JUN 28; Ley to RHG 27 JUL 28; RHG to Ley 14 AUG 28; B. O. [*sic*] to RHG 22 OCT 29; and RHG to B. O. [*sic*] 16 NOV 29, GP.

[10] Noordung, *Problem* and "Problems"; Miller, "Herman Potocnik"; NYT 10 JUN 28 and 13 OCT 29; *New York Tribune* 25 MAY 30; J. Kuhr (Consul General, Cologne) to State Department 17 APR 28, State Department to SecSI 8 MAY 28, CGA to RHG 10 MAY 28, and RHG to CGA 15 MAY 28, GP.

[11] RHG to Science Service 14 OCT 29 and RHG to E. E. Free 14 OCT 29, GP.

[12] News coverage includes *Worcester Telegram* 4 and 5 NOV 29 and *Boston Globe* 5 NOV 29; Waldemar Kaempffert to RHG 15 APR 30 and RHG to Kaempffert 18 APR 30, GP.

[13] Diary 1 APR–5 MAY 26; RHG to CGA 5 MAY 26, GP.

[14] CGA to RHG 8 MAY 26; RHG to CGA 11 and 24 MAY 26, GP.

[15] Diary JUN–JUL 26; Hamerschlag to CDW 17 JUN 26; CGA to RHG 19, 24, and 29 JUN and 1 JUL 26; RHG to CGA 22 and 29 JUN 26, GP. GP 2:596n offers ECG's comment on CGA's 24 JUN 26 letter: "The tone of this letter perhaps explains why Dr. Goddard built his next 'large' rocket twentyfold larger—a mistake he later deeply regretted." RHG had no failures except those caused by others.

[16] The history of the failed large rocket is traced in Diary 1 JUN 26–3 SEP 27, and summarized in RHG "Material." Correspondence cited includes RHG to CGA 1 OCT 26; 10 JAN, 10 FEB, 28 APR, 9 MAY, and 17 JUL 27; and CGA to RHG 14 JAN, 3 and 11 MAY 27, GP.

[17] Diary 15 JUL 27; RHG to CGA 17 JUL 27 and CGA to RHG 23 JUL 27, GP.

[18] RHG "Material"; Diary JUL 27–JUL 29; RHG to CGA 7 DEC 27, 12 and 20 FEB, 26 and 30 MAR, 30 APR, and 15 MAY 28; 3, 8, 12, 22, and 29 JAN, 2 FEB and 21 JUN 29; CGA to RHG 10 DEC 27, 15 and 23 FEB, 2 APR, 21 JUN, and 27 DEC 28; 5, 8, 19, 24, and 31 JAN 29; Rear Adm. A. L. Willard to CGA 29 MAR 28; CGA Memoranda for Files 14 JAN and 24 APR 29; and CGA to A. A. Noyes 24 JAN 29, GP. RHG to WWA 14 DEC 27 and WWA to RHG 15 DEC 27, WWAP.

[19] Diary 17–18 JUL 29; RHG excerpts from notebook 17 JUL 29, RHG to CGA 18 JUL 29, GP; Murphy Int Roope, GCR.

[20] Both from extras of 17 JUL 29.

[21] All headlines cited are from 18 JUL 29.

[22] RHG Red Notebook 17 JUL 29, RGP, excerpt in GP.

[23] NYT 19 and 29 JUL 29, *New York World* 19 JUL 29, *Washington Star* 19 JUL 29, *Washington Post* 19 JUL 29, *Boston Globe* and other papers 21 JUL 29, *Vienna Neues Wiener Journal* 23 JUL 29; *Popular Science Monthly* to RHG Tel 19 JUL 29, Gray to RHG 19 JUL 29, CGA to RHG 20 JUL 20, CGA Memorandum to AP 20 JUL 29, RHG to Blakeslee 29 JUL 29, James Stokley Science Service to RHG 25 JUL 29, RHG to Stokley 28 JUL 29, Arthur W. Ewell to RHG 2 OCT 29, and RHG to Ewell 7 OCT 29, GP.

[24] GP 2:674n. "Nell" was a stereotypical name in melodramas about unfortunate young women in the late nineteenth and early twentieth centuries. Examples included John Brougham's 1867 *Little Nell and the Marchioness*, Owen Davis's *Nellie, the Beautiful Cloak Model* (1905), and the hapless Nell rescued every week by Dudley Do-right in the *Rocky and Bullwinkle* cartoon satires on television in the 1960s. See Bordman, *Oxford Companion*, 433, and Atkinson, *Broadway* 74–78.

[25] ECG, "Our Nell."

[26] Diary 23 and 25 JUL, 1–22 AUG, 23–31 OCT and 2–11 NOV 29, and 29 NOV 29–21 APR 30; Molt to Neal 25 JUL 29; RHG to CGA 26 JUL, 9 SEP, 14 OCT, and 29 NOV 29; CGA to SecWar 27 JUL 29; CGA to RHG 7 SEP and 20 and 28 OCT 29; SecWar to CGA 3 SEP 29; H. W. Dorsey to RHG 21 OCT 29; and Text of License 25 OCT 29, GP. *Boston Herald* 25 JUL 29, *Worcester Telegram* 25 and 27 JUL 29, *Worcester Gazette* 26 JUL 29, *Boston Post* 25 JUL 29, *Boston Globe* 27 JUL 29, *Boston American* 2 AUG 29, and NYT 23 OCT 29. The Hell Pond site is now covered by development since the Devens base closed in the early 1990s.

[27] RHG to CGA 28 AUG 29 and RHG "Report on Conditions" 27 AUG 29, GP; and GP 2:688n.

[28] CGA to RHG 2 OCT 29, GP; GP 2:713n; ML Int Charles Mansur 6 JUL 56, MLP. In December Maj. C. E. Hocker of the Coast Artillery School asked about the performance of RHG's rockets and their potential as antiaircraft artillery. RHG stated that liquid-fuel rocketry was the answer to the antiaircraft problem. Hocker to RHG 6 DEC 29 and RHG to Hocker 16 DEC 29, GP.

[29] The photograph appears at GP 2:734.

[30] *Worcester Telegram* 24, 25, and 27 NOV 29; *Boston Post* 25 NOV and 18 DEC 29; *Worcester Gazette* 25 NOV 29; *Worcester Post* 26 NOV 29; *Boston Herald* 25 NOV 29; *New York Tribune* 24 NOV 29; NYT 27 NOV 29; *Boston American* 19 DEC 29; and such AP accounts as *Louisville Courier-Journal* 29 DEC 29; and *Detroit News* 26 DEC 29.

[31] This account of the Guggenheim and Lindbergh background and the meeting at Goddard's follows RHG "Material" and Diary 23 NOV 29, both sketchy; Taylor to RHG 22 NOV 29, GP; Davis, *The Guggenheims*; Cleveland, *America Fledges Wings*; Hallion, *Legacy of Flight*, 173–75 and passim; Lomask, *Seed Money*, 139–41 and passim; Pendray, *Coming Age*, 96–99; CAL, *Autobiography*, 15, 335–44, and passim; Berg, *Lindbergh*, 109, 163–69, 210–14, and passim; NYT 6 NOV 29 and 10 JUL 30; Milton, *Loss of Eden*, 184–85; Mosley, *Lindbergh*, 94, 127–28; 154–55, and passim; Ross, *Last Hero*, 243–53; ML Int CAL 27–28 JUN 56, MLP; and Anne Lindbergh, *Earth Shine*, 11–13. CAL's and HFG's books are in the bibliography. Milton, *Loss of Eden*, 184, says that RHG had applied for a Guggenheim Fund grant, but had been turned down on the grounds that he was a crackpot. I found nothing in RHG's records to support that tale. CAL told the million-dollar story to the Apollo 8 astronauts, and repeated it in his speaking engagements in the late 1960s; NYT 24 MAY 69 and Mosley, *Lindbergh*, 371.

[32] Diary 26–27 NOV 29; RHG to CGA 29 NOV 29 and CGA to RHG 30 NOV 29, GP; ML Int CAL, MLP.

[33] Diary 9–31 DEC 29; WWA to RHG 29 NOV and 5 DEC 29 and Charles H. Thurber, Board of Trustees to RHG 31 DEC 29, WWAP; RHG to CAL 2 DEC 29; CGA to RHG 2 DEC 29; Merriam to RHG 5, 19, and 28 DEC 29; RHG to CGA 14 DEC 29; RHG Memorandum on Conference 15 DEC 29; RHG to Merriam 14 and 26 DEC 29, GP.

[34] ML Int CAL, MLP.

[35] Diary 21 JAN–29 MAY 30; Merriam to RHG 25 JAN and 31 MAR 30; RHG to CGA 8 FEB, 24 APR, and 5 MAY 30; RHG to Merriam 8 FEB, 29 MAR, and 10 MAY 30; CAL to RHG Tel 12 FEB 30; CGA to RHG 28 MAR and 1 MAY 30; and DepWar to CGA 30 APR 30, GP.

[36] Hallion, *Legacy of Flight*, 175; Diary 5–24 JUN 30; RHG to CGA 28 MAY 30, CGA to RHG 2 JUN 30, WWA and RHG to members of the committee 14 JUN 30, GP; Breckinridge to WWA 9 JUN 30, Daniel Guggenheim to WWA 12 and 23 JUN 30, WWA to Guggenheim 13 JUN 30, WWA to Breckinridge 13 JUN 30, RHG to WWA 4 MAY 30, WWAP, also in RGP; NYT 27 JUL 30; ML Int Breckinridge nd 56, MLP. RGP Box 7 is stuffed with correspondence on the liability issue and establishment of the corporation.

[37] *Annual Report* of SI 1930; *Washington Post* and other papers 30 DEC 30; *Boston Transcript* 23 SEP 30. Percy Roope filled in behind RHG in math and physics, and another professor was hired on a two-year appointment.

NOTES TO CHAPTER EIGHT

[1] Lee to CGA Tel 3 JUL 30, CGA to Lee 3 JUL 30, Clark University News Release 9 JUL 30, David Lasser to RHG 10 JUL 30, H. Parrish to RHG 10 JUL 30, and RHG to Parrish 14 JUL 30, GP; *Worcester Gazette, New York Telegram, Detroit News, Worcester Telegram, Boston Post, New York American, New York Daily Mirror*, NYT, *Washington Times*, others, all 10 JUL 30. The story was in the foreign press by 11 JUL 30. Ivy Lee was image-maker for the Guggenheims, John D. Rockefeller, and a host of large corporations. DAB.

[2] *Time* 21 JUL 30.

[3] RHG to Col. D. C. Pearson 1 FEB 35, GCR; WWA to RHG 6 JUL 30 and RHG to WWA 14 JUL 30, WWAP; RHG to Merriam 14 JUL 30 and RHG to CAL 14 JUL 30, GP.

[4] RHG to Carl L. Bausch 15 and 28 JUL 30 and RHG to W. F. Clark 16 JUL 30, GP; Hagemann, "R. H. Goddard and Solar Power."

[5] Diary 25–26 JUL 30; ECG, "Life and Achievements"; Int CGA in *Worcester Telegram* 21 JUN 31.

[6] WWA to RHG 20 OCT 30, WWAP; NYT 29 SEP 30, *Chicago Tribune* 29 SEP 30, *New York Post* 30 SEP 30; RHG to Merriam 6 OCT 30 and RHG to Guggenheim Family 6 OCT 30, GP.

[7] Archuleta, *Official New Mexico Blue Book*, 7; Miller, *New Mexico*, 347–49; Fleming and Huffman, *Roundup*; Larson, *Forgotten Frontier*; Julyan, *Place Names*; Chew, *Storms*; Kelly, *History of New Mexico Military Institute*; Gibbs and Jackman, *New Mexico Military Institute*; Shinkle, *Seventy-Seven Years*; Roberts and Roberts, *History of New Mexico*. Also useful are the collections of HSSENM; its city directory collection provided census figures. Roswell is near the edge of the Permian Basin, and became an oil and gas center during the 1930s. See Rundell, *Oil in West Texas and New Mexico*.

[8] Samples of the literary output of all three are in the bibliography.

[9] Rogers's visits drew abundant notice in RDR and RMD. My thanks to Patricia Lowe, librarian of the Will Rogers Museum in Claremore, Oklahoma, who searched through the computer databases on all of Rogers's output, finding no mention of RHG. Rogers became the first honorary graduate of NMMI as tribute to his contributions to its polo and other programs. For the polo team he covered transportation, travel, and subsistence costs so that it could travel to meets. See Gibbs and Jackman, *New Mexico Military Institute*, 139. The Lindbergh-Rogers connection is detailed in Berg, *Lindbergh*.

[10] Diary AUG–SEP 30; Chaves County Historical Society *Facts and Traditions* 7 (FEB 83); ML Int Crile 5 JUL 56, MLP; RHG to Merriam 4 AUG 30, RHG to CGA 11 AUG 30, RHG to Lawrence Mansur 11 AUG 30, and RHG to Harry M. Davis of NYT, 3 JUL 37, GP.

[11] Diary SEP–OCT 30; ML Int May Corn Marley nd 56, MLP; RDR 15 DEC 63; RHG Morgue File, HSSENM.

[12] According to Charles Mansur in Mallon, *Men, Rockets*, 64.

[13] Letterhead samples in GCR; ML Int Ole Ljungquist and George Bode nd 56, MLP; Fleming and Huffman, *Roundup*, 217; Gibbs and Jackman, *New Mexico Military Institute*, 116; Lowell Randall to William D. Ebie 30 MAR 89, Randall File, GCR. Randall, who grew up in Roswell and went to work for RHG in 1941, recalled the excitement among youngsters in town.

[14] Diary 25–26 JAN and 11 FEB 31; Sachs Confidentiality Agreement 10 FEB 31, and RHG to Ljungquist 1 MAR 31, GP; Ljungquist statement in *Congressional Recognition*, 57.

[15] ML Int Ljungquist and Bode, MLP; Murphy Int Roope, GCR.

[16] Diary 1930–31 passim, esp. 8 and 19 OCT, 27 NOV, and 30 DEC 30, and 10 MAY 31; RHG to WWA 3 JAN 31, WWAP; RHG genealogy files in RGP Box 1; RDR 8 OCT 30; RMD 9 OCT 30; ML Int Charles Mansur 6 OCT 56, ML Int Ljungquist and Bode, MLP; Cobean records in RHG Morgue File, HSSENM; Mansur in Mallon, *Men, Rockets*, 62.

[17] ML Int Ljungquist and Bode, and Ljungquist to ML 27 FEB 62, MLP; Mallon, *Men, Rockets*, 61–65.

[18] ML Int Horgan, MLP.

[19] ML Int CAL, Mansur and Truax, MLP; HFG to ML 23 MAY 55, EGP; THM passim; Gartmann, *Men Behind*, 35–47, helped to establish RHG's reputation for secrecy in 1956.

[20] Millikan to RHG 9 JUL 31 and RHG to Millikan 15 JUL 31, GP. On the American Rocket (Interplanetary) Society, see Lent, *Rocket Research*, and Pendray, "Early Rocket Developments of the American Rocket Society," in Durant and James, *First Steps*. The quotation is from RHG to Pendray 29 MAY 31, GP; the society printed the letter. News about competitors includes *Worcester Gazette* 27 JAN 31, NYT 1 FEB 31, *Time* 9 FEB 31, *New York Herald* 8 APR 31, *Boston Transcript* 10 APR 31, and NYT 24 OCT 31, among others. GP includes a sampling of RHG's heavy foreign correspondence during 1930–32. On the German competition, in the 1950s aviation giant Ernst Heinkel, *Stormy Life*, 112–13, credited RHG with inventing the whole idea of rocket propulsion in 1912, and gave Oberth second place.

[21] Lasser, "Future of Rocket"; Lasser, *Conquest*; Pendray, "Giant Rockets"; MacMechen, "Rockets New Monsters"; Pendray, "Rockets to Moon"; Humphreys, "Mining the Sky"; Holt, "Amazing Turbine Rocket"; Killian, "Reaction Propulsion"; Holt, "Explore"; NYT 11 JAN, 14 JUN, 27 SEP, and 11 NOV 31; *Detroit News* 29 JAN 31; *Pittsburgh Press* 25 JAN 31; *Kansas City Times* 3 FEB 31; *Worcester Telegram* 13 JUN 31; AP reports in *Washington Post*, RMD, and other papers 14 JUN and 12 JUL 31; "Editorial" in *Scientific American* FEB 32; RHG, "Rocket Turbine for Aircraft" and "A New Turbine Rocket Plane." Much of the publicity grew out of his 1931 patent for "Propulsion of Aircraft," using a rocket to drive propellers at low altitudes, and withdrawing the propellers at high altitudes. RHG believed it the wave of the future and so persuaded many editors.

[22] RMD 16 AUG 45.

[23] RMD 6 AUG and 2, 12, and 20 DEC 30; *El Paso Evening Post* 8 AUG 30; RDR 5, 14, and 28 AUG and 12 DEC 30; *Boston Globe* 13 DEC 30; *Boston Post* 13 DEC 30; *Miami Herald* 13 DEC 30; *Chicago Tribune* 13 DEC 30. The article reviewed was Wenstrom, "Rockets to the Moon." On the Rotary speech, see also Huffman, *60 Years*, 21.

[24] RHG "Past Revisited."

[25] Merriam to RHG 27 OCT 30, RHG to Merriam 5 NOV 30, WWA to RHG 11 NOV 30, and RHG to WWA 17 NOV 30, GP.

[26] THM, xiv; RHG to Walter S. Adams 16 OCT 30, RHG to Gilbert N. Lewis 2 OCT 30, Adams to RHG 27 OCT 30, Ernest O. Lawrence to RHG 4 NOV 30, Garbedian to RHG 11 APR 32, RHG to Wells 20 APR 32, RHG to Kipling 20 APR 32, RHG to Garbedian 21 APR 32, Wells to RHG 3 MAY 32, GP.

[27] Diary 1930–32; RHG to Linde 20 APR 31, RHG to CGA 25 APR 31, RHG to Frederick G. Keyes 25 APR and 1 JUN 31, CGA to G.K. Burgess NBS 29 APR 31, Keyes to RHG 1 MAY and 8 JUN 31, Burgess to CGA 4 MAY 31, Linde to RHG 29 JUN and 21 OCT 31, GP.

[28] RHG to E.A. Varela 16 JAN 31, Fleming to RHG 17 MAR 31, RHG to CAL 27 APR 31, Willis Ford Insurance Agency to RHG 25 MAY 31, RHG to Breckinridge 2 JUN 31, Equitable Insurance to RHG 6 JUL 31, RHG to Equitable 30 JUL 31, GP; RHG to WWA 4 MAY 31, Florence Guggenheim to WWA 12 JUN 31, WWA to Florence Guggenheim 18 JUN 31, and the large file on both the credit and liability issues in RHG files, WWAP, and in RGP, Box 7.

[29] Fleming to Merriam 9 JAN 31, RHG to Fleming and accompanying statement 22 JAN 31, Brig. Gen. W.H. Tschappat to Lieut. Col. C.M. Wessen 29 JAN 31, Wessen to Tschappat 4 FEB 31, CGA to Hon. Robert Luce 7 FEB 31, and other correspondence in RGP, mostly Box 7. ECG reproduced only the 22 JAN 31 statement in GP, likely embarrassed at the absence of support for RHG's case.

[30] Diary 20–29 OCT 30; ML Int Ljungquist and Bode, MLP; Int Mansur, RDR 17 OCT 91.

[31] Diary 27 OCT 30–6 JAN 31; RHG to Merriam 3 JAN 31, GP. On Winkler's "HW-1," see Alway, *Retro Rockets*, 54–56.

[32] Diary 11–28 FEB 31; Merriam to Committee 19 JAN 31, RHG Memorandum on Advisory Committee Meeting 20 FEB 31, RHG to Florence Guggenheim 4 MAR 31, GP.

[33] Maj. K.D. Harmon to Ordnance Officer Ft. Sam Houston 23 MAR 31, Lieut. Col. W.A. Capron to Chief of Ordnance 28 MAR and 2 MAY 32, GP; RHG to WWA 22 SEP and 10 OCT 31 and WWA to RHG 6 OCT 31, WWAP.

[34] Diary 1931 passim; WWA to RHG 21 NOV 31, RHG to WWA 16 DEC 31, RHG Report on Rocket Work 15 DEC 31, and WWA to RHG 23 DEC 31, GP.

[35] Diary MAR–APR 32; RHG to Merriam 10 MAY 32, GP.

[36] ML Int CAL and Int Mansur 6 JUL 56, MLP.

[37] ML Int von Kármán 5 JUL 56, MLP.

[38] Diary MAY–JUN 32; RHG to Merriam 10, 16, and 27 MAY 32; Merriam to RHG 16 MAY 32; Merriam to Advisory Committee 17 MAY 32; Merriam memorandum on meeting 25 MAY 32; RHG to WWA 2 JUN 32; RHG to Breckinridge 12 JUN 32; WWA to Advisory Committee 14 JUN 32; and Breckinridge to RHG 16 JUN 32, GP. Berg, *Lindbergh*, ch. 10; Hallion, *Legacy of Flight*, 175–76; ML Int Breckinridge nd 56 and ML Int CAL, MLP.

[39] ML Int CAL and Int Breckinridge, MLP.

[40] Diary 16 JUN–21 JUL 32; RMD 14 JUN 32; *Worcester Telegram* 15 and 16 JUN 32; *Clark News* 5 OCT 32; ML Int Ljungquist and Bode, MLP; GP 2:833n.

[41] Diary 1932–34, passim, quotation 30 APR 33; RHG to CGA 5 AUG and 12 SEP 32, CGA to RHG 25 AUG 32, RHG Outline of Work Performed in 1932–1933 on Rocket Problems at Clark University 29 JAN 34 and RHG Summary of Work on Rocket Development 1933–1934, probably AUG 34, GP. RHG publicity during the Clark interregnum included RDR 14 JAN 33, *Worcester Telegram* 4 AUG and 13 OCT 33, and *Worcester Gazette* 15 AUG 34.

[42] RHG to CAL 8 MAY 33, GP.

[43] Diary 19–21 JUN 33; RHG to CGA 8 and 15 JUN 33, CGA to RHG 12 and 15 JUN 33, CGA to SecNav 29 JUN 33, Acting SecNav to RHG 23 JUN 33, RHG to Florence Guggenheim 24 JUN 33, HFG to RHG 5 and 15 JUL 33, and RHG to HFG 8 and 21 JUL 33, GP; Hallion, *Legacy of Flight*, 176.

[44] RHG to Roosevelt 19 JUL 33, RHG "Rocket Developments Carried on at Roswell, New Mexico" 19 JUL 33, Swanson to RHG 1 AUG 33, HFG to RHG 4 AUG 33, Swanson to HFG 1 AUG 33, HFG to Swanson 10 AUG 33, RHG to HFG 10 AUG 33, Adm. W. H. Standley to RHG 29 AUG 33, and RHG to CGA 2 SEP 33, GP.

[45] Diary 1 MAY 33; *Clark News* 14 FEB 33.

[46] RHG to WWA 15 JAN 34, GP; WWA to RHG 8 MAY 34, WWAP.

[47] Diary 29 JAN–24 AUG 34. RHG to HFG 14 and 23 JUN, 14 JUL, and 25 AUG 34; HFG to RHG 21 and 27 JUN and 3 and 11 JUL 34; RHG to WWA 14 JUL 34; and RHG to CGA 4 SEP 34, GP. *Worcester Telegram* 28 AUG 34; Hallion, *Legacy of Flight*, 176.

[48] WWA to RHG 25 SEP 34 and RHG to WWA 1 OCT 34, WWAP.

NOTES TO CHAPTER NINE

[1] Murphy Int Roope, GCR. Part of RHG's correspondence with Roope on departmental matters is preserved in the Percy Roope Papers, CUA.

[2] Diary 21 APR–5 MAY 35, 8 NOV 36; WWA to RHG 8 JAN, 4 FEB, 20 MAR, and 26 SEP 35; WWA to RHG 18 MAY and 13 OCT 36; WWA to RHG 16 JAN, 17 and 29 MAR 37, and 8 FEB 38; RHG to WWA 14 JAN, 16 FEB, and 12 MAR 35; RHG to WWA 14 MAY 36; RHG to WWA 26 JAN, 20 FEB, and 3 APR 37; RHG to WWA 15 FEB and 28 APR 38, WWAP, some also in GP. RHG to HFG 18 FEB 35;

RHG to HFG 21 JAN, 20 FEB, 16 MAR, and 3 APR 37; RHG to HFG 15 FEB and 27 and 30 APR 38; HFG to RHG 4 MAR 35; HFG to RHG 10 FEB, 5 and 16 MAR, and 3 APR 37 and 28 APR 38; RHG to CAL 21 JAN 37, GP. CAL's statement reproduced in NYT 6 JUN 37 and RDR 5 JUN 37, and editorialized in NYT 10 JUN 37.

[3] Diary 13 SEP 34, 6 FEB 37, 23 SEP 38.

[4] RHG to David Smith 16 FEB 35, GP.

[5] Diary 13–17 SEP 34; *Worcester Telegram* 28 AUG 34; NYT 21 OCT 34; RDR 14 SEP, 21 OCT, and 20 and 25 DEC 34. A selection of ECG's pictures of this joyous return is in GP 2:888–92.

[6] Diary 15–16 SEP 34; RMD 16 SEP 34, *Worcester Telegram* 17 SEP 34, RDR 17 SEP 34, NYT 17 SEP 34, *Washington Post* 23 SEP 34; James C. O'Neil (Editor *Telegram*) to RHG 17 SEP 34, RHG to O'Neil 25 SEP 34, CAL to RHG 14 OCT 35, GP; Anne Morrow Lindbergh, *Earth Shine*, 5–6.

[7] ML Int Horgan, MLP. See also Int Charles R. Brice and James B. McGhee in RDR 4 SEP 55 and Int Dwight Starr and Warren Cobean in RDR 16 MAR 76.

[8] ML Int Marley, MLP.

[9] Diary 23 SEP 34; Fisher to ECG 12 JUL 37, EGP and GP; ML Int Horgan, Ljungquist and Bode, and HAB 18 JUN 56, MLP.

[10] Diary 1934–39 passim, typical entries being 2–19 OCT 34, 28 JUL–4 AUG 35, 20 SEP 36, 28 MAR 37, 23–26 DEC 37, and 23 APR 39.

[11] Fleming and Huffman, *Roundup*, 40–41; Standhardt, *History of St. Andrews*, 72; RDR 14 and 15 MAY 35, 3 and 5 MAR 36, 2 and 3 APR 36, and 25 and 26 OCT 38. See also Diary 2 APR 36 and 20 MAR 38.

[12] ML Int Horgan, MLP.

[13] Ibid.; ML Int Crile, and several other Int Roswellians, MLP; THM, 227–28; Hurd statement in *Congressional Recognition*, 55–56.

[14] Hurd statement in *Congressional Recognition*, 55–56; Horgan, "Introduction" to Hurd, *My Land*, xxiv–xxvii; ML Int Hurd 8–9 JUL 56, MLP; THM, 227–28; Diary 29 OCT 39.

[15] Diary 22–27 SEP 35; HFG statement to the press 24 SEP 35, GP; NYT 24, 25, 26, and 29 SEP 35; RDR 23, 24, 25, and 26 SEP 35; *Worcester Telegram* 23, 24, 25, 27, and 29 SEP 35; *Washington Post* 23 and 26 SEP 35; *Worcester Gazette* 23, 25, and 26 SEP 35; *New York Sun* 25 SEP 35; *El Paso Herald-Post* 25 SEP 35; *Denver Post* 29 SEP 35; *New York Herald-Tribune* 25 and 26 SEP and 5 OCT 35; *Time* 7 OCT 35; *Christian Science Monitor* 1 NOV 35. Some of these were laudatory editorials.

[16] Berg, *Lindbergh*, 348–49; CAL, *Autobiography*, 144, 159; RHG to HFG 27 DEC 35, GP.

[17] Shirley Thomas, "Robert H. Goddard," in Gurney, *Rockets*, 72–85, quotation at 85; Diary 13 OCT 36, 23–25 MAY 37; Sunderland to RHG 25 MAY, 6 OCT, and 7 NOV 36, and 11 FEB 38; RHG to Sunderland 17 OCT and 5 NOV 36, and 19 FEB 38; Lieut. Col. J. A. Dorst to RHG 10 JUN 37; RHG to Dorst 21 JUN 37, GP.

[18] Mayes, *Wireless Communication*, 70–73, 94–101, 189–96, 223–24; Conot, *Streak of Luck*, 338–40; Aitken, *Continuous Wave*, passim, esp. 239; A. E. Anderson to ECG 10 MAR 73, Arvid E. Anderson, "Robert H. Goddard: Original Inventor-Patentee of the High Frequency Vacuum Tube Oscillator," unpublished manuscript 1976, patent assignment 7 DEC 20, and Arthur A. Collins to ECG 6 APR 54, EGP. Diary 27–31 DEC 35, 10–11 MAR and 26 APR 36. M. H. Collins to RHG 1 NOV 35; RHG to M. H. Collins 14 NOV 35; Arthur A. Collins to RHG 27 NOV 35; Arthur A. Collins to RHG 13 FEB, 13 APR, 20 MAY, 1 AUG, and 27 OCT 36 and 26 APR 37; RHG to Arthur A. Collins 2 DEC 35 and 19 FEB, 3, 8, and 17 APR, and 5 NOV 36 with Affidavit 5 NOV 36, and 2 JAN and 1 MAY 37, GP.

[19] Press accounts of Goddard during this period include NYT 21 OCT 34; 2 and 29 JUN 35; 23 SEP 36; 5 JUN, 5 and 9 SEP 37; 3 JUN, 10 JUL, and 13 OCT 38. *Literary Digest* 5 OCT 35. RMD 18 FEB, 5 and 6 AUG, 24 SEP, and 3 OCT 36; and 26 and 27 OCT 38. *Time* 2 MAR 36 and 7 FEB 38. RDR 5 AUG, 23 and 25 SEP 36, and 3 OCT 36; *New York Herald-Tribune* 4 OCT 36 and 3 JUN 38; *Worcester Telegram* 9 MAY, 3 and 6 JUN 37, and 27 FEB 38; *New York World* 5 JUN 37; *New York Sun* 5 JUN 37; *Boston Transcript* 5 JUN 37; *Worcester Post* 5 JUN 37; *Worcester Gazette* 5 JUN 37; *Albuquerque Journal* 18 JUL 37; *Los Angeles Times* 20 FEB 38; *Springfield (Mass.) Union* 3 JUL 38; *Boston Herald* 21 AUG 38; Moore, "Dr. Robert H. Goddard, Rocketeer"; Moore, "Sky-Rocketing Through Space"; Davis, "Fiery Rockets Roar"; Pendray, "Rocketry's Number One Man"; Pendray, "Number One Rocket Man"; and RHG, "Some Aspects of Rocket Engineering." See also Diary 1934–38, wherein RHG enjoys the attention. Relevant correspondence includes RHG to WWA 8 JUN 34, RHG to Albert C. Erickson and enclosures 25 SEP and 30 NOV 34, Ernie Pyle to RHG Tel 26 APR 36, RHG to Pyle Tel 27 APR 36, RHG to H. G. Wells 29 SEP 36, Pendray to RHG 29 JUN 37, RHG to Pendray 3 JUL 37, Harry M. Davis to RHG 12 JUN and 1 and 17 JUL 37, RHG to Davis 16 and 21 JUN and 21 AUG 37, RHG to HFG 5 SEP 37 and 9 JUN 38, HFG to RHG 8 SEP 37, Albert G. Ingalls to RHG 27 SEP 37, N. Laschever to RHG 1 MAR 38, RHG to Laschever 17 MAR 38, and RHG to CGA 27 JUL 38, GP.

[20] REP to RHG 17 SEP and 12 OCT 35, RHG to REP 10 OCT and 16 NOV 35, Ley to RHG 10 JAN 36, RHG to Ley 18 JAN 36, Albert Rice Leventhal to RHG 24 FEB 36, RHG to Leventhal 2 MAR 36, CAL to RHG 2 NOV 37, GP; NYT 19 JUN 38; Cleator, *Rockets Through Space*. The footnote is on p. 51.

[21] RHG to CGA 21 DEC 35, GP.

[22] Lester D. Gardner to RHG 30 MAR 36, RHG to Gardner 11 APR 36, C. W. McNash to RHG 8 JUN 36, RHG to McNash 15 JUN 36, Truax to RHG 17 MAR 36, and RHG to Truax 28 MAR 36, GP. ARS rocket experiments were reported in NYT 26 MAR, 26 AUG, and 22 OCT 35; *New York Journal-American* 26 AUG 35; and *New York Herald-Tribune* 26 AUG 35, among others. The Ley group's mail rocket tests were reported in NYT 10 NOV 35 and 6 and 10 FEB 36. On Truax, see WVB and Ordway, *History of Rocketry*, 82–83.

[23] Hallion, *Legacy of Flight,* 198–200; Millikan, *The Guggenheim Aeronautical Laboratory;* THM, 233–35; ML Int Truax, Int von Kármán, Int W.C. House nd JUL 56, MLP; Burrows, *This New Ocean,* 85–91; Malina, "On the GALCIT Rocket Research Project, 1936–38," in Durant and James, *First Steps,* 113–27; Diary 30–31 AUG 36; Millikan to RHG 15 AUG and 10 SEP 36, RHG to Millikan 1 SEP 36, Malina to RHG 1 OCT 36 and 14 OCT 37, and RHG to Malina 19 OCT 37, GP. As will be seen, von Kármán's attitude toward RHG changed considerably between his 1956 interview with ML and the publication of his memoirs, *The Wind and Beyond,* in 1967.

[24] Diary 30 OCT–2 NOV 35; RHG to CGA 28 SEP and 2 NOV 35, RHG Statement Regarding Rocket Deposited with the Smithsonian Institution 2 NOV 35, and CGA to RHG 2 OCT 35, GP; ML Int CAL, MLP.

[25] RHG to CGA 28 SEP 35 and 6 JAN 36, CGA to RHG 2 OCT 35, CAL to RHG 14 OCT 35, RHG to HFG 27 DEC 35, RHG Abstract for "Progress in the Development of Atmospheric Sounding Rockets" 31 DEC 35, RHG to WWA 6 JAN 36, and WWA to RHG 11 JAN 36, GP, which also includes Howard W. Blakeslee's AP account of 31 DEC 35; RMD 1 JAN 36, RDR 2 JAN 36, NYT 15 JAN 36.

[26] RHG, "Liquid-Propellant Rocket Development"; W.P. True to RHG 19 FEB 36, RHG to True 21 FEB 36, RHG to CGA 21 FEB 36, and CGA to Chief Justice Charles Evans Hughes and others 14 MAR 36, GP; RDR 16 MAR 36, NYT 17 and 22 MAR 36, *Worcester Gazette* 16 MAR 36, *New York Herald-Tribune* 16 MAR 36, *Time* 2 MAR 36, *American Magazine* MAR 36, *Worcester Telegram* 24 MAY 36, and Boston *Transcript* 10 AUG 36, among others; Tymeson, *Two Towers,* 171. The AP report by Stephen J. McDonough, which inspired most newspaper accounts, is reproduced in GP.

[27] HFG to RHG 7 and 21 JUL 38, RHG to HFG 14 JUL 38, and RHG Statement on Present Status of Rocket Work 14 JUL 38, GP; ML Int CAL, MLP.

[28] Diary SEP–DEC 34; CGA to RHG 17 SEP 34 and RHG to HFG 8 DEC 34, GP; ML Int Ljungquist and Bode, MLP.

[29] ML Int Ljungquist and Bode, MLP.

[30] Ibid., ML Int Truax and Int Mansur, MLP.

[31] ML Int Ljungquist and Bode, MLP.

[32] Diary SEP 34–OCT 35 (source of quotations except the one to WWA); RHG to WWA 12 MAR and 17 AUG 35; RHG to CGA 1 APR 35; CGA to RHG 6 APR 35; RHG to Charles F. Brooks 15 APR 35; RHG to HFG 2 APR, 10 JUN, 15 and 27 JUL, 17 AUG, and 29 OCT 35; HFG to RHG 11 APR, 21 JUL, 16 AUG, and 6 NOV 35; RHG Report on Rocket Development 19 APR 35; and RHG Outline of Proposed Rocket Work for 1935–1936 27 JUL 35, GP.

[33] Diary NOV 35–MAY 37; RHG to HFG 10 JAN, 24 MAR, 8 APR, 16 MAY, 1 JUN, 20 and 27 JUL, 7, 14, and 27 SEP, and 12 OCT 36; RHG to HFG 19 MAR, 3 APR, and 17 MAY 37; HFG to RHG 20 and 26 MAR, 11 MAY, 22 JUL, 16 SEP, and 5 OCT 36 and 11 MAY 37; CAL to RHG 6 APR and 29 DEC 36, 14 and 15 APR 37;

RHG Outline for Proposed Work for 1936–1937 8 APR 36; RHG to CAL 4 and 16 MAY, 7 and 15 SEP 36, 1 and 19 MAY 37; RHG Report on Rocket Work 7 SEP 36; and RHG Present Status of Rocket Work and Proposed Work for 1937–1938 19 MAR 37, GP.

[34] Diary JUN 37–SEP 38; RHG to HFG 29 AUG, 3 OCT, 15 NOV, and 4 DEC 37; RHG to HFG 3 and 7 FEB, 18 and 25 APR, 23 MAY, 17 JUN, 14 JUL, and 15 AUG 38; CAL to RHG 26 AUG and 22 SEP 37; CAL to RHG 19 JAN, 25 MAR, and 8 JUL 38; RHG to CAL 11 SEP and 8 OCT 37; RHG to CAL 3 FEB, 18 APR, 23 JUL, and 26 AUG 38; HFG to RHG 26 NOV 37, 7 APR, 13 MAY, and 10 JUN 38; RHG to NAA 22 NOV 37; William R. Enyart (Sec Contest Board NAA) to RHG 28 NOV and 21 DEC 37; Enyart to RHG 31 MAR, 6 and 12 APR, 6 JUN, and 8 SEP 38; RHG to Enyart 4 DEC 37; RHG to Enyart 3 JAN, 23 MAR, 2, 9, and 25 APR, 7, 16, 27, and 30 MAY, and 28 SEP 38; RHG to CGA 11 DEC 37; G. W. Lewis (Director Aeronautical Research NACA) to RHG 19 JAN, 3 MAR, 11 MAY, and 20 JUL 38; RHG to Lewis 24 JAN, 29 MAR, and 19 MAY 38; RHG Rocket Development, February 1937 to February 1938 7 FEB 38; RHG Proposed Sounding-rocket Casings 29 MAR 38; Enyart to Col. Pearson 6 APR 38; RHG to WWA 16 APR 38, RHG Proposed Rocket Work for 1938–1939 18 APR 38; Reports to NAA on Rocket Flight Test 20 APR 38; RHG Memorandum Regarding Rocket Using Three Types of Propulsion 26 MAY 38; C. S. Logsdon NAA to RHG 5 OCT 38; RHG to Logsdon 8 OCT 38; and RHG Outline of a Ten-Year Program on Rocket Development 15 AUG 38, GP. Correspondence and reports of H. H. Alden, Marjorie Alden, D. C. Pearson, John E. Smith, and Enyart, Pearson File, GCR. ML Int Ljungquist and Bode, Int Mansur, Int John E. Smith 10 JUL 56, MLP. CAL, *Wartime Journals*, 8. Anderson, *Orders of Magnitude*, 2–8.

[35] Diary 11 AUG–22 SEP 38; RDR 11 AUG 38.

NOTES TO CHAPTER TEN

[1] Diary 12–18 SEP 38; von Kármán, *Wind and Beyond*, 241–42; ML Int von Kármán, MLP; Hallion, *Legacy of Flight*, 198–204; Malina, "On the GALCIT Rocket Research Project"; RHG to HFG 1 NOV and 15 DEC 38, RHG to von Kármán 5 DEC 38, RHG Quarterly Report 15 DEC 38, and RHG to CAL 29 DEC 38, GP. There are conflicting accounts of what happened at the meeting, as ECG purged GP and RGP of pertinent materials, and because memories vary. In 1956 von Kármán remembered that RHG agreed that GALCIT could work on the chamber, while a decade later his memoirs said it was pumps. His memory also conflated the discussions at Falaise with later correspondence. RHG's letters show that they had agreed on chambers.

[2] Diary 10 NOV 38, 9 MAR 39.

[3] Diary 1938–42; RDR 17 OCT and 7 NOV 40; 16 and 23 JAN, 20 FEB, 6 and 20 MAR, 10 JUL, 16 and 30 OCT, 27 NOV, and 11 DEC 41; 4 JAN, 8 FEB, 22 MAR, 8 and 10 APR, 3, 4, and 20 MAY, and 21 JUN 42. Claude Simpson to RHG 14 OCT 40

and RHG to Harold Tuson 30 SEP 40, GP. George Bode recalled that he and RHG shared a passion for Laurel and Hardy movies. Int Bode, RDR 23 OCT 94. The credit report is excerpted in RHG's FBI file, AQ 105–82, Albuquerque Office, Form No. 1 dated 27 JAN 55. The file was obtained under FOIA by William P. Moyer of Portales, New Mexico, in 2001, and provided to the author; copies have since been provided to CUA and RMAC. Routine interviews with two former Soviet agents who had worked out of the main Soviet spy nest in New York, NOV–DEC 54, elicited vague recollections that they had heard of rocket experiments conducted by a "Professor Goddard" at "Rosedale," New Mexico, years earlier. They had wanted to investigate but the subject had been "dropped" by their superiors. The FBI did not know who RHG was, and the Albuquerque office sent an agent to Roswell. The FBI had no clue that the Germans were on RHG's trail in the 1930s, or there would have been a file on him earlier than 1955.

[4] RHG to WWA 20 and (Tel) 23 SEP 39 and WWA to RHG (Tel) 23 and (two letters) 25 SEP 39, WWAP; Murphy Int Roope, GCR. Melville taught at Clark from 1906 to 1948, and was registrar from 1914 to 1932. He held up Roope's Ph.D. following Webster's suicide. RHG argued that Roope should not be harmed by Webster's death, and Atwood agreed. Mott Linn, personal communication, 17 JUL 2001. Linn suggests that Melville resented taking orders from a youngster. After 1937 or so, virtually all RHG correspondence in WWAP relates to RHG's absence. Much of it was purged by ECG from GP and RGP.

[5] WWA to RHG 16 JAN and 30 MAY 39 and RHG to WWA 20 JAN 39, WWAP; RHG to HFG 20 JAN and 1 MAR 39 and HFG to RHG 24 FEB 39, GP; Diary 4–9 MAR 39.

[6] WWA to RHG 2 FEB 40 and RHG to WWA 10 FEB 40, GP.

[7] RHG to WWA 7 JUN, 10 JUL 40; RHG to WWA 17 MAR and 16 SEP 41; RHG to WWA 9 and 21 FEB and 20 and 27 JUL 42; WWA to RHG 18 JUL 40; WWA to RHG 10 MAR and 5 APR 41; WWA to RHG 16 FEB and 24 JUL 42; WWAP, some in GP. RHG to HFG 17 MAR 41 and RHG to CNH 20 JUL 42, GP. RHG to Roope 28 SEP and 19 OCT 41 and Roope to RHG 14 OCT 41, Roope Papers, CUA.

[8] ML Int Mansur, MLP; Diary SEP–OCT 38; Doolittle to RHG 11 and 25 OCT 38, 6 JUL 39 and 11 JUL 40; RHG to Doolittle 11 OCT 38, 10 APR, 6 MAY, and 27 JUN 39 and 11 JUL 40; Doolittle Notes on Visit to Mescalero Ranch 13 OCT 38; and RHG to HFG 1 NOV 38, GP.

[9] Diary 8–11 MAY 39; CAL, *Wartime Journals*, 199, 206; Berg, *Lindbergh*, 388–89; ML Int CAL, MLP; RHG to CAL 17 MAY 39, GP; RDR 10 and 11 MAY 39.

[10] Diary 20–22 JUL 39; ECG to George J. Fountaine 26 DEC 58, EGP; HFG to RHG 6 and 31 JUL 39, RHG to HFG 31 JUL 39, HFG to CAL 31 JUL 39, and HFG to WWA 31 JUL 39, GP; RDR 20 JUL 39.

[11] Diary 28 JUN 40; Mallon, *Men, Rockets*, 26–29; ML Int HAB 18 JUN 56, MLP; HAB, Address to Third Annual Goddard Memorial Dinner, Washington 17 FEB 60,

copy in EGP Box 5; HAB to RHG 17 JUN, 15 JUL, 9 AUG, 7 and 19 OCT, 14 and 18 NOV 40; RHG to HAB 22 JUN, 27 JUL, 14 and 24 OCT, 16 and 23 NOV, and 11 DEC 40, GP.

[12] RHG to CNH 21 MAY 38 and CNH to RHG 31 MAY 38, GP.

[13] Diary 1938–39; RHG to L. T. E. Thompson 17 MAR 39; RHG Quarterly Report on Rocket Work 1 JUN 39 (generator quote); RHG to HFG 1 JUN, 14 and 27 SEP, and 6 DEC 39 (launch quote); RHG to William G. Brombacher NBS 24 JUL 39; RHG to Enyart NAA 10 AUG 39; C. S. Logsdon NAA to RHG 16 AUG 39; Brombacher to RHG 11 SEP 39; RHG Quarterly Report 27 SEP 39, GP.

[14] Diary DEC 39–MAR 40; RHG Quarterly Report on Rocket Work 19 DEC 39; RHG to Logsdon 12 and 29 DEC 39; Brombacher to RHG 20 DEC 39; RHG to HFG 20 and 23 JAN, 4 and 25 MAR 40; HFG to RHG 22 and 23 JAN and 14 MAR 40; J. R. Hildebrand NGS to RHG 22 JAN 40; Gilbert Grosvener NGS to RHG 23 JAN 40; RHG to CAL 17 FEB and 4 MAR 40; CAL to RHG 27 FEB 40; RHG Quarterly Report on Rocket Work 4 MAR 40, GP.

[15] Diary 1940–41; HFG to RHG 29 MAR 40; RHG to HFG 7 APR, 19 JUN, and 23 DEC 40; RHG to HFG 3 and 12 AUG, 14 and 16 OCT 41; RHG Quarterly Report on Rocket Work 19 JUN 40; RHG to CNH 19 NOV 40; RHG Report on Rocket Work 23 DEC 40; HAB to RHG 14 and 19 JUL 40; RHG Report on Rocket Work 3 AUG 41, GP.

[16] A funding breakdown and discussion are provided at GP 3:1557n.

[17] Examples of coverage during this period include Randolph, "What Can We Expect?"; Ft. Worth Star-Telegram 5 JAN 39, Santa Fe New Mexico Sentinel 15 JAN 39, Boston Post 12 FEB 39, Science Service 2–8 OCT 39, Los Angeles Times 10 JUL 40 (weapons quote), St. Louis Globe-Democrat 27 DEC 40 (Blakeslee), RMD 27 DEC 40, and NYT 13 OCT 40 and 28 JUN 42; Rafferty, "Rockets to the Sky"; Nichols, "Flying by Fireworks"; and Wenstrom, Weather and the Ocean of Air.

[18] CAL, Wartime Journals, 210, 212; ML Int CAL, MLP; RHG to Franklin L. Fisher 26 APR 39, CAL to RHG 26 JUN 39, RHG to CAL 4 JUL 39, HFG to LaGroce 2 AUG 39, RHG to LaGroce 10 AUG 39, LaGroce to RHG 7 SEP 39, RHG to Kerbey 24 MAR 40 and 13 OCT 41, Kerbey to RHG 23 OCT 41, GP; Diary 12–13 FEB and 17–27 MAY 40. The text was published in 1996 as RHG, "Past Revisited."

[19] Diary 11–12 MAR 39; RHG to Sunderland 26 SEP 38, Sunderland to RHG 4 OCT 38, Lester P. Barlow (Martin Co.) to RHG 31 JAN and 16 and 28 FEB 39, RHG to Barlow 11 and 21 FEB and 1 MAR 39, RHG to HFG 25 MAR 39, RHG Quarterly Report on Rocket Work 25 MAR 39, RHG to CGA 31 JAN 39, CGA to RHG 10 FEB 39, Lieut. Col. J. A. Dorst (Engineers) to RHG 3 MAY 39, RHG to Dorst 6 MAY 39, and RHG to Thompson 24 FEB 39, GP.

[20] RHG to CGA 27 APR 40, CGA to RHG 30 APR 40, GP; Bush, Modern Arms.

[21] Diary 28 MAY–5 JUN 40; RHG to Thompson 12 MAY 40; RHG Memorandum on Accelerating Rocket 9 MAY 40; RHG Memorandum on Conference in Washing-

ton Regarding Rocket Developments, May 28, 1940, prepared 3 JAN 45; and Brett to RHG 24 JUN 40, GP. Hallion, *Legacy of Flight*, 176–77.

[22] RHG to CGA 7 JUN 40 and RHG to Thompson 3 SEP 40, GP; Burchard, *Rockets, Guns*, 79–103; WVB and Ordway, *History of Rocketry*, 100; Hallion, *Legacy of Flight*, 198–200. See also Millikan, *Guggenheim*; von Kármán, *Wind and Beyond*; and Malina, "On the GALCIT Rocket Research Project."

[23] RHG to Arnold, 9 JUN 40, RHG to Walsh 9 and 15 JUN 40, RHG to CAL 19 JUN 40, S.J. Cook to RHG 11 JUL 40, and RHG to Cook 17 JUL 40, GP.

[24] RHG to HFG 9 JUN 40 and HFG to RHG 21 JUN 40, GP.

[25] CNH to RHG 5 JUN, 20 and 26 JUL, and 16 AUG 40; RHG to CNH 9 JUN, 18 JUL, 13 and 25 AUG, and 9 OCT 40 and 10 JAN 42; Tolman to RHG 22 OCT 40; Bush to RHG 6 SEP and 24 OCT 40; and RHG to Tolman 30 OCT 40, GP.

[26] Fagen, *History of Engineering and Science in the Bell System: National Service*, 346–47; Green, Thomson, and Roots, *Ordnance Department Planning Munitions*, 328–30; Thomson and Mayo, *Ordnance Department Procurement and Supply*, 182–85; Brophy, Miles, and Cochrane, *Chemical Warfare Service Laboratory to Field*, 132–33, 194–95; WVB and Ordway, *History of Rocketry*, 93–96; Burchard, *Rockets, Guns*, 50–54; Hickman in *Congressional Recognition*, 50–54; Bush, *Modern Arms*, 25–31.

[27] Diary JUN–DEC 40; RHG to Capt. C.A. Ross AAC 11 JUN 40; RHG to Hap Arnold 11 JUN and 15 SEP 40; Brig. Gen. George H. Brett to RHG 24 JUN, 13 JUL, and 26 SEP 40; RHG to Brett 1 and 27 JUL and 14 OCT 40; CAL to RHG 19 JUL 40; RHG to HFG 27 JUL and 15 SEP 40; Rear Adm. W.F. Furlong BuOrd to Rear Adm. Harold G. Bowen Naval Research Laboratory 7 SEP 40; RHG to Cdr. F.W. Pennoyer BuAer 15 SEP and 9 OCT 40; RHG to CNH 15 SEP and 1 OCT 40; RHG to HAB 15 SEP 40, CNH to RHG 25 SEP and 4 OCT 40; Pennoyer to RHG 26 SEP 40; RHG to Thompson 24 OCT 40; Furlong to RHG 4 NOV 40; RHG to Furlong 10 SEP and 19 NOV 40; and RHG to Chief BuAer 14 NOV 40, GP. When Tolman again chided him for being secretive, RHG answered with his reasons "against presenting all my results to the CalTech, or to anyone else." Diary 10 SEP 40.

[28] Diary JAN–JUL 41; RHG to Thompson 7 MAR 41; HAB to RHG 27 MAR, 28 JUN, and 14 and 17 JUL 41; Lieut. Col. F.O. Carroll to RHG 28 APR 41; RHG to HFG 4 MAR, 11 and 17 MAY, and 17 and 21 JUN 41; RHG Quarterly Report on Rocket Work 4 MAR 41; RHG to HAB 12 MAY and 19 JUL 41; HFG to RHG 7 MAR, 16 MAY, 16 and 19 JUN, and 7 JUL 41; RHG to Carroll 24 JUN 41; RHG Memorandum for AAF Representatives 7 JUL 41, GP. Hay fever drove Kisk back to Massachusetts.

[29] Diary JUN–AUG 41; Truax, "Annapolis Rocket Motor Development, 1936–38," in Durant and James, *First Steps*, 295–301; RHG to George W. Lewis 23 JUN 41, Towers to RHG 14 JUL 41, RHG to HFG 19 and 25 JUL and 12 AUG 41, HFG to RHG 21 JUL 41, RHG to Towers 24 JUL 41, Capt. D.C. Ramsey BuAer to RHG 4 AUG

41, RHG to BuAer 6 and 13 AUG 41, HAB to RHG 8 AUG 41, RHG to HAB 12 AUG 41, and Fischer to RHG 25 AUG 41, GP.

[30] Diary AUG–DEC 41; HAB to RHG 31 AUG 41, RHG to HAB 15 SEP 41, RHG to CNH 7 SEP 41, Fischer to RHG 10 and 27 SEP 41, RHG to HFG 29 SEP and 5 OCT 41, RHG to Chief Experimental Engineering, Matériel Division AAF 4 OCT 41, and HFG to RHG 7 OCT 41, GP. HFG advanced RHG $3,000 in SEP and $7,000 in OCT 41. RHG repaid it to D&FGF 27 DEC 43. GP 3:1427n.

[31] HAB to RHG 25 OCT 41 and RHG to HAB 28 OCT 41, GP.

[32] Rear Adm. Ray Spear (Bu Supplies and Accounts) to RHG 28 OCT 41, Fischer to RHG 18 NOV 41, RHG to Fischer 3 NOV and 1 DEC 41, Capt. D. C. Ramsey to RHG 1 DEC 41, RHG to HFG 3 DEC 41, RHG to Ships Installation Section BuAer 24 DEC 41, GP.

[33] Diary SEP 41–MAR 42; Lowell N. Randall to William D. Ebie 30 MAR 89, Randall File, GCR; HFG to RHG 11 DEC 41, Cdr. L. C. Stevens BuAer to RHG 23 DEC 41, Fischer to RHG 5 JAN and 25 MAR 42, RHG to BuAer 24 DEC 41, RHG to Matériel Division AAF 24 DEC 41, RHG Progress Report 24 DEC 41, RHG to HFG 21 FEB 42, Capt. A. C. Miles BuAer to RHG 21 MAR 42, RHG to Miles 27 MAR 42, and RHG to Fischer 1 APR 42, GP.

[34] Diary MAR–JUN 42. Fischer to RHG 26 FEB and 7 JUN 42; RHG to Fischer 7 MAR, 27 MAY, 7 and 14 JUN 42; RHG to HFG 27 APR 42; Capt. Ralph Davison BuAer to RHG 21 MAY 42; and M. A. Norcross (Bu Supplies and Accounts) to RHG 25 JUN 42, GP.

[35] Diary JUN–JUL 42; RHG to Fischer 29 JUN 42, RHG to Curtis Hill Selective Service Board 2 JUL 42, RHG to Editor RDR 2 JUL 42, RHG to HFG 17 JUL 42, and RHG to Matériel Center AAF 17 JUL 42, GP; RDR 7 JUL 42, *Worcester Telegram* 8 JUL 42, NYT 8 JUL 42; ML Int Ljungquist and Bode, MLP; Marjorie Alden to ECG 1 NOV 42 and ECG to Marjorie Alden 9 NOV 42, EGP. Those making the move were Ljungquist, Glenn E. Loughner, Mansur, Arthur P. Freund, A. S. Campbell, Randall, Bode, and Alden. The latter returned to NMMI by 1 SEP 42, and rejoined RHG at Annapolis 24 DEC 42.

NOTES TO CHAPTER ELEVEN

[1] RHG to Hawley 4 AUG 42, GP; ECG to Alden 9 NOV 42, EGP.

[2] ML Int Mansur, Int William C. House nd JUL 56, Int Truax, Int Riffolt, Int Ljungquist and Bode, MLP; Heppenheimer Int Truax 1989, in Heppenheimer, *Countdown*, 32.

[3] Rocket work at Annapolis is summarized in the Diary, along with progress reports and correspondence in GP and RGP. On the AAF unit, see RHG to Lieut. Col. P. H. Dane 2 OCT and 3 DEC 42 and 23 MAR 43 and RHG to Col. Paul H. Kemmer, 3 FEB 43, GP. The manual is in RGP.

[4] Diary JUL–DEC 42; RHG Report on Assisted-Takeoff Tests of PBY 23 SEP 42, GP; ECG to Marjorie Alden 9 NOV 42, EGP; Gibbs and Jackman, *NMMI Centennial,* 150–51.

[5] Diary 1942–45; RHG to Warfel 4 JAN 43; RHG Memorandum on Pump-type Assist Units 3 NOV 44; RHG Shop Memorandum 20 DEC 44; RHG Performance Data on Gas-generator-driven Turbine-pump Unit Developed for BuAer 17 JAN 45; RHG Progress Report on Pump and Turbine Propulsion 31 JAN 45; RHG to Robert L. Earle 18 FEB 45; RHG Progress Reports on Pump and Turbine Propulsion 28 FEB, 31 MAR, 30 APR, and 31 MAY 45; GP. Ljungquist to ML 27 FEB 62, ML Int Truax, ML Int Mansur, and ML Int Ljungquist and Bode, MLP. Mallon, *Men, Rockets,* 21.

[6] RHG to Bleecker 23 JUL and 4 AUG 42 and RHG to HFG 7 AUG 42, GP; Diary 21 AUG 42.

[7] RHG to Hawley 11 SEP 42 and W. B. Woodson (Judge Advocate General of the Navy) to RHG 22 SEP 42, GP.

[8] Diary 4–29 OCT 42; Fisher to Bleecker 30 OCT 42 and Bleecker to Fisher 3 NOV 42, GP.

[9] RHG to Fisher 9 DEC 42, Fisher to RHG 16 DEC 42, RHG to HFG 4 and 19 JAN 43, HFG to RHG 15 JAN 43, RHG to WWA 9 JAN 43, RHG to Bleecker 21 JAN 43, and RHG to J. J. Murphy (Linde) 21 JAN 43, GP.

[10] RHG to HFG 15 FEB 43, Bleecker to HFG 22 FEB 43; Fisher to RHG 16 APR 43, and RHG to Breckinridge 9 JUN 43, GP; Diary 8–10 APR and 23 JUN 43.

[11] Rear Adm. E. M. Pace to RHG 26 JUN 43; Fisher to RHG 14 JUL and 6 AUG 43, and 29 MAY 44; H. B. Mallett to RHG 26 JUL and 9 OCT 43; RHG to Mallett 28 JUL, 28 SEP, 11 and 22 OCT, and 13 DEC 43; RHG to Bleecker 13 AUG, 30 SEP, and 3 NOV 43; F. A. Collins to RHG 26 NOV 43; RHG to HFG 27 DEC 43, GP. Diary 20 JUL, 28 SEP, and 11–22 OCT 43.

[12] Earle to RHG 7 JUN 44 and RHG to Earle 10 JUN 44, GP.

[13] F. A. McConnell (TIAA) to RHG 19 JAN 43, RHG to WWA 21 JAN 43, and WWA to RHG 25 JAN 43, WWAP.

[14] WWA to RHG 23 JAN and 10 JUL 43, RHG to WWA 1 FEB and 10 JUL 43, WWAP; RHG to HFG 25 JAN 43, GP.

[15] RHG to WWA 9 JAN, 21 APR, and 3 MAY 43 and WWA to RHG 16 and 27 APR 43, WWAP. ECG censored these letters for GP and RGP.

[16] WWA to RHG 2 and 11 AUG and RHG to WWA 4, 5, and 11 AUG 43, WWAP and GP.

[17] Koelsch, *Clark University,* 143–45; WWA to RHG 9 MAR and 16 APR 45, and RHG to WWA 15 MAR and 26 APR 45, GP; Murphy Int Roope, GCR. Diary 30 MAY–3 JUN 45 is devoid of any but a terse mention of the ceremony. Marigold's remarks appeared in *Worcester Telegram* 3 JUN 45, with other details on 6 MAY and 2 JUN. The citation is also reproduced in GP.

¹⁸ Diary MAR–NOV 44; HAB to RHG 4 FEB 44, HFG to RHG 19 FEB 44, RHG to HAB 15 MAR and 17 JUL 44, RHG to Branley Seeley (Patent Department C-W) and attachments 13 APR 44, RHG to Hawley 9 MAY 44, and RHG to F. A. Collins (D&FGF) 28 JUL 44, GP.

¹⁹ Diary DEC 44–APR 45. RHG to HAB 21 DEC 44; RHG to Robert L. Earle C-W 3 JAN, 8 and 28 FEB, 7, 10, and 26 APR 45; RHG to Col. G. W. Trichel 29 JAN 45; RHG to R. W. Porter GE 8 FEB 45, Ray Stearns GE to RHG 6 APR 45; A. W. Robinson (Project HERMES, GE) to RHG 7 APR 45, GP. Heppenheimer, *Countdown*, 33–37.

²⁰ Diary APR–JUN 45; RHG to Robert L. Earle C-W 27 APR 45, RHG to J. W. Long C-W 6 JUN 45, and RHG to G. W. Brady C-W 17 JUN 45, GP; Ljungquist to ML 27 FEB 62, ML Int Mansur, and ML Int Ljungquist and Bode, MLP. Those going to New Jersey included research engineer Alden; supervisor of construction Ljungquist; construction and design engineers A. P. Freund, Albert Campbell, and Henry Sachs; and test engineers Lawrence Randall, Charles Mansur, and George Bode; all had been with Goddard since before Pearl Harbor. Alden went to Wyoming, then on to Ohio State University. Ljungquist and Bode spent the rest of their careers at C-W, while Mansur and Randall went on to White Sands Missile Range, New Mexico. Freund, Campbell, and Sachs retired after the war.

²¹ RHG to Wilson 10 JUN 44, GP; Ley, *Rockets*. On the German rocket programs, see Neufeld, *Rocket and Reich*; King and Kutta, *Impact*; and sources cited below.

²² Vernon A. McLaskey to RHG 16 JUL 44, RHG to McLaskey 19 JUL 44, Theodore Adams to RHG 6 AUG 44, RHG to Adams 10 AUG 44, John N. Wheeler NANA to RHG 22 AUG 44, and RHG to Wheeler 24 AUG 44, GP. The patent was No. 1,980,266, issued 13 NOV 34.

²³ Neufeld, *Rocket and Reich*, 43–44, 147–49, and Dornberger, *V-2*, 94. See also the interrogation of Schmidt in Zwicky, *Report*, 15.

²⁴ RHG to HFG 15 DEC 44 and HFG to RHG 16 DEC 44, GP.

²⁵ Diary 28 DEC 44 and 4 JAN 45; RHG to HFG 28 DEC 44 and 9 FEB 45, RHG Comparison of the German V-2 Rocket and the Rocket Developed by R. H. Goddard in New Mexico 28 DEC 44, RHG to Kerbey 8 JAN 45, RHG to Gilbert Grosvenor 26 JAN 45, Henry C. Parker to RHG 1 MAR 45, RHG to Parker 5 MAR 45 and RHG to Warren S. Orton (Patent Attorney C-W) 14 MAR 45, GP; Kerbey, "Germany's Flying Vengeance Bomb." Kerbey's piece was widely reported, beginning with *Washington Times-Herald* and *Worcester Telegram* 19 JAN 45.

²⁶ ML Int Ljungquist and Bode, Int Mansur, Int CAL, MLP; Murphy Int Roope, GCR; and the literature cited below.

²⁷ *Science News Letter* 18 MAY and 24 AUG 46; *Time*, 20 MAY and 2 SEP 46; *Fortune* SEP 46; *New York Herald-Tribune* 6 OCT 46; NYT 14 FEB 47; *Science Digest* JUL 47; *Wall Street Journal* 24 JUN 47; Ross, "Where Were Our Rockets?"; *Saturday Evening Post* 18 OCT 47; *Worcester Telegram* 1 NOV 47 and 9 MAR 48; Fuller, "Uncle Sam

'Discovers' the Rocket"; *Readers Digest* APR 48; *New York Herald-Tribune* 22 APR 48; RMD 24 and 25 APR 48; D&FGF news release 14 DEC 48, reported in *Worcester Telegram* 14 and 15 DEC 48, *New York World-Telegram* 14 DEC 48, and NYT 14 DEC 48, among others; Thompson speech "Robert Goddard and the Weapons Program" to the 23 MAR 49 dinner of the fifth Joint Army-Navy Solid Propellant Conference, copy in EGP; RDR 1 FEB 49; *Christian Science Monitor* 9 APR 49; *St. Louis Post-Dispatch* 24 APR 49; *New York Journal-American* 24 APR 49; *El Paso Times* 25 APR 49; Thompson paper "Review of Rocket Systems, Outline of Discussions, Instrumentation Laboratory, MIT," 12 MAY 52, copy in EGP; Lehman, "Strange Story"; Grosvenor "Introduction," v–vi, to Abbot, *Adventures*, 101; Dornberger in NYT 6 JUN 58; Reinhardt and Kintner, *Haphazard Years*, 130, 159–60, 171; Boushey Address to Third Annual Goddard Memorial Dinner 17 FEB 60, copy in EGP; Davis, "Father of Rocketry"; Lehman, "Father of Modern Rocketry"; Parson, *Missiles*, 25; Caidin, *Overture to Space*, 44.

[28] Neufeld, *Rocket and Reich*; King and Kutta, *Impact*; Zwicky, *Report*; Kennedy, *Vengeance Weapons*; Brodie, *From Crossbow*, 230–32; Pocock, *German Guided Missiles*; Ordway and Sharpe, *Rocket Team*. On psychological and social aspects, see Speer, *Inside*, 362–76; Keegan, *Mask of Command*, 302; and Knightley, *First Casualty*, 327.

[29] Neufeld, *Rocket and Reich*, 267–79; Crouch, *Aiming for Stars*, 101–02; Ordway and Sharpe, *Rocket Team*, 271–90; Lasby, *Project Paperclip*; McGovern, *Crossbow and Overcast*, 145–50; Bar-Zora, *Hunt for German Scientists*; Hunt, *Secret Agenda*; Piszkiewicz, *Nazi Rocketeers*; Bowe, *Paperclip Conspiracy*; Zwicky, *Report*, which contains interrogation reports; File "Peenemunde East Through the Eyes of 500 Detained at Garmisch," copy in NASA History Office Archives; Simon to ECG 25 MAR 57 and Hamill to ECG 10 APR 57, EGP.

[30] ML Int WVB 30 JUL 56, MLP.

[31] William W. Cook to ECG 1 APR 58, EGP.

[32] ML Int WVB, Int CAL, Int Mansur, Int Ljungquist and Bode, Int Truax, ML to ECG 31 MAR 57, MLP; rocket literature collection in EGP Box 55; THM, 389–90.

[33] Dornberger, *V-2*, 273; Neufeld, *Rocket and Reich*, 267–79; Michel, *Dora*, esp. ch. 14.

[34] Ley, "Introduction" to Dornberger, *V-2*, x; Ley, *Rockets*, 104, 109; Dornberger quoted in Caidin, *Overture to Space*, 44–45; Dornberger, "The V-2 Rocket"; Dornberger, "The German V-2"; Bergaust, *Rocket City*, 11–12. On later admissions, see Bergaust, *Reaching*, 44; and NASM Oral History interviews of surviving German rocketeers; as examples Int Konrad K. Dannenberg 7 NOV 89, Int Gerhard Reisig 27 JUN 85, and Int Arthur Rudolph 4 AUG 89, available in summary at the website www.nasm.edu/nasm/dsh/peenintro.html. American historians have generally swallowed the German story whole. See, e.g., Neufeld, *Rocket and Reich*, 7; Swenson, Grimword, and Alexander, *This New Ocean*, 15–16; Williams and Epstein, *Rocket Pioneers*, 204–31; Ley, *Rockets, Missiles and Space Travel*, 202–31.

[35] Bergaust, *Wernher von Braun*, 518–20; Piszkiewicz, *Wernher von Braun*, 29, 66–67; WVB and Ordway, *History*, 44–56; WVB, "Reminiscences of German Rocketry," in Clarke, *Coming of Space Age*, 33–55; L. T. E. Thompson to ECG 7 MAR 57, EGP; ML Int WVB and WVB to ML 8 MAY 63, MLP. Interestingly the format of WVB, *Mars Project*, follows that of RHG's 1919 SI publication.

[36] Neufeld, *Rocket and Reich*, 53, 67, 129–30; Kahn, *Hitler's Spies*, 77; Farago, *Game of Foxes*, 39–41; Nash, *Spies*, 131. Surviving documents on this espionage are on National Archives Microfilm T-78 and NASM Archives FE Microfilm FE-349-366; thanks to Mike Neufeld, NASM, for providing copies. WVB denied, ML Int WVB, MLP, knowing of any spying, but his initials are on many of the documents.

[37] Diary 1942–45, esp. 15 OCT 42 and 10 OCT 44; ECG to Marjorie Alden 9 NOV 42, EGP; RHG to Harold Tuson 2 MAR and 28 JUN 43 and 2 MAR 45, GP.

[38] Diary 24–25 DEC 42, 4 APR, 23 MAY, 23 JUN, 19–26 SEP, and 24–25 DEC 43; 22–25 DEC 44; and 1–5 APR 45. RHG to CGA 10 AUG 44, GP. Lindbergh performed a number of jobs for the army and navy during the war and after, including unofficial combat missions. Berg, *Lindbergh*, ch. 15 and 16.

[39] Diary 1942–45; NYT 8 and 9 JAN 44, *Time* 4 SEP 44, *Worcester Telegram* 25 SEP 44, *Worcester Gazette* 13 NOV 44.

[40] GEP to SecSI 24 AUG 44; A. Wetmore to GEP 31 AUG 44; RHG to GEP 4 and 13 SEP, 31 OCT, 14 NOV, and 11 DEC 44 and 8 MAR 45; GEP to RHG 5 SEP, 9 NOV, and 7 and 14 DEC 44; RHG to Wetmore 23 OCT 44; RHG to Roope 11 DEC 44; Roope to RHG 14 DEC 44, GP. RHG, *Rockets*.

[41] Diary FEB 43–17 JUN 45; ML Int Ljungquist and Bode and Int Mansur, MLP; RHG to Harold Tuson 28 JUN 43, GP.

[42] HFG to RHG 23 JUN and 27 JUL 45 and WWA to RHG 6 JUL 43, GP.

[43] ML Int Ljungquist and Bode, MLP.

[44] *Worcester Telegram* 11, 12, 13, and 15 AUG 45; *New York Sun* 11 AUG 45; *New York Journal-American* 11 AUG 45; *Philadelphia Bulletin* 11 AUG 45; *Baltimore Sun* 11 AUG 45; *Kansas City Star* 11 AUG 45; NYT 11 and 12 AUG 45; *New York World-Telegram* 11 AUG 45; *New York Herald-Tribune* 12 AUG 45; *Newsweek* 20 AUG 45; *Time* 20 AUG 45; RDR 12 AUG 45; RMD 15 AUG 45; *WPI Journal* SEP 45; *Science* 23 NOV 45; and many others.

[45] RMD 16 AUG 45.

NOTES TO CHAPTER TWELVE

[1] Leckie, *Elizabeth Bacon Custer*.

[2] She evidently sent typescripts of every speech to RMAC, and GCR houses them; they are all virtually identical. On the speeches in New York, see *Worcester Telegram* 1 and 16 DEC 46.

[3] HFG to ECG 28 AUG and 8 NOV 45, ECG to HFG 11 SEP 45 and 11 OCT 46, and ECG to F. A. Collins (Sec D&FGF) 13 and 24 NOV 45, EGP. See also Dill, "God-

dard's Home on the Range." Mescalero Ranch still exists, extensively remodeled on the interior but on the exterior much as the Goddards left it, except that white has replaced the brown paint of the 1930s.

[4] Berg, *Lindbergh*, 471–72; Milton, *Loss of Eden*, 417; CAL, *Autobiography*, 349–50; Mosley, *Lindbergh*, 343–44, 352–53; *Science News Letter* 11 MAY 46; NYT 23 FEB and 2 MAR 46.

[5] ECG to Hawley 17 JAN 46, ECG to HAB 2 APR 46, and HFG to ECG 20 NOV 46 and 8 APR 47, EGP.

[6] HFG to GEP 8 APR 47, HFG to ECG 8 APR and 10 JUL 47, Hawley to ECG 10 APR 47, and Robert L. Earle C-W to HFG 21 JUN 47, EGP.

[7] HFG Memorandum to Directors 14 JUL 47 and ECG to HFG 23 JUL 47, RGP.

[8] ECG Memorandum of Conversations Concerning the Goddard Patents nd DEC 47, McCarthy to HFG 17 DEC 47, and GEP Memorandum to Directors 6 JUN 48, EGP.

[9] HFG introduction to RHG, *Rocket Development*, xi–xii.

[10] HFG to ECG 18 DEC 50 and the heavy correspondence between them in EGP.

[11] WVB and Ordway, *History*, 44–56; Bergaust, *Wernher von Braun*, 518–20; Piszkiewicz, *Wernher von Braun*, 67. On the government using secrecy classification to hamper the case, see GEP Report on the Goddard Patent Situation and Related Matters 16 APR 52, EGP.

[12] ECG to HAB 11 MAR 51, ECG Report on Goddard Patent Exploration 23–26 APR 51, HFG to Donal F. McCarthy 7 SEP 51, ECG to F. A. Collins 23 OCT 51, Murchison to HFG 29 OCT 51, HFG to ECG 3 DEC 51, ECG to HFG 7 DEC 51, ECG to George Fountaine Sec D&FGF 22 FEB 56, GEP Report to Foundation on the Goddard Patent Situation 16 APR 52, EGP. Murchison and Beauregard were of the firm of Stockton, Ulmer and Murchison, Jacksonville, Florida, and Washington, D.C. If successful they would receive 25 percent of any net profits up to $200,000, 10 percent above that, for ten years.

[13] HFG to ECG 23 JAN 56, Beauregard to HFG 26 JAN 56, ECG to HFG 15 FEB 56, ECG to Fountaine 22 FEB 56, Fountaine to ECG 28 FEB 56, ECG to HAB 1 JUL 57, HAB to ECG 17 AUG 57, and ECG's quarterly reports, EGP.

[14] Hawley to ECG 24 AUG 57 and ECG to Hawley 20 AUG 57, ECG.

[15] ECG to HAB 9 JAN and 25 JUN 60, EGP.

[16] Fountaine to ECG 29 JUN 60; General Councel DepNav to Beauregard 22 JUN 60; Beauregard to Merritt H. Steger 24 JUN 60; Patent License and Release Contract Executed Between ECG and D&FGF and the U.S. Government Represented by the USA, USN, USAF, and NASA 24 JUN 60; ECG to HAB 9 JUL 60; Beauregard to ECG 15 and 19 JUL 60; NASA News Release 5 AUG 60; UPI Release 5 AUG 60; EGP. *Boston Traveler* 5 AUG 60; *Boston American* 5 AUG 60; *New York Journal-American* 5 AUG 60; *Boston Herald* 5 AUG 60; AP accounts in *Newsday* 5 AUG 60, *New York Herald-Tribune* 5 AUG 60, and *Washington Star* 5 AUG 60; Davis, "Father of

Rocketry." The government acknowledged infringement of Patents No. 2,397,657, "Control Mechanism for Rocket Apparatus," filed 23 JUN 41, issued 2 APR 46, and No. 2,397,659, also "Control Mechanism for Rocket Apparatus," filed 29 JUL 42, issued 1 APR 46. The former described a method for operating a pump-and-turbine rocket motor for a single run. The latter is an improvement on that, by controls for successive runs—a throttle-controlled rocket motor.

[17] ECG to Hawley 23 JUL 60, EGP.

[18] ECG to HFG 1 MAY 62, EGP.

[19] ECG to HAB 8 FEB 46 and ECG's extensive reports to HFG, EGP; ECG "Introduction" to GP, 1:xv–xvi; RHG, *Rocket Development*.

[20] ECG to HFG 4 MAY and 3 JUL 53, HFG to ECG 29 JUN 53, Price to Roger W. Straus, Jr., nd, HFG to Straus 29 JUN 53, ECG to HAB 15 MAY 55, and extensive files on the Price and Lehman projects in EGP.

[21] HFG to CNH 30 AUG 56, ECG to HFG 18 MAY 55 and 17 NOV 56, HFG to ML 23 MAY 55, ECG Notes for Lunch with HFG 5 NOV 56, ECG Quarterly Reports to HFG 1955–58, EGP; ML Int CAL, and CAL file, MLP; ML, "Secrets of White Sands" and "Strange Story." EGP has literally reams of ECG's notes and memoranda on the Lehman project.

[22] CAL to ML 16 and 20 APR and 23 NOV 58, MLP; ECG to HFG 5 AUG 58, EGP.

[23] ECG Memorandum of Phone Call from CAL 11 OCT 59; ECG to CAL 24 OCT 59; ECG to CAL 30 APR, 11 AUG, and 8 DEC 62; CAL to ECG 1 DEC 60; CAL to ECG 17 JUL 61; CAL to ECG 25 APR, 5 MAY, 25 AUG, and 17 DEC 62; CAL to ECG 11 FEB 63; and ECG to HFG 11 AUG 62, EGP. CAL to ML 12 APR 60; CAL to ML, 30 SEP 62; CAL to ML 20 FEB, 24 JUN, and 17 NOV 63; HFG to ML, 4 MAR 63, MLP.

[24] THM, xiii.

[25] ECG to HFG 7 MAY 58 and 14 JUL 59 and HFG to ECG 6 JUL 59, EGP; GEP, "Introduction," GP 1:ix.

[26] ECG Quarterly Reports to HFG 1958–70, ECG to HFG 4 MAY 60, 11 AUG 62, 23 AUG 63, 30 MAY 64; ECG to HFG 20 FEB, 15 MAY, 2 JUL, 14 AUG, and 18 NOV 65; ECG to HFG 18 FEB and 2 JUN 67; ECG to HFG 2 FEB, 7 MAY, and 6 NOV 68; ECG to HFG 5 FEB, 27 MAY, and 4 NOV 69; ECG to HFG 12 FEB, 27 JUL, and 20 NOV 70; ECG to Fountaine 2 FEB 63 and 13 JUN 70; HFG to ECG 4 JUN 64; and ECG to Henry C. Beauregard 20 NOV 65, EGP.

[27] CAL to ECG 30 MAY 70, EGP; NYT *Book Review* 2 AUG 70.

[28] RDR 15 DEC 63.

[29] D&FGF Press Release on Exhibit Opening 21 APR 48, Guggenheim File, GCR. *Worcester Telegram* 3, 15, and 21 APR 48 and 22 MAR 49; NYT 22 APR and 14 DEC 48; ECG to HFG 29 SEP 49; ECG to CAL 5 AUG 50; and Jenne Biltz (CAL Sec) to ECG 18 SEP 50, EGP. Hallion, *Legacy of Flight*, 177–79, 183. Mosley, *Lindbergh*, 345.

[30] NYT 2 DEC 50 (quotation); "Goddard Memorial Exhibit Opens"; "Robert H. Goddard Memorial Exhibit"; and the large AIAA File, GCR.

[31] ECG to HAB 6 SEP 49, EGP.

[32] ECG to HAB 11 MAR 51, EGP. *Worcester Telegram* 2 JAN, 18 FEB, and 11 APR 49; *Worcester Gazette* 3 JAN 49; *Boston Traveler* 3 JAN 49; RDR 1, 13, 17, 20, and 22 FEB, 11 and 29 MAR, 10 and 11 APR 49, 4 SEP 50, 9 JAN 53, and 4 SEP 55; *Boston Globe* 18 FEB 49; RMD 18 and 22 FEB, 29 MAR, and 3 APR 49; *El Paso Times* 18 FEB and 7 APR 49; *Albuquerque Journal* 18 FEB 49; ECG to Tom Messer RMAC 7 APR and 25 MAY 49; Messer to ECG 21 MAY 49 and extensive correspondence in ECG 1945–49 File and ECG 1950–54 File, GCR; RMAC *Bulletin* Winter 1953–54.

[33] ECG to HAB 10 FEB 58, HFG to ECG 23 APR 58, and other correspondence in EGP; extensive correspondence in ECG 1955–59 File and Guggenheim File, GCR; Hallion, *Legacy of Flight*, 183; Hurd to Paul Horgan 21 MAR 59 in Hurd, *My Land*, 381. In 1954 the Roswell Chamber of Congress asked ECG to intercede with CAL to have Roswell selected for the USAF Academy. She did. ECG to CAL 21 APR 54, EGP. CAL was on the site-selection committee, but the nod went to Colorado Springs. Berg, *Lindbergh*, 496. In 1968 the American Institute for Aeronautics and Astronautics (AIAA), successor to IAS and ARS, formally transferred title to the artifacts to RMAC. Correspondence and other documentation in AIAA File, GCR.

[34] Heavy local newspaper coverage, including RDR 26 APR 59, in RHG Morgue File, HSSENM; Exhibit History, Dedication File, GCR; HFG to David Gebhard RMAC 10 FEB 59, Guggenheim File, GCR; "Robert Hutchings Goddard Space and Rocket Collection"; "Von Braun Tribute to Goddard"; "Rocket Museum"; and NYT 14 SEP 59 for another instance of WVB's "boyhood hero" speech. On German audiences, see Rauschenbach, *Hermann Oberth*. A Paul Horgan wing was dedicated in the same ceremonies.

[35] ECG to HAB 15 FEB 62 and ECG to HFG 10 AUG and 7 NOV 64, EGP; RDR 18 FEB 62, 15 DEC 63, 3–10 OCT 64, 20 and 26 SEP 65. The eight-cent airmail stamp was designed by Robert J. Jones of the Bureau of Engraving and Printing, and the design was displayed for a time at Clark University. It showed Goddard's portrait to the left, with an Atlas taking off to the right. NYT 27 SEP 64. First day of issue was 8 OCT 64.

[36] Exhibit History Miscellaneous File, ECG Files 1960–79, and Guggenheim File, GCR; Huffman, *60 Years*, 75–77; Fleming and Huffman, *Roundup*, 90, 192. The real angel of the project was local contractor and engineer Harold E. DeShurley, who did the research and design, supervised the construction, and paid for much of it.

[37] ECG File 1970–79, GCR; RDR 19 AUG 73 and 11 MAR 75. The display was still in place in 2002, although Schmitt's coveralls replaced the spacesuit in 2000.

[38] NYT 10 AUG 55; ECG to HFG 14 AUG 55, EGP.

[39] Hill Award documentation in AIAA File, GCR; NYT *Magazine* 2 FEB 58; Eisenberg, "Tragic Case" (for which ECG even provided NGS photographs); Rothman,

"Father of Rocket"; Dryden to ECG 31 JUL 59 and ECG to HFG 11 AUG 59, EGP.

[40] ECG to HAB 11 JUL and 4 AUG 59, HAB to ECG 24 JUL 59, and ECG to HFG 14 AUG 65 and 14 NOV 66, EGP; NYT 29 JUN 60, 27 JUL 64 and 14 MAR 65; Hallion, *Legacy of Flight*, 165–66.

[41] *Tech News* 29 APR 59; *New York World-Telegram* 2 JUL 60; *Worcester Telegram* 6 JUN 64; NYT 8 SEP and 14 OCT 65, 20 MAY and 24 AUG 69; ECG to HAB 20 MAR 61 and ECG's quarterly reports to D&FGF through the 1960s, EGP. Details of the various dedications in AIAA File and ECG Files, GCR.

[42] ECG to CAL 24 MAR 65 and CAL to ECG 1 JAN 69, EGP; Berg, *Lindbergh*, 537; Mosley, *Lindbergh*, 371; Anne Lindbergh, *Earth Shine*, 13; Borman, *Countdown*, 196–97; NYT 24 MAY 69.

[43] Examples include Dewey, *Robert Goddard*; Moore, *Robert Goddard: Pioneer Rocket Boy*; Lomask, *Robert H. Goddard: Space Pioneer*; Quackenbush, *The Boy Who Dreamed of Rockets: How Robert Goddard Became the Father of the Space Age*; Farley, *Robert H. Goddard*; Coil, *Robert Hutchings Goddard: Pioneer of Rocketry and Space Flight*; Maurer, *Rocket! How a Toy Launched the Space Age*; Streissguth, *Rocket Man: The Story of Robert Goddard*; and Schaefer, *Robert Goddard*.

[44] Von Kármán, *Wind and Beyond*, 240–42.

[45] NYT 17 MAY 64; "New Trustees Elected"; E&FGF Files 1970–76, EGP; RDR 16 MAR 76, 7 JUN 82, 16 MAR 86, 7 APR 83, and 6 and 16 MAR 2001; *Worcester Telegram* 6 JUN 82; *Boston Globe* 6 JUN 82; and *Worcester Gazette* 6 JUN 82. She left RMAC $5,000 in her will. Robert M. Flint to RMAC 6 JUN 83, ECG Miscellaneous File, GCR. She left the house to both WPI and Clark University, but neither wanted it. As of 2001, the house was undergoing nomination to the National Register of Historic Places. Barbara S. Berka to the author 17 APR 2001.

NOTES TO EPILOGUE

[1] A contemporaneous example of this is Smith, *We Came in Peace*, published immediately after the first moon landing, and which includes a chronology that gives each his due.

[2] CAL to ECG 15 NOV 63, EGP, reproduced in "Our Professor Goddard."

BIBLIOGRAPHY

NOTE: NEWSPAPER AND NEWSLETTER ARTICLES ARE CITED IN THE CHAPTER NOTES ONLY,
WITH A FEW EXCEPTIONS; THE SAME IS TRUE OF ARTICLES INCLUDED IN BOOKS CITED IN
THE BIBLIOGRAPHY.

ARCHIVAL AND MUSEUM RESOURCES

Archives, libraries, and other institutions consulted during research for this volume are listed in the Acknowledgments. Of individual collections, the most productive were the Robert Goddard Papers, Esther Goddard Papers, Wallace W. Atwood Papers, Milton Lehman Papers, Percy W. Roope Papers, and photograph collection, Clark University Archives; the Robert H. Goddard Collection, Exhibit History Files, Esther Goddard Correspondence, and photograph collections, Roswell Museum and Art Center; the General Collections, Morgue Files, and photograph collections, Historical Society for Southeastern New Mexico; the Archives of the History Office and the Photograph Index Files, including the NIX and GRIN databases, of the News and Imaging Branch, Headquarters, National Aeronautics and Space Administration. The largest and most instructive collections of Goddard rockets and other artifacts are housed at the Roswell Museum and Art Center and the National Air and Space Museum.

PUBLICATIONS OF ROBERT H. GODDARD

"Bachelet's Frictionless Railway at Basis a Tech Idea." *Journal of the Worcester Polytechnic Institute*, November 1914.

"The Development of the Rocket for the Investigation of the Upper Atmosphere." Institute of Radio Engineers, Philadelphia Section, *Proceedings*, February 1930.

"The High-Altitude Rocket." *Monthly Weather Review*, February 1924.

"High-Velocity Jets of Gas Applicable to Rockets." Paper presented to the Annual Meeting of the American Physical Society, Washington, D.C., 26 April 1919.

"How My Speed Rocket Can Propel Itself in a Vacuum." *Popular Science Monthly*, September 1924.

"In Memoriam—Arthur G. Webster." *Clark University Library Publications*, March 1924.

"The Limit of Rapid Transit." *Scientific American*, 20 November 1909.

"Liquid-Propellant Rocket Development." *Smithsonian Miscellaneous Collections* 95, no. 3, 1936.

"A Method of Reaching Extreme Altitudes." *Smithsonian Miscellaneous Collections* 71, no. 2, 1919.

"A Method of Reaching Extreme Altitudes." *Nature*, 26 August 1920.

"A New Invention to Harness the Sun." *Popular Science Monthly*, November 1929.

"A New Rocket Plane for Stratosphere." *The New York Times*, 27 September 1931.

"A New Turbine Rocket Plane for the Upper Atmosphere." *Scientific American*, March 1932.

"On High-Altitude Research." *Science*, 6 February 1920.

"On Interference Colors in Clouds." *Science*, 19 December 1913.

"On Mechanical Force from the Magnetic Field of a Displacement Current." Paper presented at the Annual Meeting of the American Physical Society, Cambridge, Massachusetts, 27 April 1912.

"On Ponderomotive Force Upon a Dialectric Which Carries a Displacement Current in a Magnetic Field." *Physical Review*, August 1914.

"On Recording Apparatus for Meteorological Research with Rockets." *Science*, 10 December 1920.

"On Some Anomalous Electrical Conductors." *Journal of the Worcester Polytechnic Institute*, November 1908.

"On Some Peculiarities of Electrical Conductivity Exhibited by Powders and a Few Solid Substances." *Physical Review*, June 1909.

On the Conduction of Electricity at Contacts of Dissimilar Solids. Ph.D. dissertation (Physics), Clark University, 1911. Abstracted in *Physical Review*, June 1912, and *Electrician*, 16 August 1912.

"On the Efficient Utilization of Solar Energy." *Journal of the Optical Society of America*, July 1929.

"On the Present Status of the High-Altitude Rocket." Paper presented at the Annual Meeting of the American Physical Society, Cincinnati, Ohio, 27–30 December 1923.

"On the Production of Rare Gases in Vacuum Tubes." *Science*, 7 May 1915.

"The Past Revisited: An Unpublished Account." Draft for *National Geographic Magazine*, 1939, published in *Quest, the Magazine of Spaceflight*, no. 2, 1996.

"The Possibilities of the Rocket in Weather Forecasting." National Academy of Sciences *Proceedings*, August 1920.

"Progress in the Development of Atmospheric Sounding Rockets." *Bulletin of the American Meteorological Society*, February 1936.

"Robert H. Goddard: An Autobiography." *Astronautics*, April 1959.

"The Rocket Method." *Journal of the Worcester Polytechnic Institute*, April 1920.

"Some Aspects of Rocket Engineering." *Tech Engineering News*, April 1938.

"A Study of Crystal Rectifiers." Paper presented at the Annual Meeting of the American Physical Society, Washington, D.C., 27–30 December 1911.

"A Study of Crystal Rectifiers." *Physical Review*, February 1912.

"That Moon Rocket Proposition: Its Proponent Says a Few Words in Refutation of Some Popular Fallacies." *Scientific American*, 26 February 1921.

"Theory of Diffraction." M.A. thesis (Physics), Clark University, 1910.

"The Use of the Gyroscope in the Balancing and Steering of Airplanes." *Journal of the Worcester Polytechnic Institute*, November 1907.

"The Use of the Gyroscope in the Balancing and Steering of Airplanes." *Scientific American Supplement*, 29 June 1907.

"Whither Go the Rockets?" *Journal of the Worcester Polytechnic Institute*, March 1936.

Robert H. Goddard and Albert C. Erickson. "Periodically Interrupted Flow through Air Passages." *Journal of the Aeronautical Sciences*, May 1935.

POSTHUMOUS COMPILATIONS OF WRITINGS BY ROBERT H. GODDARD

The Papers of Robert H. Goddard, Including the Reports to the Smithsonian Institution and the Daniel and Florence Guggenheim Foundation. Edited by Esther C. Goddard and G. Edward Pendray. 3 volumes. New York: McGraw-Hill, 1970.

Rocket Development: Liquid-Fuel Rocket Research, 1929–1941. Edited by Esther C. Goddard and G. Edward Pendray. New York: Prentice-Hall, 1948, 1961.

Rockets. New York: American Rocket Society, 1946.

BOOKS

Abbot, Charles G. *Adventures in the World of Science*. Washington, D.C.: Public Affairs Press, 1958.

———. *Great Inventions*. Washington, D.C.: Smithsonian Institution, 1932.

———. *The Sun and the Welfare of Man*. Washington, D.C.: Smithsonian Institution, 1929.

Adams, Carsbie C. *Space Flight: Satellites, Spaceships, Space Stations, and Space Travel*. New York: McGraw-Hill, 1958.

Aftermath, 1908. Worcester, Massachusetts: Worcester Polytechnic Institute, 1908.

Aitken, Hugh G.J. *The Continuous Wave: Technology and American Radio, 1900–1932*. Princeton: Princeton University Press, 1985.

Alway, Peter. *Retro Rockets: Experimental Rockets, 1926–1941*. Ann Arbor, Michigan: Saturn Press, 1996.

———. *Rockets of the World*. 3rd Edition. Ann Arbor, Michigan: Saturn Press, 1999.

Anderson, Frank W., Jr., *Orders of Magnitude: A History of NACA and NASA, 1915–1976*. NASA History Series, SP-4403. Washington, D.C.: National Aeronautics and Space Administration, 1976.

Archuleta, Edward J., editor. *Official New Mexico Blue Book, 1997–1998*. Santa Fe: Office of the Secretary of State, 1997.

Atkinson, Brooks. *Broadway*. New York: Macmillan, 1970.

Atwood, Wallace W. *The Physiographic Provinces of North America*. Boston: Ginn, 1940. Reprinted New York: Blaisdell, 1964.

———. *The Rocky Mountains*. New York: Vanguard, 1945.

Atwood, Wallace W., and Harold C. Bryant. *Research and Education in the National Parks*. Washington, D.C.: Government Printing Office, 1932.

Bainbridge, William Sims. *The Spaceflight Revolution: A Sociological Study*. Seattle: University of Washington Press, 1976.

Baker, David. *The History of Manned Space Flight*. New York: Crown, 1981.

———. *The Rocket*. New York: Crown, 1978.

————. *Spaceflight and Rocketry: A Chronology*. New York: Facts on File, 1996.

Bar-Zorah, Michel. *The Hunt for the German Scientists*. New York: Hawthorn, 1967.

Barzun, Jacques. *From Dawn to Decadence: 500 Years of Western Cultural Life, 1500 to the Present*. New York: HarperCollins, 2000.

Berg, A. Scott. *Lindbergh*. New York: Berkley, 1998.

Bergaust, Erik. *Reaching for the Stars*. Garden City, New York: Doubleday, 1960.

————. *Rocket City U.S.A.: From Huntsville, Alabama, to the Moon*. New York: Macmillan, 1963.

————. *Wernher von Braun*. Washington: National Space Institute, 1976.

The Bhagavad-Gita: Krishna's Counsel in Time of War. Translated by Barbara Stoler Miller. New York: Bantam, 1986.

Bilstein, Roger E. *Flight in America, 1900–1983: From the Wrights to the Astronauts*. Baltimore: Johns Hopkins University Press, 1984.

————. *Orders of Magnitude: A History of NACA and NASA, 1915–1990*. NASA History Series, SP-4406. Washington, D.C.: National Aeronautics and Space Administration, 1989.

Bordman, Gerald. *The Oxford Companion to American Theatre*. New York: Oxford University Press, 1984.

Borman, Frank, with Robert J. Serling. *Countdown: An Autobiography*. New York: Morrow, 1988.

Bowe, Tom. *The Paperclip Conspiracy*. Boston: Little, Brown, 1987.

Brodie, Bernard, and Fawn M. Brodie. *From Crossbow to H-Bomb*. Revised and enlarged edition. Bloomington: Indiana University Press, 1973.

Brooke, John L. *The Heart of the Commonwealth: Society and Political Culture in Worcester County, Massachusetts, 1713–1864*. New York: Cambridge University Press, 1989.

Brophy, Leo P., Wyndham D. Miles, and Rexmond C. Cochrane. *The Chemical Warfare Service: From Laboratory to Field*. United States Army in World War II: The Technical Services. Washington, D.C.: Department of the Army, 1959.

Brown, Richard D., and Jack Tager. *Massachusetts: A Concise History*. Amherst: University of Massachusetts Press, 2000.

Bubnov, I. N. *Robert Goddard, 1882–1945*. Moscow: Nauka, 1978.

Burchard, John E., editor. *Rockets, Guns, and Targets: Rockets, Target Information, Erosion Information, and Hypervelocity Guns Developed During World War II by the Office of Scientific Research and Development*. Boston: Little, Brown, 1948.

Burgess, Eric. *Frontier to Space*. New York: Macmillan, 1955.

Burrows, William E. *This New Ocean: The Story of the First Space Age*. New York: Random House, 1998.

Bush, Vannevar. *Modern Arms and Free Men: A Discussion of the Role of Scientists in Preserving Democracy*. New York: Simon and Schuster, 1949.

Butterfield, Herbert. *The Origins of Modern Science, 1300–1800*. Revised edition. New York: Free Press, 1957.

Caidin, Martin. *Countdown for Tomorrow: The Inside Story of Earth Satellites, Rockets and Missiles and the Race Between American and Soviet Science*. New York: Dutton, 1958.

———. *Overture to Space.* New York: Duell, Sloan and Pearce, 1963.

———. *Rockets and Missiles, Past and Future.* New York: McBride, 1954.

———. *Vanguard! The Story of the First Man-Made Satellite.* New York: Dutton, 1957.

Carter, Paul A. *Politics, Religion, and Rockets: Essays in Twentieth-Century American History.* Tucson: University of Arizona Press, 1991.

Chew, Joe. *Storms Above the Desert: Atmospheric Research in New Mexico, 1935–1985.* Albuquerque: University of New Mexico Press, 1987.

Clarke, Arthur C. *The Exploration of Space.* New York: Harper and Brothers, 1951.

———. *The Making of a Moon: The Story of the Earth Satellite Program.* New York: Harper and Brothers, 1957.

———. *The Promise of Space.* New York: Harper and Row, 1968.

———, editor. *The Coming of the Space Age.* New York: Meredith, 1967.

Cleator, P. E. *An Introduction to Space Travel.* New York: Pitman, 1961.

———. *Rockets Through Space: On the Dawn of Interplanetary Travel.* London: Allen and Unwin, 1936.

Cleveland, Reginald M. *America Fledges Wings: The History of the Daniel Guggenheim Fund for the Promotion of Aeronautics.* New York: Pitman, 1942.

Cohen, Morris H. *Worcester's Ethnic Groups: A Bicentennial View.* Worcester: Worcester Bicentennial Commission, 1970.

Coil, Suzanne M. *Robert Hutchings Goddard: Pioneer of Rocketry and Space Flight.* New York: Facts on File, 1992.

Collins, Michael. *Liftoff: The Story of America's Adventure in Space.* New York: Grove, 1988.

Conot, Robert. *A Streak of Luck.* New York: Seaview, 1979.

Crane, Ellery Bicknell. *Historic Homes and Institutions and Genealogical and Personal Memoirs of Worcester County, Massachusetts.* 4 volumes. New York: Lewis, 1907.

Crane, W. S., compiler. *The Roswell Museum and Art Center Robert H. Goddard Collection.* Roswell, New Mexico: Roswell Museum and Art Center, 1994.

Crouch, Tom D. *Aiming for the Stars: The Dreamers and Doers of the Space Age.* Washington, D.C.: Smithsonian Institution Press, 1999.

———. *The Bishop's Boys: A Life of Wilbur and Orville Wright.* New York: Norton, 1989.

Davis, Heather M. *Wernher von Braun.* New York: Putnam, 1967.

Davis, John H. *The Guggenheims: An American Epic.* New York: Morrow, 1978.

Davis, Kenneth S. *The Hero: Charles A. Lindbergh and the American Dream.* New York: Doubleday, 1959.

Dewey, Anne Perkins. *Robert Goddard: Space Pioneer.* Boston: Little, Brown, 1962.

Dornberger, Walter. *V-2.* Translated by James Cleugh and Geoffrey Halliday. New York: Viking, 1954.

Duff, A. Wilmer. *A Text-Book of Physics.* 5th edition. Philadelphia: P. Blakiston's Sons, 1921.

Duff, A. Wilmer, and Henry T. Weed. *Elements of Physics.* New York: Longmans, Green, 1928.

Durant, Frederick C., III, and George S. James. *First Steps Toward Space: Proceedings of the First and Second History Symposia of the International Academy of Astronautics, at Belgrade, Yugoslavia, 26 September 1967, and New York, USA, 16 October 1968.* Washington, D.C.: Smithsonian Institution Press, 1974.

Erskine, Margaret A. *Worcester: An Illustrated History.* Woodland Hills, California: Windsor, 1981.

Ethell, Jeffrey L. *Frontiers of Flight.* Washington: Smithsonian Institution Press, 1992.

Ewing, Edgar G., and Roy A. Squires. *Comparison of the Achievements of Two American Rocket Pioneers: A Report . . . Comparing the Contributions to World Astronautics by the Two Foremost American Space Pioneers—Robert H. Goddard and Edmund V. Sawyer.* Napa, California: Crescent Engineering and Research, 1965, 1991.

Fagen, M. D., editor. *A History of Engineering and Science in the Bell System: National Service in War and Peace (1925–1975).* N.P.: Bell Telephone Laboratories, 1978.

Farago, Ladislas. *The Game of the Foxes: The Untold Story of German Espionage in the United States and Great Britain During World War II.* New York: David McKay, 1971.

Farley, Kevin Clafford. *Robert H. Goddard.* Englewood Cliffs, New Jersey: Silver Burdett Press, 1991.

Farnsworth, Albert, and George B. O'Flynn. *The Story of Worcester, Massachusetts.* Worcester: Davis, 1934.

Feynman, Richard P., as told to Ralph Leighton. *Surely You're Joking, Mr. Feynman! Adventures of a Curious Character.* New York: Norton, 1985.

———. *What Do You Care What Other People Think? Further Adventures of a Curious Character.* New York: Norton, 1988.

Fleming, Elvis E., and Minor S. Huffman, editors. *Roundup on the Pecos.* Roswell, New Mexico: Chaves County Historical Society, 1978.

Ford, Brian. *German Secret Weapons: Blueprint for Mars.* New York: Ballantine, 1969.

Freeman, Marsha. *How We Got to the Moon: The Story of the German Space Pioneers.* Washington, D.C.: 21st Century Associates, 1993.

Garbedian, H. Gordon. *Major Mysteries of Science.* New York: Covici-Friede, 1933.

Gartmann, Heinz. *The Men Behind the Space Rockets.* New York: McKay, 1956.

Gibbs, William E., and Eugene T. Jackman. *New Mexico Military Institute: A Centennial History.* Roswell, New Mexico: NMMI Centennial Commission, 1991.

Goddard, Charles Austin, compiler. *The Samuel Goddard Families: Descendants of William Goddard, New England Immigrant, Who Settled at Watertown, Massachusetts, in 1665.* Fayetteville, West Virginia: Privately printed, 1935.

Goddard, William Austin. *A Genealogy of the Descendants of Edward Goddard.* Worcester: Spooner, 1833.

Green, Constance McLaughlin, Harry C. Thomson, and Peter C. Roots. *The Ordnance Department: Planning Munitions for War.* United States Army in World War II: The Technical Services. Washington, D.C.: Department of the Army, 1955.

Guggenheim, Harry F. *The Seven Skies.* New York: Putnam, 1930.

———. *The United States and Cuba: A Study in International Relations.* New York: Macmillan, 1934.

Gunston, Bill. *The Illustrated Encyclopedia of the World's Rockets and Missiles.* London: Salamander Books, 1979.

Gurney, Gene, editor. *Rockets and Missile Technology.* New York: Franklin Watts, 1964.

Hall, G. Stanley. *Aspects of Child Life and Education.* Boston: Ginn, 1907.

———. *Adolescence.* New York: Arno, 1969.

———. *Senescence: The Last Half of Life.* New York: Appleton, 1923.

Hallenbeck, Cleve. *Land of the Conquistadores.* Caldwell, Idaho: Caxton, 1950.

———. *Spanish Missions of the Old Southwest.* Garden City, New York: Doubleday, 1926.

Hallenbeck, Cleve, and Juanita H. Williams. *Legends of the Spanish Southwest.* Glendale, California: Arthur H. Clark, 1938.

Hallion, Richard P. *Legacy of Flight: The Guggenheim Contribution to American Aviation.* Seattle: University of Washington Press, 1977.

Harford, James. *Korolev: How One Man Masterminded the Soviet Drive to Beat America to the Moon.* New York: Wiley, 1997.

Heiman, Grover. *Jet Pioneers.* New York: Duel, Sloan and Pearce, 1963.

Heinkel, Ernst. *Stormy Life: Memoirs of a Pioneer of the Air Age.* Edited by Jurgen Thorwald. New York: Dutton, 1956.

Heppenheimer, T. A. *Countdown: A History of Space Flight.* New York: Wiley, 1997.

Hertog, Susan. *Anne Morrow Lindbergh: Her Life.* New York: Doubleday, 1999.

Hoffman, Paul. *The Man Who Loved Only Numbers: The Story of Paul Erdös and the Search for Mathematical Truth.* New York: Hyperion, 1998.

Holme, Bryan. *Bullfinch's Mythology.* New York: Viking, 1979.

Horgan, Paul. *The Centuries of Santa Fe.* New York: Dutton, 1956.

———. *A Distant Trumpet.* New York: Farrar, Straus and Cudahy, 1960.

———. *Great River: The Rio Grande in North American History.* 2 volumes. New York: Holt, Rinehart and Winston, 1954.

———. *Lamb of God.* Roswell, New Mexico: Privately printed, 1927.

———. *Lamy of Santa Fe: His Life and Times.* New York: Farrar, Straus and Giroux, 1975.

———. *Main Line West.* New York: Harper and Brothers, 1936.

———. *No Quarter Given.* New York: Harper and Brothers, 1935.

Howard, Fred. *Wilbur and Orville: A Biography of the Wright Brothers.* New York: Knopf, 1987.

Huffman, Minor. *60 Years of Roswell Rotary.* Roswell, New Mexico: Privately printed, 1980.

Humphries, John. *Rockets and Guided Missiles.* New York: Macmillan, 1956.

Hunt, Linda. *Secret Agenda: The United States Government, Nazi Scientists, and Project Paperclip, 1945 to 1990.* New York: St. Martin's, 1991.

Hurd, Peter. *My Land Is the Southwest: Peter Hurd Letters and Journals.* Edited by Robert Metzger. College Station: Texas A & M Press, 1983.

James, Peter, and Nick Thorpe. *Ancient Inventions.* New York: Ballantine, 1994.

Jardine, Lisa. *Ingenious Pursuits: Building the Scientific Revolution.* New York: Doubleday, 1999.

Julyan, Robert. *The Place Names of New Mexico.* Albuquerque: University of New Mexico Press, 1996.

Kahn, David. *Hitler's Spies: German Military Intelligence in World War II.* London: Hodder and Stoughton, 1978.

Keegan, John. *The Mask of Command.* New York: Viking, 1987.

Kelly, J. R. *A History of New Mexico Military Institute, 1891–1941.* Roswell: The Author, 1953.

Kennedy, Gregory P. *Rockets, Missiles, and Spacecraft of the National Air and Space Museum.* Washington, D.C.: Smithsonian Institution Press, 1983.

———. *Vengeance Weapon 2: The V-2 Guided Missile.* Washington, D.C.: Smithsonian Institution Press, 1983.

King, Benjamin, and Timothy Kutta. *Impact: The History of Germany's V-Weapons in World War II.* Rockville Centre, New York: Sarpedon, 1998.

Kirkpatrick, J. E. *The American College and Its Rulers.* New York: New Republic, 1926.

Knightley, Paul. *The First Casualty: From the Crimea to Vietnam: The War Correspondent as Hero, Propagandist, and Myth Maker.* New York: Harcourt Brace Jovanovich, 1975.

Koelsch, William A. *Clark University, 1887–1987: A Narrative History.* Worcester, Massachusetts: Clark University Press, 1987.

Koppes, Clayton R. *JPL and the American Space Program.* New Haven: Yale University Press, 1982.

Krieger, F. J. *Behind the Sputniks: A Survey of Soviet Space Science.* Washington: Public Affairs Press, 1958.

Larson, Carole. *Forgotten Frontier: The Story of Southeastern New Mexico.* Albuquerque: University of New Mexico Press, 1993.

Lasby, Clarence. *Project Paperclip: German Scientists and the Cold War.* New York: Atheneum, 1971.

Lasser, David. *The Conquest of Space.* New York: Penguin, 1931.

Leckie, Shirley A. *Elizabeth Bacon Custer and the Making of a Myth.* Norman: University of Oklahoma Press, 1993.

Lent, Constantin Paul. *Rocket Research History and Handbook.* New edition. New York: Pen-Ink Publishing, 1945.

———. *Rocketry, Jets, and Rockets: The Science of the Reaction Motor and Its Practical Application for Aircraft and Space Travel.* New York: Pen-Ink Publishing, 1947.

Ley, Willy. *Rockets, Missiles, and Men in Space.* New York: Viking, 1968.

———. *Rockets, Missiles, and Space Travel: Revised Edition.* New York: Viking, 1957.

———. *Rockets: The Future of Travel Beyond the Stratosphere.* New York: Viking, 1944.

Lincoln, William. *History of Worcester, Massachusetts, from Its Earliest Settlement to September, 1836.* Worcester: Charles Hasey, 1862.

Lindbergh, Anne Morrow. *Earth Shine.* New York: Harcourt, Brace and World, 1969.

Lindbergh, Charles A. *Autobiography of Values.* New York: Harcourt Brace Jovanovich, 1978.

———. *Of Flight and Life.* New York: Charles Scribner's Sons, 1948.

———. *The Wartime Journals of Charles A. Lindbergh.* New York: Harcourt Brace Jovanovich, 1970.

———. *"We."* New York: G. P. Putnam's Sons, 1927.

Lomask, Milton. *Seed Money: The Guggenheim Story.* New York: Farrar, Straus and Giroux, 1964.

———. *Robert H. Goddard: Space Pioneer.* Champaign, Illinois: Garrard, 1972.

Lowell, Percival. *Mars and Its Canals.* New York: Macmillan, 1906.

Mackenzie, Catherine. *Alexander Graham Bell: The Man Who Contracted Space.* Boston: Houghton Mifflin, 1928.

Magill, Frank N., editor. *The Great Scientists.* Danbury, Connecticut: Grolier, 1989.

Mallan, Lloyd. *Men, Rockets, and Space Rats.* New York: Julian Messner, 1955.

Massachusetts: A Guide to Its Places and People. American Guide Series. Boston: Houghton Mifflin, 1937.

Maurer, Richard. *Rocket! How a Toy Launched the Space Age.* New York: Crown, 1995.

Mayes, Thorn L. *Wireless Communication in the United States: The Early Development of American Radio Operating Companies.* East Greenwich, Rhode Island: New England Wireless and Steam Museum, 1989.

McCurdy, Howard E. *Space and the American Imagination.* Washington, D.C.: Smithsonian Institution Press, 1997.

McDougall, Walter A. *The Heavens and the Earth: A Political History of the Space Age.* New York: Basic Books, 1985.

McGovern, James. *Crossbow and Overcast.* New York: Morrow, 1964.

Michel, Jean, in association with Louis Nucera. *Dora.* Translated by Jennifer Kidd. New York: Holt, Rinehart and Winston, 1980.

Miller, Francis Trevelyan. *The World in the Air: The Story of Flying in Pictures.* 2 volumes. New York: G. P. Putnam's Sons, 1930.

Miller, Joseph. *New Mexico: A Guide to the Colorful State.* American Guide Series. Revised edition. New York: Hastings House, 1953.

Miller, Ron. *The Dream Machines: An Illustrated History of the Spaceship in Art, Science, and Literature.* Malabar, Florida: Krieger, 1993.

Millikin, Robert. *The Guggenheim Aeronautical Laboratory of the California Institute of Technology: The First Twenty-Five Years.* Pasadena: CalTech Press, 1954.

Milton, Joyce. *Loss of Eden: A Biography of Charles and Anne Morrow Lindbergh.* New York: HarperCollins, 1993.

Milton, Lehman. *This High Man: The Life of Robert H. Goddard.* New York: Farrar, Straus and Company, 1963. Reissued as second edition, *Robert H. Goddard: Pioneer of Space Research.* New York: DaCapo, 1988.

Moore, Clyde B. *Robert Goddard: Pioneer Rocket Boy.* Indianapolis: Bobbs-Merrill, 1966.

Mosley, Leonard. *Lindbergh: A Biography*. Mineola, New York: Dover, 1976.

Nash, Jay Robert. *Spies: A Narrative Encyclopedia of Dirty Deeds and Double Dealing from Biblical Times to Today*. New York: M. Evans, 1997.

Needell, Allan A., editor. *The First 25 Years in Space: A Symposium*. Washington, D.C.: Smithsonian Institution Press, 1983.

Nelson, John. *Worcester County: A Narrative History*. 3 volumes. New York: American Historical Society, 1934.

Neufeld, Michael J. *The Rocket and the Reich: Peenemünde and the Coming of the Ballistic Missile Era*. Cambridge: Harvard University Press, 1995.

Newell, Homer E. *Beyond the Atmosphere: Early Years of Space Science*. NASA History Series SP-4211. Washington, D.C.: National Aeronautics and Space Administration, 1980.

———. *Express to the Stars: Rockets in Action*. New York: McGraw-Hill, 1961.

Noordung, Hermann. *Das Problem der Befahrung des Weltraums: Der Raketen Motor*. Berlin: Carl Schmidt, 1929. Reissued as *The Problem of Space Travel: The Rocket Motor*. Translated and edited by Ernst Stuhlinger, J.D. Hunley, and Jennifer Garland. NASA History Series, SP-4025. Washington, D.C.: National Aeronautics and Space Administration History Office, 1995.

Nutt, Charles. *History of Worcester and Its People*. 3 volumes. New York: Lewis Historical Publishing, 1919.

Oberth, Hermann. *Die Rakete zu den Planetenräumen*. Munich: Oldenbourg, 1923.

———. *Primer for Those Who Would Govern . . . in a World Parliament*. New York: West-Art, 1987.

Ordway, Frederick I., III, and Mitchell R. Sharpe. *The Rocket Team: From the V-2 to the Saturn Moon Rocket—The Inside Story of How a Small Group of Engineers Changed World History*. New York: Crowell, 1979.

Parson, Nels A., Jr. *Missiles and the Revolution in Warfare*. Cambridge: Harvard University Press, 1962.

Pendray, G. Edward. *The Coming Age of Rocket Power*. New York: Harper and Brothers, 1945.

Philp, Charles G. *Stratosphere and Rocket Flight*. London: Pitman, 1935.

Piszkiewicz, Dennis. *The Nazi Rocketeers: Dreams of Space and Crimes of War*. New York: Praeger, 1995.

———. *Wernher von Braun: The Man Who Sold the Moon*. Westport, Connecticut: Praeger, 1998.

Pocock, Rowland F. *German Guided Missiles of the Second World War*. New York: Arco, 1967.

Poe, Edgar Allan. *The Complete Tales and Poems of Edgar Allan Poe*. New York: Vintage, 1975.

Pope, Loren. *Colleges That Change Lives*. New York: Penguin, 1996.

Proctor, Mary. *The Romance of the Moon*. New York: Harper and Brothers, 1928.

Pruette, Lorine. *G. Stanley Hall: A Biography of a Mind*. New York: Appleton, 1926.

Pursell, Carroll W., Jr., editor. *Technology in America: A History of Individuals and Ideas*. Cambridge, Massachusetts: MIT Press, 1981.

Quackenbush, Robert M. *The Boy Who Dreamed of Rockets: How Robert Goddard Became the Father of the Space Age*. New York: Parents Magazine Press, 1978.

Rabelais, François. *Gargantua and Pantagruel.* Translated by Sir Thomas Urquhart and Peter Motteux. Great Books of the Western World, volume 24. Chicago: University of Chicago Press, 1952.

Rauschenbach, Boris V. *Hermann Oberth: The Father of Space Flight.* New York: West-Art, 1994.

Reinhardt, George C., and William R. Kintner. *The Haphazard Years: How America Has Gone to War.* Garden City, New York: Doubleday, 1960.

Riabchikov, Evgeny. *Russians in Space.* Translated by Guy Daniels. Garden City, New York: Doubleday, 1971.

Roberts, Susan A., and Calvin A. Roberts. *A History of New Mexico.* 2nd edition. Albuquerque: University of New Mexico Press, 1998.

Rosen, Milton W. *The Viking Rocket Story.* New York: Harper and Brothers, 1955.

Rosenzweig, Roy. *Eight Hours for What We Will: Workers and Leisure in an Industrial City, 1870–1920.* Cambridge: Harvard University Press, 1983.

Rosenzweig, Saul. *Freud, Jung, and Hall the King-Maker: The Historic Expedition to America (1909) with G. Stanley Hall as Host and William James as Guest.* St. Louis: Rana House, 1992.

Ross, Dorothy. *G. Stanley Hall: The Psychologist as Prophet.* Chicago: University of Chicago Press, 1972.

Ross, Walter S. *The Last Hero: Charles A. Lindbergh.* New York: Harper and Row, 1964, 1968.

Rundell, Walter, Jr. *Oil in West Texas and New Mexico: A Pictorial History of the Permian Basin.* College Station: Texas A & M University Press, 1982.

Rush, N. Orwin, editor. *Letters of G. Stanley Hall to Jonas Gilman Clark.* Worcester, Massachusetts: Clark University Library, 1948.

Ryan, W. Carson. *Studies in Early Graduate Education: The Johns Hopkins, Clark University, the University of Chicago.* New York: Carnegie Foundation for the Advancement of Teaching, 1939.

Scamehorn, Howard L. *Balloons to Jets.* Chicago: Regnery, 1957.

Schaefer, Lola M. *Robert Goddard.* Mankato, Minnesota: Pebble Books, 2000.

Schauer, William H. *The Politics of Space: A Comparison of the Soviet and American Space Programs.* New York: Holmes and Meier, 1976.

Shinkle, James D. *Seventy-seven Years of Roswell School History.* Roswell, New Mexico: Hall-Poorbaugh, 1958.

Smith, LeRoi, editor. *We Came in Peace.* San Rafael, California: Classic Press, 1969.

Speer, Albert. *Inside the Third Reich.* New York: Macmillan, 1970.

Standhardt, Edith Wolf. *A History of St. Andrews Church: The Establishment and Development of the Protestant Episcopal Church in Roswell, New Mexico, 1891–1966.* Roswell, New Mexico: Frank M. Standhardt, 1966.

Stoiko, Michael. *Pioneers of Rocketry.* New York: Hawthorn, 1974.

———. *Soviet Rocketry: Past, Present, and Future.* New York: Holt, Rinehart and Winston, 1970.

Streissguth, Thomas. *Rocket Man: The Story of Robert Goddard.* Minneapolis: Lerner, 1995.

Stuhlinger, Ernst, and Frederick I. Ordway, III. *Wernher von Braun: Crusader for Space.* Melbourne, Florida: Krieger, 1994.

Swenson, Loyd S., Jr., James M. Grimword, and Charles C. Alexander. *This New Ocean: A History of Project Mercury.* NASA History Series, SP-4201. Washington, D.C.: National Aeronautics and Space Administration History Office, 1998.

Thomas, Shirley. *Men of Space.* Philadelphia: Chilton, 1960.

Thomson, Harry C., and Lida Mayo. *The Ordnance Department: Procurement and Supply.* United States Army in World War II: The Technical Services. Washington, D.C.: Department of the Army, 1960.

Top, Franklin H., and others. *Communicable and Infectious Diseases: Diagnosis, Prevention, Treatment.* 5th edition. St. Louis: C. V. Mosby, 1964.

Tripp, Edward. *The Meridian Handbook of Classical Mythology.* New York: New American Library, 1970.

True, Michael. *Worcester Area Writers, 1680–1980.* Worcester, Massachusetts: Worcester Public Library, 1987.

Tsiolkovsky, K. E. *Works on Rocket Technology.* Moscow: Publishing House of The Defense Industry, 1947. NASA Technical Translation NASA TT F-243. Washington, D.C.: National Aeronautics and Space Administration, 1965.

Tymeson, Mildred McClary. *Two Towers: The Story of Worcester Tech, 1865–1965.* Worcester, Massachusetts: Worcester Polytechnic Institute, 1965.

Vaeth, J. Gordon. *200 Miles Up: The Conquest of the Upper Air.* 2nd edition. New York: Ronald Press, 1955.

Verne, Jules. *From the Earth to the Moon.* Paris, 1865. New York: Dodd, Mead, 1962.

———. *Round the Moon.* Paris, 1870. New York: Dodd, Mead, 1962.

Verral, Charles Spain. *Robert Goddard: Father of the Space Age.* Englewood Cliffs, New Jersey: Prentice-Hall, 1963.

Veysey, Laurence R. *The Emergence of the American University.* Chicago: University of Chicago Press, 1965.

Von Braun, Wernher. *The Mars Project.* Urbana: University of Illinois Press, 1953.

———. *Space Frontier.* Revised edition. New York: Holt, Rinehart and Winston, 1971.

Von Braun, Wernher, and Frederick I. Ordway, III. *History of Rocketry and Space Travel.* New York: Crowell, 1966.

Von Braun, Wernher, Frederick I. Ordway, III, and Dave Dooling. *Space Travel: A History.* New York: Harper and Row, 1985.

Von Kármán, Theodore, with Lee Edson. *The Wind and Beyond: Theodore von Kármán, Pioneer in Aviation and Pathfinder in Space.* Boston: Little, Brown, 1967.

Webster, Arthur Gordon. *The Dynamics of Particles and of Rigid, Elastic, and Fluid Bodies: Being Lectures on Mathematical Physics by Arthur Gordon Webster.* Leipzig: Teubner, 1904. Reprinted New York: Dover, 1959.

————. *Partial Differential Equations of Mathematical Physics*. Edited by Samuel J. Plimpton. New York: Stechert, and Leipzig: Teubner, 1927. Reprinted New York: Dover, 1966.

————. *The Theory of Electricity and Magnetism: Being Lectures on Mathematical Physics by Arthur Gordon Webster*. London and New York: Macmillan, 1897.

Wedlake, G. E. C. *SOS: The Story of Radio Communication*. New York: Crane, Russak, 1973.

Weeks, John. *Men Against Tanks: A History of Anti-Tank Warfare*. New York: Mason/Charter, 1975.

Wells, H. G. *Seven Science Fiction Novels of H. G. Wells*. New York: Dover, n.d.

Wells, Ralph G., and John S. Perkins, compilers. *New England Community Statistical Abstracts*. 3rd edition. Boston: Boston University College of Business Administration, 1942.

Wenstrom, William H. *Weather and the Ocean of Air*. Boston: Houghton Mifflin, 1942.

Williams, Beryl, and Samuel Epstein. *The Rocket Pioneers: On the Road to Space*. New York: Julian Messner, 1955, 1958.

Williams, Ernestine Chesser. *Echoes Break the Silence: A Collection of Articles from New Mexico and Arkansas*. Roswell, New Mexico: Old-Time Publishers, 1987.

Winter, Frank H. *The First Golden Age of Rocketry: Congreve and Hale Rockets of the Nineteenth Century*. Washington, D.C.: Smithsonian Institution Press, 1990.

————. *Prelude to the Space Age: The Rocket Societies, 1924–1940*. Washington, D.C.: Smithsonian Institution Press, 1983.

————. *Rockets into Space*. Cambridge, Massachusetts: Harvard University Press, 1990.

Wolko, Howard S. *In the Cause of Flight: Technologists of Aeronautics and Astronautics*. Smithsonian Studies in Air and Space, Number Four. Washington: Smithsonian Institution Press, 1981.

Woolf, Virginia. *A Room of One's Own*. Reprint. New York: Cambridge University Press, 1995.

Wulforst, Harry. *The Rocketmakers*. New York: Orion, 1990.

Zaehringer, Alfred. *Soviet Space Technology*. New York: Harper and Brothers, 1961.

Zwicky, Fritz. *Report on Certain Phases of War Research in Germany, Volume I*. Pasadena, California: Aerojet Engineering, 1945.

ARTICLES

Armagnac, Alden P. "Aims Rocket at Roof of Sky." *Popular Science Monthly*, October 1929.

Bailey, Arnold C. "Prophet of the Space Age." *Electronic Age*, Winter 1969–70.

Banks, Phyllis Eileen. "Robert H. Goddard, Space Pioneer." *The Roswell/Chaves County Way* 78, no. 1, 1998.

Bode, George. "Museum Tower Used for Rockets by Goddard." *Roswell Daily Record*, 31 January 1994.

Boushey, Homer A. "Space Weaponry." *Ordnance*, July–August 1958.

————. "Who Controls the Moon Controls the Earth." *U.S. News & World Report*, 7 February 1958.

Clark, Tim. "The Angel Shooter." *Yankee*, October 1982.

Clarkson, Paul S. "Ah, Distinctly I Remember." *Goddard Biblio Log*, Spring 1979.

Dajer, Tony. "The Sleeping Giant." *Discover*, May 2001.

Davis, Harry M. "Fiery Rockets Roar, Forerunners of Swift Ships of Space." *The New York Times Sunday Magazine*, 5 September 1937.

Davis, Kenneth S. "Father of Rocketry." *The New York Times Magazine*, 23 October 1960.

Davis, L.J. "Worlds Enough: Travels in an Adolescent Genre." *Harper's Magazine*, January 2002.

Deisch, Noel. "How High Up?" *Scientific American*, July 1926.

Dill, Gailanne Teresa. "Goddard's Home on the Range." *Roswell Daily Record Vision Magazine*, 12 March 1999.

Dornberger, Walter R. "The V-2 Rocket." Paper presented at the Fifth Annual Meeting of the Society for the History of Technology, Philadelphia, 28 December 1962.

Durant, Frederick C., III. "Robert H. Goddard: Accomplishments of the Roswell Years (1930–1941)." Paper presented at the Seventh International History of Astronautics Symposium of the International Academy of Astronautics, Baku, USSR, October 1973.

———. "Robert H. Goddard and the Smithsonian Institution." Paper presented at the Second Symposium on the History of Astronautics, Nineteenth Congress of the International Astronautical Federation, 16 October 1968.

Eisenberg, Arlene and Howard. "The Tragic Case of Robert Goddard." *Saga*, November 1961.

Emme, Eugene M. "Yesterday's Dream ... Today's Reality: A Biographical Sketch of the American Rocket Pioneer, Dr. Robert H. Goddard." *The Aerospace Historian*, October 1960.

England, Terry. "Robert H. Goddard." *New Mexico Magazine*, February 1999.

Fairman, Milton. "The Race to Explore Outer Space." *Popular Mechanics*, March 1930.

Fleming, Elvis E. "Charles Lindbergh: Advisor to Dr. Robert H. Goddard." *Roswell Daily Record Vision Magazine*, 12 March 1999.

Fox, Dorothea Magdalene. "Dr. Goddard." *New Mexico Magazine*, September 1960.

Fuller, Curtis. "Uncle Sam 'Discovers' the Rocket." *Flying*, March 1947.

Goddard, Esther C. "The Life and Achievements of Dr. Robert H. Goddard." Speech at Special Evening Colloquium, NASA Goddard Space Flight Center, Greenbelt, Maryland, 5 October 1973.

———. "Our Nell." *Aeronautica*, October–December 1950.

"Goddard Memorial Exhibit Opens." *Aeronautica*, October–December 1950.

Gray, George W. "Speed." *The Atlantic Monthly*, March 1930.

Hagemann, E.R. "R.H. Goddard and Solar Power, 1924–1934." *Solar Energy*, April–June 1962.

Hendrickson, Nancy L. "Roswell's Real Flying Objects." *Astronomy*, February 1998.

Hersey, Irwin. "It's a Goddard Year." *Astronautics*, June 1959.

Hold, Lew. "Amazing Turbine-Rocket to Explore Outer Space." *Modern Mechanics and Inventions*, December 1931.

Hull, Richard. "A Trail to Space." *The Bee-Hive*, Spring 1958.

Humphreys, W. J. "Mining the Sky for Scientific Knowledge." *Scientific American*, January 1931.

Hunley, J. D. "The Enigma of Robert H. Goddard." *Technology and Culture*, April 1995.

"Jet Propulsion: The U.S. Is Behind." *Fortune*, September 1946.

Kaempffert, Waldemar. "Rocket Ships and a Visit to Mars." *Forum*, October 1930.

Kerbey, McFall. "Germany's Flying Vengeance Bomb V-2 Follows American Prewar Rocket Design." *National Geographic News Bulletin*, 19 January 1945.

Kernan, Michael. "50th Anniversary of Step Toward Space." *Smithsonian*, March 1976.

Killian, J. Rhyne, Jr. "Reaction Propulsion: Adventures of the Rocketeers, Both Fanciful and Scientific." *Technology Review*, October 1931.

Kluger, Jeffrey. "Robert Goddard, Rocket Scientist." *Time*, 29 March 1999.

Kriegbaum, H. "Beyond the Stratosphere." *Future*, April 1940.

Lasser, David. "The Future of the Rocket." *Scientific American*, March 1931.

Lehman, Milton. "The Father of Modern Rocketry." *The Sunday New York Times*, 20 November 1966.

———. "The Secrets of White Sands." *Saturday Evening Post*, 27 March 1954.

———. "The Strange Story of Dr. Goddard." *Reader's Digest*, November 1955.

Lindbergh, Charles A. "The Rocket Offers Freedom from the Air." *Astronautics Journal*, July 1937.

MacMechen, Rutherford. "Rockets—the New Monsters of Doom." *Liberty*, 19 September 1931.

Magee, Richard E., Jr. "The Two Goddards of Southern New Mexico." *Southern New Mexico Historical Review*, January 1998.

Masius, Morton. "A. Wilmer Duff, Scientist and Educator." *Journal of the Worcester Polytechnic Institute*, March 1951.

Meigs, John. "Roswell Museum: Experience in Excellence." *New Mexico Magazine*, May–June 1970.

Miller, Ron. "Herman Potocnik *alias* Hermann Noordung." *Journal of the British Interplanetary Society*, July 1992.

Moore, Alvin Edward. "Dr. Robert H. Goddard, Rocketeer." *Popular Aviation*, March 1937.

———. "Sky-Rocketing Through Space." *Popular Aviation*, April 1937.

"New Trustees Elected." *Clark University Bulletin Alumni Magazine*, July 1964.

Nichols, Herbert B. "Flying by Fireworks." *Christian Science Monitor*, 7 December 1940.

Noordung, Hermann. "The Problems of Space Flying." Translated by Francis M. Currier. *Science Wonder Stories*, July, August, September 1929.

"The Object at Hand." *Scientific American*, November 1989.

Ordway, Frederick I., III. "Applications of the Rocket Engine." *Sky and Telescope*, January 1955.

"Our Professor Goddard." *Clark Now: The Magazine of Clark University,* Fall–Winter, 1982.

Page, Brian R. "The Rocket Experiments of Robert H. Goddard, 1911 to 1930." *Physics Teacher,* November 1991.

Pearson, Drew, and Jack Anderson. "Wernher von Braun: Columbus of Space." *True,* February 1959.

Pendray, G. Edward. "Developing Rockets to Explore Stratosphere." *Literary Digest,* 29 September 1934.

———. "Giant Rockets." *Popular Science Monthly,* August 1931.

———. "The Man Who Ushered in the Space Age." *American Legion Magazine,* September 1960.

———. "Number One Rocket Man." *Scientific American,* May 1938.

———. "Pipe Lines to the Heavens." *Colliers,* 26 April 1947.

———. "Obituary—Robert H. Goddard." *Science,* 23 November 1945.

———. "Rocketry's Number One Man." *Astronautics,* July 1937.

———. "Rockets to the Moon." *Elks Magazine,* October 1931.

Pollard, Hugh. "Professor Goddard's Rocket to the Moon." *Review of Reviews,* June 1924.

"Push into Space." *Time,* 19 January 1959.

Rafferty, Keen A. "Rockets to the Sky." *New Mexico Magazine,* March 1939.

Randolph, Major James R., USAR. "What Can We Expect of Rockets?" *Army Ordnance,* January–February 1939.

Rhodes, Richard. "The Ordeal of Robert Hutchings Goddard: 'God Pity a One-Dream Man.'" *American Heritage,* June–July 1980.

"Robert Hutchings Goddard Space and Rocket Collection at the Roswell Museum and Art Center, Roswell, New Mexico." *New Mexico Professional Engineer,* April 1959.

"The Robert H. Goddard Memorial Exhibit." *Aeronautical Engineering Review,* February 1951.

"Rocket Museum." *New Mexico Magazine,* April 1959.

"Rockets." *Time,* 2 March 1936.

Ross, John C. "Where Were Our Rockets?" *Esquire,* September 1947.

"Roswell Museum to Open Goddard Wing This Month." *Astronautics,* April 1959.

Rothman, Milton A. "Father of the Rocket." *Boy's Life,* June 1962.

Sagan, Carl. "Boyhood Dreams Helped Launch Flight to Mars." *Sunday Los Angeles Times,* 17 October 1976.

Schuessler, Raymond. "How America Delayed Space Research for 20 Years." *The Retired Officer,* January 1973.

"700 Miles per Hour: Guggenheims Back Rocket Developed by Goddard." *Literary Digest,* 5 October 1935.

Shepperd, J. McLaughlin. "What of the Rocket Plane?" *Air Travel News,* March 1930.

Shiers, George. "The First Electron Tube." *Scientific American,* March 1969.

Thompson, L. T. E. "Robert Goddard and the Weapons Program." Remarks delivered at dinner of the Fifth Joint Army-Navy Solid Propellant Conference, 23 March 1949.

"Von Braun Tribute to Goddard." *Astronautics*, June 1959.

Warren, David M. "The Forgotten Father of Rocketry." *Family Weekly*, 19 April 1959.

Wenstrom, William H. "Rockets to the Moon." *Review of Reviews*, September 1930.

———. "Sounding the Sky's Blue Depths." *New York Times Magazine*, 7 August 1932.

Winter, Frank H. "Who First Flew in a Rocket?" *Journal of the British Interplanetary Society*, July 1992.

Wright, Robert E. "Rocketeering." *Wisconsin Engineer*, November 1939.

OTHER

Annual Report of the Board of Regents of the Smithsonian Institution ... 1930. Washington, D.C.: Government Printing Office, 1931.

Chasan, Joshua S. *Civilizing Worcester: The Creation of Institutional and Cultural Order, Worcester, Massachusetts, 1848–1876*. Ph.D. dissertation (History), University of Pittsburgh, 1974.

Kolesar, Robert J. *Politics and Policy in a Developing City: Worcester, Massachusetts, in the Late Nineteenth Century*. Ph.D. dissertation (History), Clark University, 1987.

Manual for the City Council of the City of Worcester, MA, 1998–99, Worcester Public Library Reference Desk.

Nowlan, Philip Francis. "The Airlords of Han." *Amazing Stories*, March 1929.

———. "Armageddon—2419 A.D." *Amazing Stories*, August 1928.

Roberge, Roger A. "The Three-Decker: Structural Correlate of Worcester's Industrial Revolution." M.A. thesis (Geography), Clark University, 1965.

United States. Congress. House. Committee on Science. *Memorial to Dr. Robert H. Goddard. Hearing, Eighty-ninth Congress, First Session, on H. J. Res. 597, September 7, 1965*. Washington, D.C.: Government Printing Office, 1965.

United States. Congress. Senate. Committee on Aeronautical and Space Sciences. *Congressional Recognition of Goddard Rocket and Space Museum, Roswell, New Mexico, with Tributes to Dr. Robert H. Goddard, Space Pioneer, 1882–1945*. Washington, D.C.: Government Printing Office, 1970.

United States. Department of Commerce and Labor. Bureau of the Census. *A Century of Population Growth: From the First Census of the United States to the Twelfth, 1790–1900*. Washington: Government Printing Office, 1909.

———. *Thirteenth Census of the United States Taken in the Year 1910*. Washington, D.C.: Government Printing Office, 1912.

INDEX